Flood Risk Management and Response

Flood Risk Management and Response

Editors

D. Proverbs
Birmingham City University, UK

C.A. Brebbia
Wessex Institute, UK

WIT*PRESS* Southampton, Boston

Editors:

D. Proverbs
Birmingham City University, UK

C.A. Brebbia
Wessex Institute, UK

Published by

WIT Press
Ashurst Lodge, Ashurst, Southampton, SO40 7AA, UK
Tel: 44 (0) 238 029 3223; Fax: 44 (0) 238 029 2853
E-Mail: witpress@witpress.com
http://www.witpress.com

For USA, Canada and Mexico

Computational Mechanics International Inc
25 Bridge Street, Billerica, MA 01821, USA
Tel: 978 667 5841; Fax: 978 667 7582
E-Mail: infousa@witpress.com
http://www.witpress.com

British Library Cataloguing-in-Publication Data

A Catalogue record for this book is available
from the British Library.

Library of Congress Catalog Card Number: 2016917267

ISBN: 978-1-78466-241-7
eISBN: 978-1-78466-242-4

The text of the papers in this volume were set individually by the authors or under their supervision. Only minor corrections to the text may have been carried out by the publisher.

Printed in Great Britain by Lightning Source, UK.

The material contained herein is reprinted from a special edition of the International Journal of Safety and Security Engineering, Volume 6, Number 3, published by WIT Press.

Preface

This Volume contains a selection of papers presented at the International Conference on Flood Risk Management and Response which took place on San Servolo Island, Venice, organised by the Wessex Institute of Technology and Birmingham City University of the UK.

Flooding continues to be a major environmental hazard that affects all world regions. This has been demonstrated in recent months with flash floods in China, US, France, Germany and the UK bringing sever disruption and damage to communities. It is important that these communities become more resilient to flooding as it is inevitable that these events will become more prevalent in the future. The research findings published in these papers provide new insights into how this might be achieved, drawing on evidence from a range of international contexts and flood risk challenges.

A key feature of emerging flood risk management approaches is the need to accept that some flooding is inevitable and that communities need to become more resilient. This demands that community members take some responsibility and ownership of 'living with water'. For home owners and businesses this means taking steps to make their property more resilient to the effects of flooding, so that they can quickly return to normal in the aftermath of a flood. Indeed, retrofit and adaptation of the existing building stock represents a vitally important part of a long term integrated approach to flood risk management.

The papers in this Volume discuss all these topics. They contribute a valuable source of up to date information regarding the understanding of flood events and the appropriate response, together with the more recent studies on flood recovery and resilience.

The Editors
2017

Contents

ESTIMATION OF FLOOD RISK MANAGEMENT IN 17TH CENTURY ON OKAYAMA ALLUVIAL PLAIN, JAPAN, BY NUMERICAL FLOW SIMULATION

TADAHARU ISHIKAWA[1] & RYOSUKE AKOH[2]
[1]Dept. of Civil and Environmental Engineering, Hosei University, Japan.
[2]*Graduate School of Environmental and Life Science,* Okayama University, Japan.

ABSTRACT

In this study, the hydraulic function of the Hyakken-gawa Floodway, which was constructed in the 17th century to reduce the flood risk to Okayama Castle City, was evaluated by numerical flow simulation. The calculation conditions were determined by referring to the records about the floodway in old documents as well as from the numerical data based on the present conditions. The alluvial plain topography used for the inundation calculation was obtained from recent GIS data. The computation results showed that the flood control function of the floodway was composed of two stages: (1) Just before inundation occurred in the castle city, river water flowed into the floodway by collapsing the earthen dike located at the floodway head; this limited the increase in the river flow rate in the city area. (2) Just before the floodway capacity was exhausted, the backwater generated by the transverse masonry dikes in the floodway induced inundation to the paddy field region, which was on the side opposite to the castle city; this reduced the flood damage in the city area. The results also suggested that the civil engineers in the 17th century, who had neither the knowledge of modern hydraulics (including the concept of river flow rates) nor the advantage of using machinery for construction, presumably developed flood risk management strategies by combining the measures possible in those days. Finally, the results indicate that the floodway design used to distinguish the paddy fields from the urban areas for flood acceptance can be utilized for risk management against extraordinary floods even in present times.
Keywords: 17th century, distinction of land use, floodway function, historic flood risk management, numerical flow simulation, transfer of flood risk.

1 INTRODUCTION

In the 17th century, a long floodway named Hyakken-gawa was constructed to reduce the flood risk from the Asahi River to Okayama Castle City located on an alluvial plain in the mid-west part of Japan. Because the civil engineers of the day did not have the knowledge of modern hydraulics or the advantage of using large construction machinery, the planning and design was entirely empirical, and the dimensions of embankments were limited because only manual labor was employed. However, the main structure of the floodway survived until the end of the 19th century without much modification. Although the banks and the structures providing flow divergence were enlarged and reinforced with concrete in the 20th century, the design concept and framework of the system are still valid.

Taniguchi *et al.* [1] summarized the background and process of constructing Hyakken-gawa based on old documents and drawings from the standpoint of historical science. According to them, because of the lack of knowledge about velocity vectors, the engineers of the day wrongly believed that the flow divergence was controlled by the ratio of the cross-sectional

DOI: 10.2495/SAFE-V6-N3-455-465

areas of the channel. Imamoto *et al.* [2] assumed the flood hydrograph for the floodway design by comparing the flooding records for the 17th and 18th centuries with recent hydrological data and performed a hydraulic model test to investigate the flow diverging from the Asahi River to the floodway. However, the reliability of their experimental result was questionable because the river reach of the model was too short [3]. Therefore, the actual hydraulic function of the system for the reduction of flood risk in the castle city has not been fully understood yet.

In this study, a series of numerical flow simulations were performed to investigate the capacity of the floodway, the function of structures at the floodway head, the reduction of flood risk in the castle city, and the manner in which the civil engineers of the day managed floods larger than the channel capacity. A two dimensional shallow water equation formulated on an unstructured triangular mesh system was adopted for the numerical simulation. The calculation conditions were assumed by referring to the description and data on floods available in the literature [1, 2] as well as recent hydrological and surveying data. The simulation results were collected and arranged to estimate the flood risk management in the 17th century when the construction scale was limited and to discuss the possibility of applying these results to reduce the risk of floods exceeding the channel capacity, which may happen occasionally even in present times.

2 STUDY SITE

Figure 1 shows the location, an old map drawn in the late 17th century, and a recent aerial photograph of the study site. The white rectangle in Fig. 1c represents the area shown in Fig. 1b. The castle city was developed on the west side of the Asahi River, considering the bending part of the channel as a moat of the castle. However, because flooding often occurred at the bending, the local government planned the development of a floodway in the farmland on the east side of the river, which had wide paddy fields on low land and villages distributed at relatively high spots. The floodway was enclosed by banks located on the original ground, and these banks had some openings for irrigation channels and roads. During flooding, wooden blocks were inserted into the slots on the side face of the openings to close them.

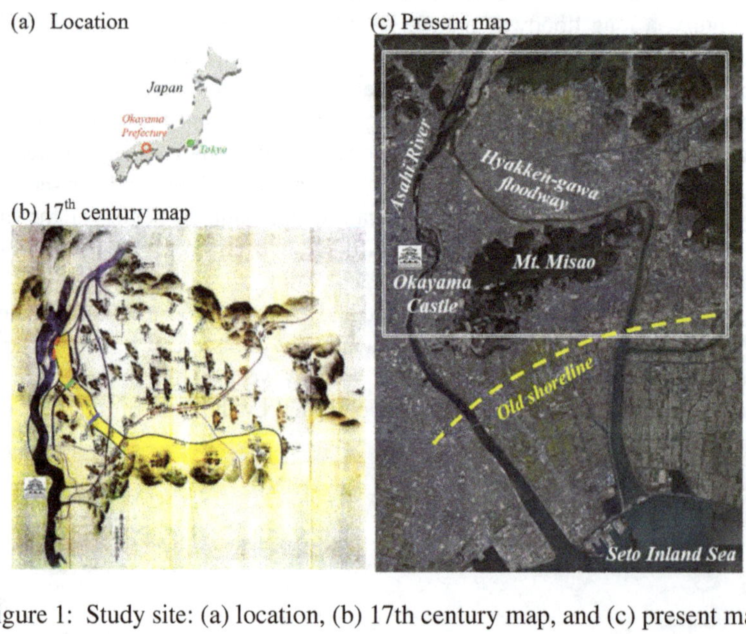

Figure 1: Study site: (a) location, (b) 17th century map, and (c) present map.

At present, the urban area extends to the east side of Hyakken-gawa. The floodway has a composite cross section to ensure large flow capacity; the low water channel is always filled with water, but the level of the high water channel is almost equal to the level of the outside ground, as was the case in the 17th century. The course and width of the river and floodway have not changed much since the 17th century. However, the shoreline has moved from the yellow broken line to the present position because of reclamation in the Seto Inland Sea.

The three thick colored lines marked on the floodway in Fig. 1b show the locations of transverse dikes that control the flow into the floodway. The first dike (red) was made of earth, and it was lower than the neighboring banks; hence, it was broken in a short time when the river water overtopped it. The main body of the second dike (green) was masonry, but it had a 30-cm-thick cap made of earth. The third dike was entirely masonry. The approximate heights of the first and second dikes were recorded in the early 20th century [4], but that of the third dike is not known because it was destroyed by a large flood in the 19th century. In this study, the height of the third dike was assumed to be the same as that of the second dike.

3 NUMERICAL MODEL

A set of shallow water equations was adopted for the numerical simulation model.

$$\frac{\partial h}{\partial t}+\frac{\partial (Uh)}{\partial x}+\frac{\partial (Vh)}{\partial y}=0 \tag{1}$$

$$\frac{\partial (Uh)}{\partial t}+\frac{\partial (UUh)}{\partial x}+\frac{\partial (UVh)}{\partial y} = -gh\frac{\partial H}{\partial x}+\frac{\partial (h\tau_{UU})}{\partial x}+\frac{\partial (h\tau_{UV})}{\partial y}-\frac{\tau_0}{\rho}\frac{U}{\sqrt{U^2+V^2}} \tag{2}$$

$$\frac{\partial (Vh)}{\partial t}+\frac{\partial (UVh)}{\partial x}+\frac{\partial (VVh)}{\partial y} = -gh\frac{\partial H}{\partial x}+\frac{\partial (h\tau_{UV})}{\partial x}+\frac{\partial (h\tau_{VV})}{\partial y}-\frac{\tau_0}{\rho}\frac{V}{\sqrt{U^2+V^2}} \tag{3}$$

where (U, V) are the velocity components in (x, y) coordinates, h is the water depth, H is the water surface level (= h + ground level), ρ is the water density, and g is the gravitational acceleration. τ_0 is the bed friction force, and τ_{UU}, τ_{UV} and τ_{VV} are the horizontal shear stresses, which are expressed by the following equations:

$$\tau_0=\rho U_f^2=n^2\frac{\rho g(U^2+V^2)}{h^{1/3}}$$

$$\tau_{UU}=2\varepsilon\frac{\partial U}{\partial x}-\frac{2}{3}k \quad \tau_{UV}=\varepsilon\frac{\partial U}{\partial y}+\varepsilon\frac{\partial V}{\partial x} \quad \tau_{VV}=2\varepsilon\frac{\partial V}{\partial y}-\frac{2}{3}k \tag{4}$$

$$\varepsilon=\frac{1}{6}kU_f h \qquad k=2.07U_f^2$$

where U_f is the friction velocity, n is Manning's roughness coefficient, ε is the vertically averaged eddy viscosity, k is the turbulent kinetic energy, and κ is the Karman constant (0.41).

The differential equations were converted to difference equations by the finite volume method on the unstructured triangular mesh system. The equation forms and solving process are described in a previous paper [4]. The triangular meshes were generated from the bank lines and the outline of the inundation area using the software ANSYS.

The flow rate over the dikes and banks was calculated using the following formula [5]:

$$q = \begin{cases} 0.35h_1\sqrt{2gh_1} & \textit{if } h_2/\mathrm{h}_1 \leq 2/3 \\ 0.91h_2\sqrt{2g(h_1 - h_2)} & \textit{otherwise} \end{cases} \tag{5}$$

where q is the flow rate over the unit length, and h_1 and h_2 are the water depths on the upstream and downstream sides, respectively. The height reduction rate of the earthen first dike and the earthen cap of the second dike was assumed to be 0.36 m/min based on an existing study [6].

4 CALCULATION CONDITION

4.1 Topography

A report on river improvement works since the 19th century [7] suggests that the bank lines of the Asahi River and Hyakken-gawa in the 17th century were almost the same as the present ones. However, because no information was available about the channel bed elevation of Asahi River in the 17th century, we adopted the recent surveying data of channel cross sections for calculation. On the other hand, the channel bed elevation of Hyakken-gawa was considered to be almost the same as that of the present high water channel, and the channel cross sections were obtained by removing the data for the low water channel from the recent surveying data. Manning's roughness was assumed to be 0.28 and 0.35 for the Asahi River and Hyakken-gawa Floodway, respectively, based on existing studies [8, 9].

According to the modern policy for flood risk management in Japan, the safety level should be kept equal for each river watershed, except for special cases such as the inside of flood-retarding basins. In the 17th century, on the other hand, uneven treatment depending on the land use was common because of poor flood control facilities. Actually, the Hyakken-gawa Floodway itself was expected to transfer the flood risk from the castle city to the farming area on the east. Therefore, we supposed that the east bank of Hyakken-gawa was a little lower than the west bank. The height of the east bank from the high water channel bed was assumed to be 3 m by measuring the remaining fragments of old banks (**Picture. 1**), whereas the overflow from the west bank was not considered. The topography of the alluvial plain was obtained from the GIS data (2001), which is shown in Fig. 2 along with the boundary for the numerical flow simulation. The total number of grid points for the calculation was 106,252.

4.2 Hydraulic condition

The highest water level at each big flood event since the 17th century was measured at the Kyobashi Station in the castle, and the critical water level for inundation was 4.5 *T.P.m* (*T.P.m*; standard unit of measurement for the elevation above the average sea level in Japan) [1]. A

Picture 1: Old bank

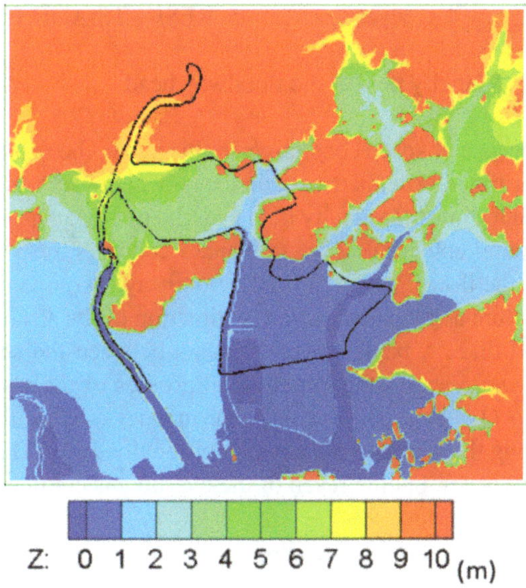

Figure 2: Topography of alluvial plain.

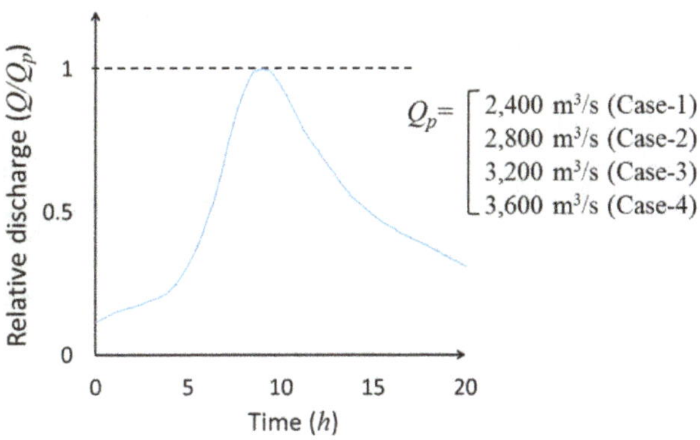

Figure 3: Flood hydrograph used for numerical simulation.

preliminary test using one-dimensional flow calculation with $n = 0.028$ for the Asahi River showed that the flow divergence to Hyakken-gawa started before the water level reached the critical line, as was mentioned in the old document; the preliminary test also showed that the flood discharge at the critical water level was 2,000 m^3/s [3]. Therefore, flood events larger than this level were considered in this study. A series of assumed flood hydrographs were obtained by changing the scale of the hydrograph that is adopted by present flood prevention works (Fig. 3).

The downstream ends considered in the calculation were the river mouths at the old shoreline shown in Fig. 1c, where the water level was assumed to be 0.25 *T.P.m*, which is the average tidal level recorded at the nearest tidal station.

5 RESULTS AND DISCUSSION

5.1 Velocity field at the floodway head

Because of space constraints in this paper, the results from only Case-3 (Q_p = 3,200 m^3/s) are illustrated in detail to explain the typical flow characteristics in the floodway system. Figure 4a and b show the velocity field near the floodway head (arrows) and the ground elevation (background colors) at Q = 2,600 and 3,200 m^3/s, respectively. The first dike was already broken by the flow intruding into the floodway (two openings along the separating bank). The flow vectors are changed from the direction of the Asahi River to that of the floodway by the second dike. A bank overflow occurs between the second and third dikes at Q = 3,200 m^3/s. Figure 5 shows the longitudinal profiles of the water surface near the floodway head at the two Q values. The second dike induces a hydraulic jump, reducing the energy of the flow along the downstream channel of the floodway. The third dike induces a side overflow in the backwater, but the reach of the overflow is confined to the section between the two dikes.

5.2 Inundation on the alluvial plain

Figure 6a shows the hydrograph for the floodway (red line) along with the flood hydrograph for the region upstream of the Asahi River (blue line) in Case-3 (Q_p = 3,200 m^3/s). Flow

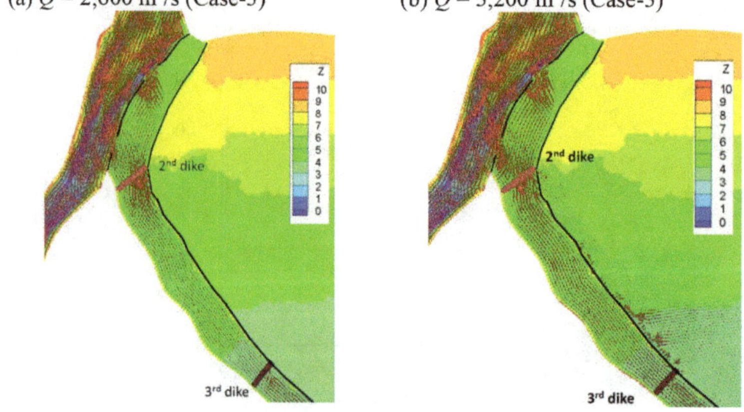

Figure 4: Velocity field near floodway head (Case-3, Q_p = 3,200 m^3/s).

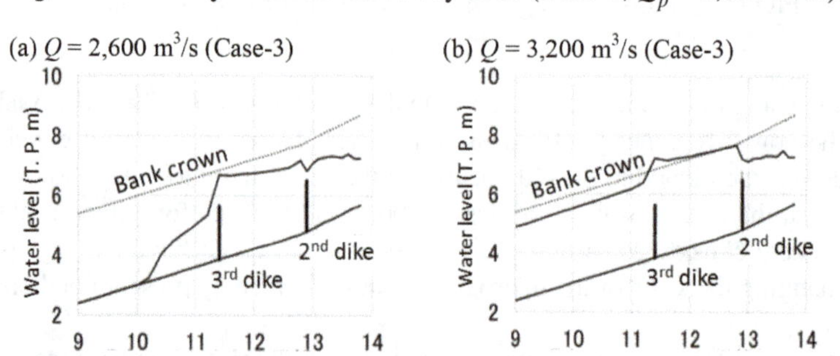

Figure 5: Longitudinal water surface profiles (Case-3, Q_p = 3,200 m^3/s).

divergence started when the river discharge exceeded 2,000 m^3/s, and a maximum divergence of 800 m^3/s was attained at the flood peak. Bank overflow started when the floodway discharge reached 500 m^3/s. Figure 6b and c show the inundation progress on the alluvial plain for the two stages shown in Fig. 6a and d shows the maximum inundation depth during the flood event. The flood water flowed through the section between the two dikes along the channel bank following the ground elevation. The maximum water depth was less than 40 cm except in the area close to the third dike, which is acceptable for a paddy field if the inundation does not last very long.

It should be noted that a famous big shrine is located near the floodway head, which was built by the feudal lord of this district before the construction of Hyakken-gawa. The governing classes offered worship at the shrine for a long time. Although the shrine was located on the alluvial plain, it was not hit by the flood water discharged between the two dikes because of the mild ground slope from the north to the south.

5.3 Reduction of flood risk in the castle city

Figure 7 shows the calculated water level at the Kyobashi Station in the castle city for the peak discharge (Q_p) of each flood. The horizontal dotted line (4.5 *T.P.m*) is the critical water level above which the city would be inundated. The blue line shows the result for the condition before the construction of the Hyakken-gawa Floodway. The red line shows the result for the condition

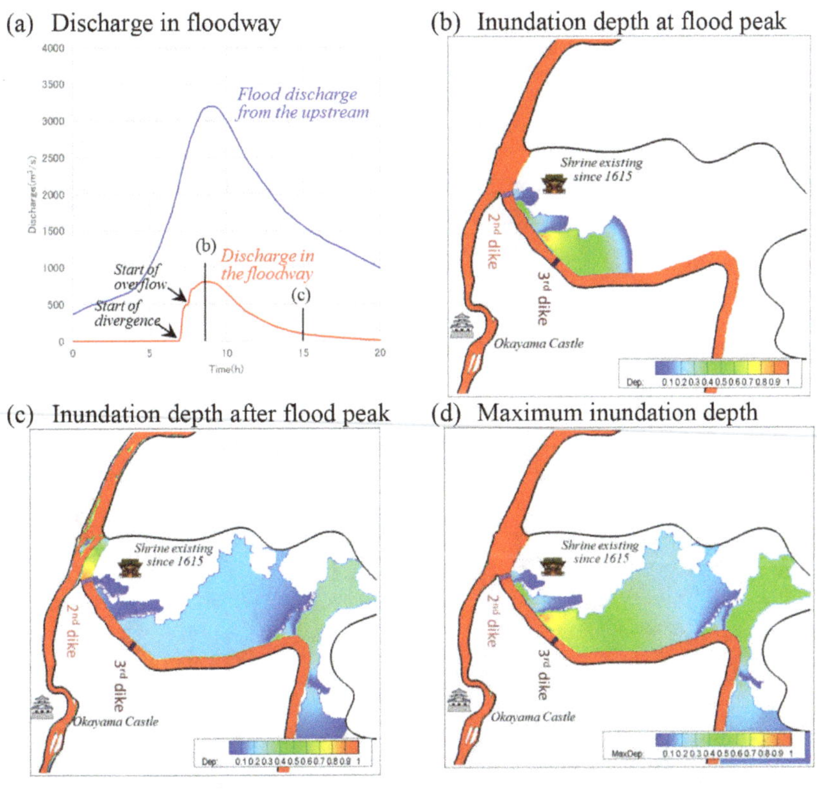

Figure 6: Inundation due to overflow of Hyakken-gawa.

in which a part of the flood is diverted to Hyakken-gawa, but bank overflow is not considered in the calculation. The green line shows the case in which the bank overflow is considered.

At Q_p = 2,400 m^3/s, the highest water level before the floodway construction was very close to the critical line, and the use of the floodway decreased the water level by approximately 50 cm, which ensured that the city was safe. At Q_p = 2,800 m^3/s, the highest water level without the floodway exceeded the critical line, but it fell below the critical line because of the effect of the floodway and further decreased by approximately 50 cm when bank overflow occurred in the floodway. At Q_p = 3,200 m^3/s, the water level exceeded the critical line even with the discharge to the floodway, but it fell below the critical line with the bank overflow in the floodway. At Q_p = 3,600 m^3/s, flooding occurred in the castle city, but the volume of inundation (Fig. 7d, color shading) was quite small, and the rate of decrease in the water surface was higher because of the bank overflow in the floodway.

Figure 8 shows the dependence of the highest water level at the Kyobashi Station on the peak discharge from the upstream region (Q_p). The original channel capacity was about 2,400 m^3/s. Flow divergence to the floodway started at $Q_p \sim$ 2,000 m^3/s when the water level was still 50 cm below the critical line. The Hyakken-gawa Floodway was effective in preventing inundation in the castle city only when Q_p was below 2,800 m^3/s. However, a part of the flow through Hyakken-gawa was released to the east side farmland by bank overflow at $Q_p \sim$ 2,600 m^3/s when the water level at the Kyobashi Station was still 50 cm below the critical line. The overall effectiveness of the floodway including the effectiveness attributed to the effect of the bank overflow continued until Q_p was around 3,200 m^3/s. The result suggests that the flood risk reduction was composed of two stages; the flow divergence to the floodway which was

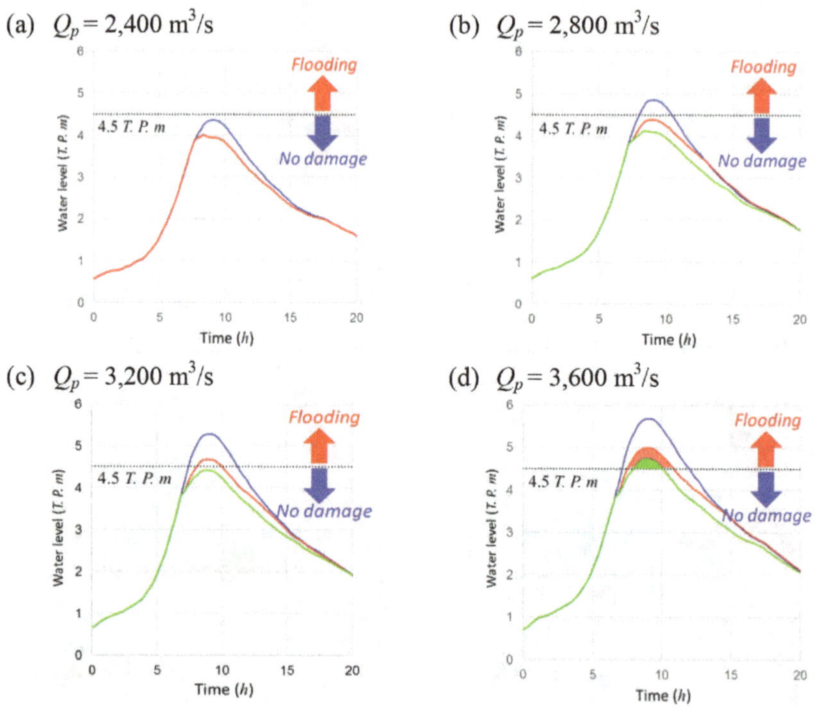

Figure 7: Calculated hydrographs at Kyobashi station.

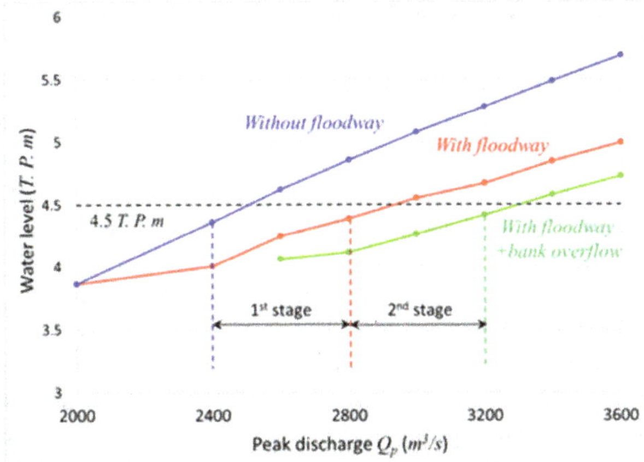

Figure 8: Dependence of highest water level on flood scale at Kyobashi Station.

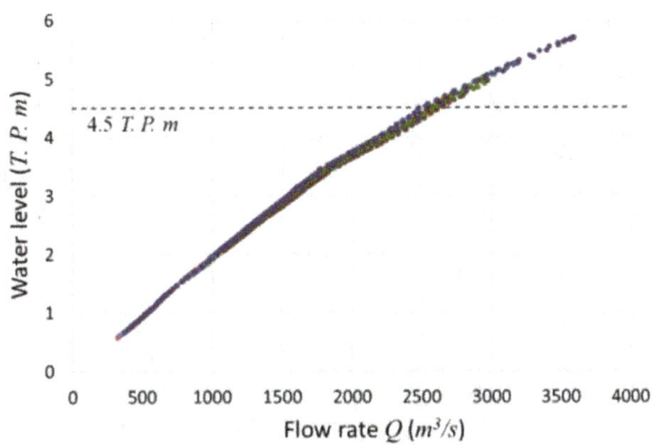

Figure 9: Dependence of water level on flow rate at Kyobashi Station.

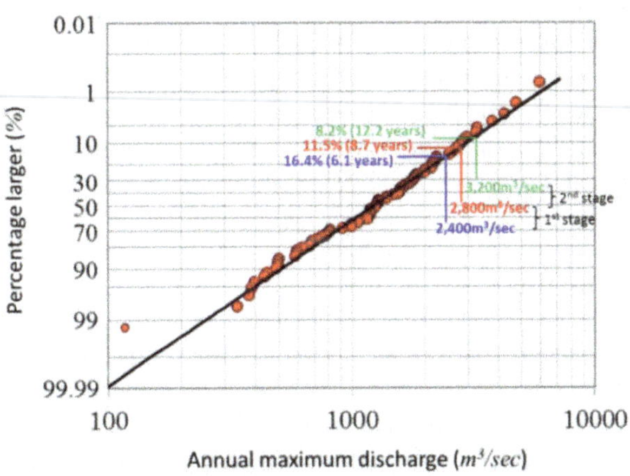

Figure 10: Improvement of safety level.

effective to Q_p ~ 2,800 m^3/s and the bank overflow to the east side farmland which was effective to Qp ~ 3,200 m^3/s.

Figure 9 shows the relationship between the water level and channel discharge recorded at the Kyobashi Station; in this figure, the entire calculation results during all the flood events are plotted together. All the data points approximately lie along a single smooth curve, and the hysteresis due to the unsteady variation in the flow discharge is very weak. This fact suggests that the result described in Fig. 8 will be valid irrespective of the shape of the hydrograph.

The probability plot of the annual maximum discharge in the Asahi River for the last 70 years is shown in Fig. 10 in the form of a Hazen Plot. The data approximately follow the log-normal distribution. Assuming that the probability in the 17th century followed the regression line shown in the figure for the present data, the return period of flooding in the castle city increased from 6.1 to 8.7 years because of the first stage and from 8.7 to 12.2 years because of the second stage.

6 CONCLUSION

In this study, the function of the Hyakken-gawa Floodway in reducing the flood risk in Okayama Castle City in the 17th century was investigated by performing a series of numerical flow simulations using a shallow water model. The major conclusions are as follows:

1. The flood risk reduction using the floodway was composed of two stages. The first stage was the flow divergence to the Hyakken-gawa channel, which was induced by the premeditated collapse of the first dike made of earth at the floodway head. The second stage was the inundation of the east side farmland, which was induced by the bank overflow due to the backwater from the third dike. Note that the function of the second dike was to induce a hydraulic jump for flow energy dissipation and to smoothly change the flow direction toward the floodway axis.
2. The first stage of flood control increased the flood discharge safety level of the castle city against inundation from 2,400 to 2,800 m^3/s, which corresponds to an increase in the return period from 6.1 to 8.7 years. The second stage of flood control involving the inundation of the east side farmland further increased the flood discharge safety level to 3,200 m^3/s and the return period to 12.2 years.
3. The second stage of flood control was realized by suffering the loss of harvest in the farmland roughly once in ten years. Old documents indicate that farmers occasionally petitioned the local government through the regional magistrate, to reinforce the Hyakken-gawa banks, and that the local government gave tax remissions to the 12 farming villages on the east side alluvial plains [1]. These facts suggest that the regional administration gave comprehensive considerations to the farmers and followed suitable procedures while undertaking the transfer of flood risk from the castle city to the farmland through the floodway system.
4. According to the modern policy for flood risk management in Japan, the safety level is kept constant for each river basin, although the land condition changes along the channel. However, recent changes in the global climate might increase the frequency of floods that are larger than the channel capacity under the present regulation for flood risk management. Hence, to reduce the total flood damage on floodplains inundation should be controlled by considering the distribution of land cover and properties on such floodplains. The revaluation of past civil engineering works that used only manual labor would contribute to the creation of a new phase of flood risk management in the near future.

ACKNOWLEDGEMENTS

We would like to thank Emeritus Professor Hideo Kikkawa of Tokyo Institute of Technology for his valuable suggestions pertaining to this study. We would also like to thank Editage (www.editage.jp) for providing English language editing. This study was supported by a grant from the RIVER FOUNDATION (Number 27-1212-006).

REFERENCES

[1] Taniguchi, S. et al.: The history of Hyakken-gawa, Office for construction and administration of Asahi River, Ministry of Construction, 1978.

[2] Imamoto, H., Ishigak, T., Baba, Y. & Tanaka, N., Experimental study on river improvement for the Asahi River in Edo-era. *Historical Studies in Civil Engineering, JSCE*, **15**, pp. 119–128, 1995.
http://dx.doi.org/10.2208/journalhs1990.15.119

[3] Ishikawa, T., Akoh, R., Kikkawa, H. & Kobayashi, Y., A hydraulic study on the design of Hyakken River Construction in the Edo Period. *Journal of Hydraulic Engineering, JSCE* ser. B1, **71**(4), pp. 613–618, 2015.
http://dx.doi.org/10.2208/jscejhe.71.I_613

[4] Zeng, X., Akon, R., Ishikawa, T. & Nakamura, T., On the efficiency of GPGPU acceleration of Tsunami simulation using an unstructured triangular mesh system. *Journal of Hydraulic Engineering, JSCE ser. B1*, **69**(4), pp. 619–624, 2013.
http://dx.doi.org/10.2208/jscejhe.69.I_619

[5] Honma, H., Coefficient of flow volume on low overflow weir. *Proceeding of JSCE*, **26**(6), pp. 635–645, 1940.

[6] Suga, K, Ishikawa, T. & Kasai, T., Report of hydraulic test on bank collapse by overtopping flow, part of data analysis. PWRI Report, Ministry of Construction, 1760, 1982.

[7] Okayama Office for River Improvement Works: Records of the Past Improvement of Hyakken-gawa, Ministry of Construction, 1983.

[8] Maeno, S., Miyauchi, Y. & Mori, T., Effect of vegetation on flood flow in the asahi river. *Journal of Hydraulic Engineering, JSCE*, **48**, pp. 757–762, 2004.
http://dx.doi.org/10.2208/prohe.48.757

[9] Ishikawa, T., Yoshida, K. & Minoura, Y., Application of a wuasi-3d numerical model to a flood flow observed in the lower tone river in september. 1998, *Journal of Hydraulic Engineering JSCE ser. B1*, **69**(4), pp. 619–624, 2013.

SLOWING THE FLOW IN PICKERING: QUANTIFYING THE EFFECT OF CATCHMENT WOODLAND PLANTING ON FLOODING USING THE SOIL CONSERVATION SERVICE CURVE NUMBER METHOD

H. THOMAS & T.R. NISBET
Centre for Ecosystems, Society & Biosecurity, Forest Research, UK.

ABSTRACT
The Soil Conservation Service (SCS) Runoff Curve Number method has been successfully applied to the Pickering Beck catchment at Pickering in North Yorkshire to assess the impact of land use change on flood flows. While limited-scale woodland creation (3% of the catchment) was predicted to have a small effect on the range of peak flows studied (<1% to 4% reduction), in line with previous model applications in the catchment, the conversion of the 25% cover of existing woodland to improved grassland produced a large increase in peak flow, up to 41% for a 1 in 100-year event. These numbers need to be treated with particular caution since the SCS method remains to be validated for UK conditions, however, they support growing evidence that woodland creation and management could have a significant role to play in flood risk management. The SCS method provides a potentially powerful tool for evaluating the impact of land-use change and management on flood runoff, as well as for identifying areas where such measures could be most effective.
Keywords: curve number method, flooding, land-use change, natural flood management, rainfall-runoff modelling.

1 INTRODUCTION

Recent flooding incidents in the UK have put immense pressure on existing flood defences, and have strengthened the case for the implementation of Natural Flood Management (NFM) measures in catchments to complement and reduce the strain on these hard-engineered forms of protection.

To develop a better understanding of how flood flows are generated across catchments and where NFM measures, in particular woodland planting, would be most effective, a hydrological assessment has been undertaken using the Soil Conservation Service Runoff Curve Number **(SCS CN)** methodology reported in Urban Hydrology for Small Watersheds TR-55 [1]. This method was originally developed to facilitate the hydrological assessment of catchments in North America subject to land use change, principally urbanisation.

The principles of the approach were adapted to quantify the effects of woodland planting measures in the Pickering Beck catchment in North Yorkshire, UK, as part of the "Slowing the Flow at Pickering" project [2]. A spatial application using a Geographical Information System (GIS) was used to identify key areas of runoff generation within the catchment. This offers a useful tool to help target woodland planting to where it can be most effective at alleviating downstream flood risk.

DOI: 10.2495/SAFE-V6-N3-466-474

2 MATERIALS AND METHODS

2.1 Study area and the slowing the flow at pickering project

The 'Slowing the Flow at Pickering' project was established in April 2009 to look at how changes in land use and land management can help to reduce flood risk in the town of Pickering in North Yorkshire. Pickering has a long history of flooding, with four floods in the last fifteen years (1999, 2000, 2002 and 2007). The 2007 flood was the most serious to date, causing an estimated £7 million of damage to residential and commercial properties. Whilst a flood alleviation capital scheme had been proposed to alleviate the problem, a cost-benefit analysis showed this to be unaffordable when set against national cost-benefit thresholds and other priorities. The use of high flood walls was also not favoured by the local community due to aesthetic impacts on the town. The project represented a new approach to flood management that sought to work with natural processes to help reduce the risk of flooding for affected communities.

A crucial element of the project was to understand better how floods are generated in the catchment and how the way the land is used and managed affects the speed and volume of flood flows. The combined effects of past land management practices by humans over centuries are thought to have increased flood risk by promoting rapid runoff and increasing siltation within river channels, although more recent changes, for example afforestation since 1945, may have partly offset this trend. There are four principal land uses in the c. 68 km^2 catchment of the Pickering Beck that drains to Pickering, comprising forest, arable, heather moorland and improved grassland. The overall aim of the project was to demonstrate how land-use change and improvements in management practices could help to restore the catchment's natural flood attenuation capacity.

2.2 Data sources

Land use/cover data were extracted from the LCM2000 Land Capability Map [3] dataset and merged with the Hydrology of Soil Types (HOST) dataset [4], which was then used to assign appropriate runoff curve numbers according to the SCS CN method (detailed in Section 2.3 below). A 1:50,000 Digital Elevation Model was used to delineate the catchment boundary for modelling purposes.

Rainfall and flow data were obtained from two Environment Agency gauges situated within the catchment whilst design rainfall events were generated using the Revitalised Flood Estimation Handbook (ReFH) methodology using appropriate catchment parameters.

2.3 Soil conservation service runoff curve number model

The SCS Curve Number model has been widely used internationally for water resources management and planning [5–9]. Halcrow [10] applied the SCS CN method within a UK context to establish baseline hydrological conditions in the Allan Water catchment in Scotland.

The model estimates precipitation excess as a function of cumulative precipitation, soil cover, land use, and antecedent moisture, using the following equation:

$$Q = \frac{(P - Ia)^2}{(P - Ia) + S} \tag{1}$$

Where

Q = runoff

P = rainfall

Ia = the initial abstraction (initial loss)

S = potential maximum retention, a measure of the ability of a catchment to abstract and retain storm precipitation.

Initial abstraction (Ia) comprises all water losses before runoff begins. It includes water retained in surface depressions, water intercepted by vegetation, evaporation, and soil infiltration. Ia is highly variable but generally well correlated with soil and land cover parameters. Through studies of many small agricultural watersheds in North America, Ia was found to be approximated by the following empirical equation:

$$Ia = 0.2S \tag{2}$$

S is related to the soil and land cover conditions of the catchment through the CN. CN has a range of 0 to 100, and S is related to CN by:

$$S = \left(\frac{25400 - 254CN}{CN}\right) \text{ (SI units)} \tag{3}$$

CN values range from 100 (open water) to approximately 30 for permeable soils with high infiltration rates.

2.4 Estimating curve number

The CN for a catchment can be estimated as a function of soil type and land cover using tables published by the SCS in the United States Department of Agriculture, Urban Hydrology for Small Watersheds TR 55 (1986) report.

For a catchment that consists of several soil types and land uses, a composite CN is calculated as:

$$CN_{composite} = \frac{\sum AiCNi}{\sum Ai} \tag{4}$$

In which CNcomposite = the composite CN used for runoff volume computations; i = an index of catchment subdivisions of uniform land use and soil type; CNi = the CN for subdivision i; and Ai = the drainage area of subdivision i.

The method classifies soils into one of four hydrologic soil groups (HSGs; A, B, C and D) based on their hydrological characteristics. This includes the infiltration rate (the rate at which water enters the soil at the surface) and the transmission rate (the rate at which water moves within/through the soil). Approximate numerical ranges for both rates were first published by Musgrave [11]. The four groups are defined as follows:

1. Group A (low runoff potential): soils with low runoff potential and high infiltration rates even when thoroughly wetted. High rates of water transmission (greater than ~8 mm/h).
2. Group B: soils with moderate infiltration rates when thoroughly wetted. These soils have a moderate rate of water transmission (~4 to 8 mm/h).

Table 1: Conversion used for HOST soil classes to SCS soil classes.

Standard Percentage Runoff (HOST) (%)	SCS Soil Class	HOST soil class
<10	A	1, 2, 4, 11, 13
10–20	B	3, 5
20–40	C	6, 9, 10, 14, 16, 17, 18, 24
>40	D	7, 8, 12, 15, 19, 20, 21, 22, 23, 25, 26, 27, 28, 29

Table 2: Curve numbers assigned to land use and soil classes using the SCS land use descriptions and cross referencing these to UK land cover types present in the catchment.

LCM2000 Description	SCS Land Use Description	Soil A	Soil B	Soil C	Soil D
Woodland	Fair, woods	36	60	73	79
Arable & horticulture	Good, fallow ground	74	83	88	90
Grassland	Poor, pasture	68	79	86	89
Set-a-side grassland	Meadow	30	58	71	78
Bracken	Fair, brush	35	56	70	77
Dwarf shrub heath	Good, brush	48	67	77	83
Inland water	Open water	100	100	100	100
Bare ground	Fallow, bare soil	77	86	91	94
Suburban	Residential	61	75	83	87
Urban	Urban	89	92	94	95

3. Group C: soils with low infiltration rates when thoroughly wetted. These soils have a low rate of water transmission (~1 to 4 mm/h).
4. Group D: soils with high runoff potential. They have very low infiltration rates when thoroughly. These soils have a very low rate of water transmission (<1 mm/h).

The HSGs were converted to fit the UK's HOST classification system by Halcrow [10] based on Standard Percentage Runoff values, as shown in Table 1.
Curve numbers were assigned to different combinations of land use and soil classes using the SCS land use descriptions and cross referencing these to UK land cover types present in the catchment (Table 2). Different values are available for each land cover type depending on general condition, ranging from poor, fair to good. Numbers were selected based on local judgement. This allowed a SCS Curve Number grid to be generated for the catchment and for weighted average curve numbers to be determined for each sub-catchment represented in the model.

2.5 Rainfall-runoff modelling using HEC-HMS and the SCS CN method

The Hydrologic Modeling System (HEC-HMS) is designed to simulate the rainfall-runoff processes within catchments. It is applicable to diverse geographic areas for solving a wide range of problems, including managing flood flows draining small urban and headwater catchments to large river basins.

HEC-HMS is a numerical model that can be used to simulate catchment and channel behaviour and thereby predict river flows and response times. It was applied in this study to run the SCS CN method to investigate the impact of land use change on flood runoff within the Pickering Beck catchment.

2.6 Calibration of the SCS CN method to the revitalised flood estimation (ReFH) method and measured June 2007 flood event

The SCS Runoff Curve Number method and HEC-HMS model were used to generate a set of predicted flood growth curves based on the existing mix of soil types and land covers in the catchment. The results are compared with the growth curve using the observed gauged flow data along with those derived from the FEH Revitalised Flood Hydrograph Method and FEH Pooling Group analysis in Fig. 1. This shows that the method provides a reasonable fit in the middle of the flow range but both under-predicts smaller flood events and over-predicts larger ones.

These discrepancies could be due to the following factors:

1. The CN values are derived from US studies and catchment conditions. While an attempt was made to translate values to UK soil and land cover types, as well as general management condition, this is based on judgement rather than calibrated values.
2. The use of the HOST data set and mapped Soil Associations to assign catchment soils to HSGs within the SCS model introduces significant uncertainties linked to the scale of mapping and soil variability. This could introduce marked local discontinuities in runoff conditions that are not reflected by the model.

Early development of the SCS CN method established that there could be significant variability in the curve number, and that the same catchment could have more than one curve number, or indeed, a set of curve numbers [12]. Ponce & Hawkins [13] stated that the likely

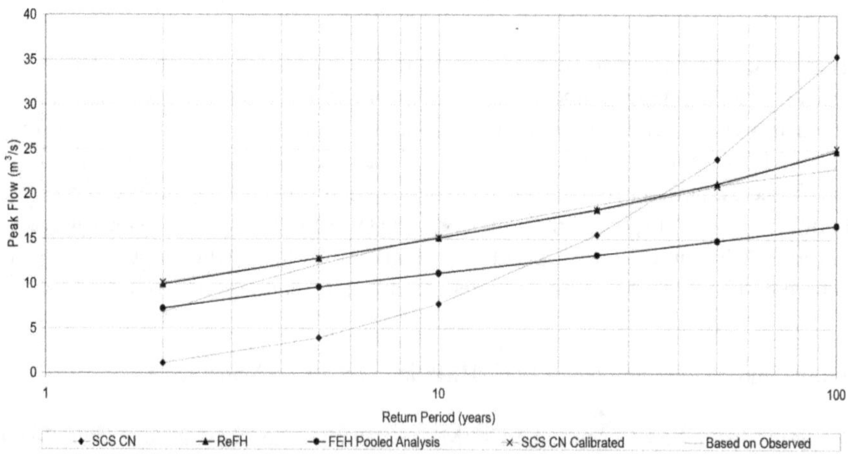

Figure 1: Comparison of the flood growth curves of the Pickering Beck catchment from five different analyses.

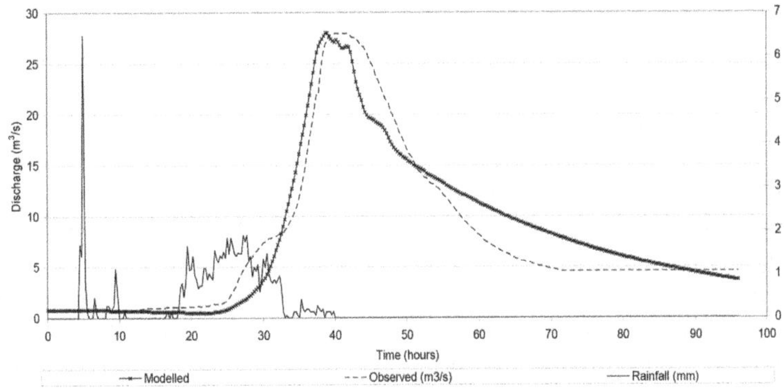

Figure 2: Modelled versus observed hydrograph for the June 2007 flood event at Pickering Beck gauging station.

sources of this variability are the spatial variability of storm and catchment properties, temporal variability of storm rainfall, i.e. storm intensity, and, differing antecedent rainfall and associated soil moisture conditions.

In order to address this variability and improve model fit, a calibration exercise was undertaken to scale the original CN values for each land cover and soil type combination so that the modelled peak flows better matched those of the ReFH growth curve. This involved applying a separate scaling factor for each return period until a satisfactory peak flow was achieved.

The adjusted CN values were applied to the June 2007 flood event, the highest recorded flow on record at the Pickering Beck gauging station, to compare the modelled versus observed hydrograph (Fig. 2) This gave an acceptable fit, with peak flows of 28 and 27.9 m3/s and total hydrograph volumes of 2,177,603 and 2,253,960 m^3 for predicted and observed conditions, respectively.

3 RESULTS AND DISCUSSION

3.1 The effect of land use change on runoff generation

The SCS CN methodology was applied to the Pickering Beck catchment to assess the effect of woodland planting measures in the catchment. The following scenarios were applied:

1. Baseline land use scenario using the existing land cover map, showing 16.2 km^2 of woodland in the catchment
2. An additional 7 km^2 of woodland planting in the catchment, assuming all previously mapped opportunities were achievable.
3. An additional 0.2 km^2 of woodland representing that achieved in the catchment since the Slowing the Flow project commenced in 2009.
4. Conversion of all existing woodland (16.2 km^2) to improved grassland
4. Conversion of all existing woodland to Dwarf Shrub Heath.

The FEH was used to generate design 24-hour rainfall events for given return periods as shown in Table 3.

Figure 4 shows the effect of the two woodland planting scenarios on peak flows for the range of return periods. The results for Scenario 2 (opportunity mapping) predicted a 14% decrease in peak flow for more frequent events (e.g. 1 in 2 year) and a 6% decrease for the less frequent 1 in 100-year flood. The small areas of woodland planted under Scenario 3 produced a 4% decrease for predicted peak flows to 1 in 25 years and a <1% decrease for larger events. These numbers are relatively similar to those generated by a separate modelling study by Odoni and Lane [14] using a different, combined hydrology-hydraulic model (OVERFLOW).

Existing improved grassland amounts to ~12 km^2 of the catchment (18%), while dwarf shrub heath covers ~3.4 km^2 (5%). Scenarios 4 and 5 involved converting all 16.2 km^2 (~25%) of existing woodland to each of these land covers.

As expected, converting all woodland in the catchment to improved grassland would have a much larger effect on peak flows, increasing more frequent floods by as much as 87% (1 in 2 year flood) and up to 41% for less frequent events (1 in 100 year). Conversion of woodland to dwarf shrub heath had a smaller effect, increasing peak flows for a 1 in 2-year flood by 27% and by 14% for a 1 in 100-year event. These represent large reductions and are above or at the top end of ranges reported by other UK modelling studies based on the scale of land cover change involved [15]. In view of the scaling required for the calibration exercise and the fact that the method remains to be validated under UK conditions, these results need to be treated with caution. However, they do add further support to the growing evidence that woodland creation and management could have a significant role to play in flood risk management.

Table 3: FEH-generated design 24-hour rainfall events.

Return period	Rainfall depth (mm)
2	30.2
5	38.7
10	45.7
25	56.3
50	65.7
100	76.5

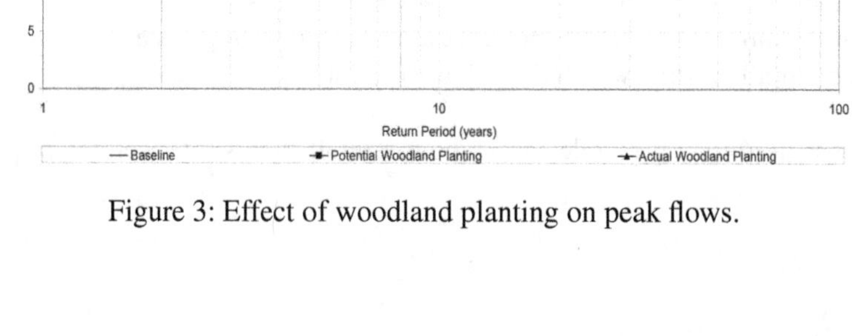

Figure 3: Effect of woodland planting on peak flows.

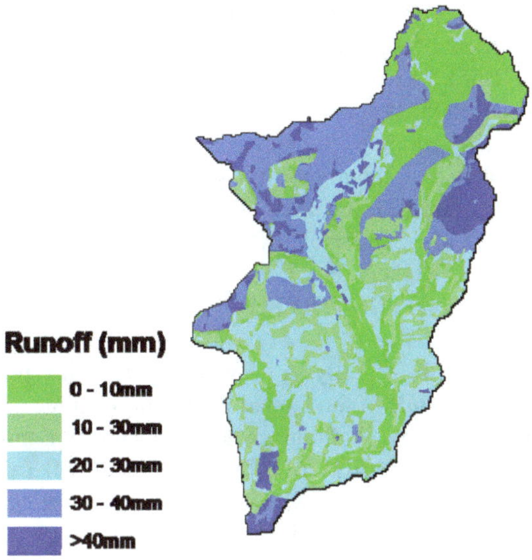

Figure 4: Runoff from June 2007 rainfall event (total rainfall 85.5 mm).

3.2 Spatial representation of the SCS CN method

The SCS CN method can be used to identify areas within catchments that generate the greatest volumes of runoff. This could provide a very useful tool for targeting NFM measures, such as woodland planting, to where they would be most effective, as well as allow different options to be quantified and compared.

As an example, Fig. 4 shows the spatial distribution of runoff within the Pickering Beck catchment for the June 2007 rainfall event, based on the modelled runoff from each 50 m by 50 m grid cell. The greatest runoff (>40 mm in some areas) for this 85.5 mm storm event occurred on areas predominantly covered with improved grassland and heath in the northeast of the catchment, but also from existing woodland in the northwest (as well as built up areas to the south).

4 CONCLUSIONS

The Soil Conservation Service (SCS) Runoff Curve Number method has been successfully applied to the Pickering Beck catchment at Pickering in North Yorkshire to assess the impact of land use change on flood flows. While limited-scale woodland creation (3% of the catchment) was predicted to have a small effect on the range of peak flows studied (<1% to 4% reduction), in line with previous model applications in the catchment, the conversion of the 25% cover of existing woodland to improved grassland produced a large increase in peak flow, up to 41% for a 1 in 100 year event. These numbers need to be treated with particular caution since the SCS method remains to be validated for UK conditions, however, they support growing evidence that woodland creation and management could have a significant role to play in flood risk management. The SCS method provides a potentially powerful tool for evaluating the impact of land use change and management on flood runoff, as well as for identifying areas where such measures could be most effective.

REFERENCES

[1] United States Department of Agriculture Urban hydrology for small watersheds, Technical Release 55 (TR-55), 2nd edn., Natural Resources Conservation Service, Conservation Engineering Division, 1986.

[2] Nisbet, T.R., Roe, P., Marrington, S., Thomas, H., Broadmeadow, S. & Valatin, G., Slowing the flow at pickering, final report phase II for the department of environment, food and rural affairs (Defra), Project RMP5455, Defra, London, 2015.

[3] Fuller, R.M., Smith, G.M., Sanderson, J.M., Hill, R.A., Thomson, A.G., Cox, R., Brown, N.J., Clarke, R.T., Rothery, P. & Gerard, F.F., Countryside Survey 2000 Module 7 LAND COVER MAP 2000 FINAL REPORT CSLCM/Final, Centre for Ecology and Hydrology, 2002.

[4] Boorman, D.B., Hollis, J.M. & Lilly, A., Hydrology of soil types: a hydrologically based classification of the soils of the United Kingdom, Institute of Hydrology Report No.126, Institute of Hydrology, Wallingford, 1995, available at http://www.ceh.ac.uk/products/publications/documents/IH126HYDROLOGYOFSOILTYPES.pdf

[5] Hawkins, R.H., Runoff curve numbers with varying site moisture. *Journal of the Irrigation and Drainage Division, ASCE,* **10**(4), pp. 389–398, 1978.

[6] Ragan, R.M. & Jackson, T.J., Runoff synthesis using Landsat and SCS model. *Journal of Hydrologics Division, ASCE*, **106**, pp. 667–678, 1980.
http://dx.doi.org/10.1111/j.1752-1688.1980.tb02504.x

[7] Slack, R.B. & Welch, R., Soil conservations service runoff curve number estimates from Landsat data. *Water Resources Bulletin*, **16**(5), pp. 887–893, 1980.

[8] Hawkins, R.H., Asymptotic determinations of runoff curve numbers from data. *Journal of Irrigation and Drainage Engineering*, **119**, pp. 334–345, 1993.
http://dx.doi.org/10.1061/(ASCE)0733-9437(1993)119:2(334)

[9] Lewis, M.J., Singer, M.J. & Tate, K.W., Applicability of SCS curve number method for a California Oak Woodlands watershed. *Journal of Soil Water Conservation*, **55**(2), pp. 226–230, 2000.

[10] Halcrow, Allan water natural flood management techniques and scoping study, Report to Scottish environment agency, 2011.

[11] Musgrave, G.W., How much of the rain enters the soil? *Water Yearbook of Agriculture*, U.S. Department of Agriculture: Washington, DC. pp. 151–159, 1955.

[12] Hjelmfelt, A.T., Investigation of curve number procedure. *Journal of Hydraulic Engineering. ASCE*, **117**(6), pp. 725–737, 1991.
http://dx.doi.org/10.1061/(ASCE)0733-9429(1991)117:6(725)

[13] Ponce, V.M. & Hawkins, R.H., Runoff curve number: has it reached maturity? *Journal of Hydrologic Engineering*, **1**, pp. 11–19, 1996.
http://dx.doi.org/10.1061/(ASCE)1084-0699(1996)1:1(11)

[14] Odoni, N.A. & Lane, S.N., Assessment of the impact of upstream land management measures on flood flows in Pickering Beck using Overflow. Forestry Commission, [Online], 2010, available at http://www.forestry.gov.uk/pdf/Pickering_crim_report_April_2010.pdf/$FILE/Pickering_crim_report_April_2010.pdf

[15] McIntyre, N.L. & Thorne, C.R. (eds), Land use management effects on flood flows and sediments – guidance on prediction CIRIA report C719 construction industry research and information association, London, 2013.

FLOOD RISK ASSESSMENT AND PRIORITISATION OF MEASURES: TWO KEY TOOLS IN THE DEVELOPMENT OF A NATIONAL PROGRAMME OF FLOOD RISK MANAGEMENT MEASURES IN MOLDOVA

E. FRANK[1], D. RAMSBOTTOM[2] & A. AVANZI[1]
[1]BETA Studio srl, Italy.
[2]HR Wallingford Ltd, UK.

ABSTRACT

Following severe floods in 2008 and 2010, the Government of Moldova requested assistance to improve flood protection throughout the country. The European Investment Bank has funded a Technical Assistance project to develop a programme of flood risk management measures. The project included the detailed 2D hydraulic modelling of the high-risk rivers (about 3400 km) to produce flood hazard and flood risk maps, the identification of measures to reduce the flood risk, the prioritisation of measures and the development of a phased investment programme and a Short-Term Investment Plan. Flood risk was assessed using 12 flood risk indicators representing social, economic and environmental impacts of flooding. Prioritisation of measures took account of: (i) the urgency of the measure; (ii) the magnitude of the risk that can be reduced with the measure; (iii) the benefit-cost ratio of the measure. The approaches used and in particular the methodologies implemented and the results obtained for flood risk assessment and for prioritisation of measures proved to be valuable tools to reach the objective of the study and, in particular, to facilitate the discussion with the stakeholders and the decision-making process.

Keywords: flood management, flood risk assessment, hydraulic modelling, measures, national programme.

1 INTRODUCTION

Flooding of river systems is one of the most frequent and costly natural hazard affecting the majority of the world's countries on a regular basis [1].

Analysis of past flood events shows that the majority of losses arise in urban areas, due to damages of structures, costs of business shut-down and failure of infrastructure [2, 3].

Consequences of flooding, or damages, are usually divided in two classes, tangible and intangible damages, depending on the possibility of assessing them in monetary terms. A further classification distinguishes losses in direct and indirect damages, with the first resulting from the physical contact of water with exposed elements and the second induced by the flood, but occurring, in space and time, outside the event [4–6]. Even though it is acknowledged that intangible and indirect damages can have an important contribution in total flood damage, a large part of the literature is focused only on direct tangible damages.

Insurance companies and research institutions in many countries develop and use flood damage models to assess the expected flood impact in economic terms.

DOI: 10.2495/SAFE-V6-N3-475-484

The contribution of damage modelling is becoming of essential importance in the field of flood risk management, for decision-making processes and for developing flood control policies and strategies. This role has been emphasised by the European Floods Directive [7], which requires Member States to prepare floods risk maps and to establish flood risk management plans focused on prevention, protection and preparedness, taking also into account cost-benefit aspects (e.g. related to the implementation of flood mitigation measures). This means that the 'classical' concept of flood protection, based on design standards related to predefined return periods, is increasingly being replaced by more comprehensive understanding of the risk over a certain period [8, 9].

This paper describes the approach used in the development of a programme of flood risk management measures in Moldova and shows not only the importance of flood potential damages assessments and cost-benefit evaluations but also the need to summarise the results of such assessments and evaluations using few parameters (e.g. 'total risk', 'priority of measure') to assist the decision-making processes.

2 MASTER PLAN FOR FLOOD RISK MANAGEMENT IN MOLDOVA

In the framework of European Neighbourhood Policy (ENP) the European Investment Bank, following a request for support from the Ministry of Environment following the 2010 floods, decided to develop a countrywide flood management Master Plan and an Investment Programme, funded by EPTATF (Eastern Partnership Technical Assistance Trust Fund). In order to define this investment programme, BETA Studio srl (Italy) and HR Wallingford Ltd (UK) were appointed to prepare a Master Plan for flood risk management in Moldova. This Master Plan provides a phased investment programme of flood management measures for rivers in Moldova covering a 20-year period and a Short-Term Investment Plan (STIP) covering the next 7 years.

A large proportion of the flood risk in Moldova occurs on the floodplains of the two main rivers (the Prut and the Dniester). There are systems of flood defence dykes on these rivers and some on the tributaries. These provide flood protection but there is a concern about their condition. Flooding also occurs on the tributaries and failure of dams is regarded as a serious potential cause of flooding. Several studies have been carried out in the past related to hydrology and flooding, but there was a lack of a comprehensive study for flood risk assessment (FRA) for the whole country and for the identification of measures to reduce such risk.

The Master Plan was developed in the following main stages: (i) Preliminary Flood Risk Assessment (PFRA) covering all 12,000 km of rivers in Moldova and identification of high-risk areas; (ii) detailed fully two-dimensional hydraulic modelling of the high-risk rivers to produce flood hazard and flood risk maps; (iii) identification of measures to reduce the flood risk; (iv) prioritisation of measures and development of a phased investment programme and an STIP.

3 FLOOD RISK ASSESSMENT

The purpose of FRA is to assess the potential adverse consequences of flood scenarios. FRA requires knowledge of the probability of flooding and the consequences of flooding.

The probability of flooding was obtained from the results of a detailed fully two-dimensional hydraulic modelling of the studied rivers (about 3,400 km in the whole of Moldova) for floods with annual exceedance probabilities of 1, 0.5 and 0.1%. The results of the hydraulic modelling included flood hazard maps showing the flood extent and flood depth for each of the modelled flood events.

The consequences of flooding consist of the impacts of flooding on the 'receptors' (people, buildings, agriculture, infrastructure, etc.). These consequences have been estimated as described in the following sections.

3.1 Criteria for FRA

Different consequences of flooding should be taken into account in FRA. The impacts to be considered in this study have been agreed with the local stakeholders. In particular, the following consequences of flood events were discussed: loss of lives and property, loss of livelihoods, environmental damage, health impacts, resettlement away from flood risk areas, psychosocial effects, hindering of economic growth and development and political implications. An open discussion was held with the stakeholders on the basis of these concepts and the chosen criteria agreed with the stakeholders can be grouped into: (i) human impacts; (ii) economic impacts; (iii) environmental impacts.

3.2 Risk evaluation methodology

In the project, the main objectives of the risk evaluation methodology were: (i) to identify and map areas of high, medium or low risk; (ii) to calculate damages for the appraisal of flood risk management options.

To do this, some indicators for FRA had to be chosen: these indicators on one hand should be representative of the most important flood consequences and, on the other hand, should be sufficiently simple to be used in an efficient way in the study.

The developed methodology took into account the criteria agreed with the stakeholders, the Floods Directive [7] and a guideline [10] prepared by the European Commission. As a result, 12 different flood risk indicators were considered, grouped in 3 categories of impact as shown in Table 1.

Using the results of hydraulic modelling (flood outlines for the three return periods considered) the value of each flood risk indicator is calculated using raster square grid discretisation (200 m) of the whole Moldova.

The risk indicator HU1, number of affected people, is calculated by intersecting the flooded area with a GIS layer file of the urbanised area, which includes information on population density for each city, town and village in Moldova.

The risk indicators HU2 and HU3, that represent the number of severely and very severely affected people during a flood, respectively, are calculated in a similar way as HU1, but only considering the flooded area where the combination of flood depth (d) and velocity (v) is greater than certain threshold values that can constitute a risk to people, according to the standard matrix used in the UK (Defra [11]).

The last two flood risk indicators for the category 'human impact' represent the impact due to the interruption of critical services: number of water supply points flooded (HU4) and length of key infrastructure flooded (main roads/railways) (HU5).

Economic impacts are calculated using the internationally accepted and most common method for the estimation of direct flood damage that consists of the application of depth–damage functions [12, 13]. Depth–damage functions represent relationships between flood depth and the resulting monetary damage: for a given flood depth the function gives expected losses to a specific property or land use type, either as a percentage of a pre-defined asset value (relative function) or directly in financial terms (absolute function). The following procedure

Table 1: Flood risk indicators used in the project.

Categories of impacts	Symbol	Flood risk indicators	Unit
Human impacts	HU1	Number of affected people	-
	HU2	Number of severely affected people ($d\cdot(v+0.5)$)>1.5) where d is depth of flooding and v is flow velocity	-
	HU3	Number of very severely affected people ($d\cdot(v+0.5)$)>2.5)	-
	HU4	Number of water supply points flooded	-
	HU5	Length of key infrastructure flooded (main roads, railways)	M
Economic impacts	EC1	Damages for residential area (cities, towns, villages) [€]	€
	EC2	Damages for non-residential area (industrial and commercial) [€]	€
	EC3	Agriculture damages [€]	€
Environmental impacts	EN1	Extent of environmental sites flooded	m^2
	EN2	n° of heritage sites flooded	-
	EN3	n° of pollution sources (WWTP, oil stations, waste disposal sites …) flooded	-
	EN4	Extent of diffuse source of pollution	m^2

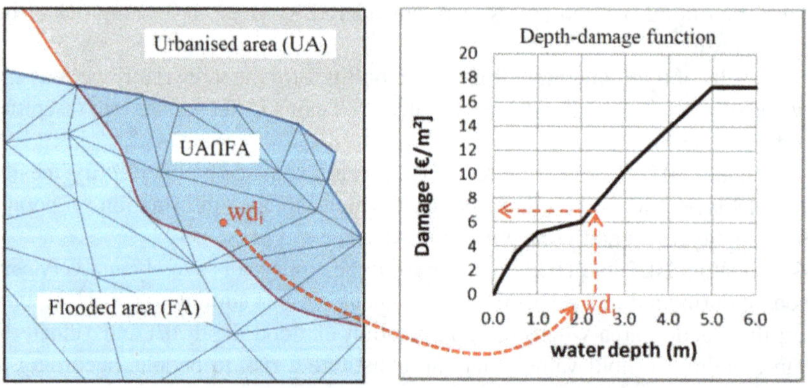

Figure 1: Calculation of damages for residential area.

was applied for EC1 (damages for residential area), for each square grid of the domain (see Fig. 1):

- The flooded area (FA) is intersected with the urbanised area (UA) to obtain the polygon UAՈFA. Note that the polygon UAՈFA consists of different triangular cells (the hydraulic model mesh) and that three different types of urban area are considered: city, town and village, each with a different damage function;

- For each triangular cell of UA∩FA, the damage value (€/m²) is calculated applying the relevant damage function for the particular urbanised area to the maximum water depth (wd_i) in the cell as calculated by the hydraulic model;
- The damage for each residential area in the grid square [€] is calculated summing, for each triangular cell, the product of the above value and the extent of the triangular cell.

The flood risk indicators EC2 – damages for non-residential area (industrial and commercial) – and EC3 - agriculture damages - have been calculated taking into account the relevant depth-damage functions for those categories. As suggested in studies that compare post-flood damage data with predictions [14, 15] caution in transferring and applying damage curves from one country to another should always be used, as transferability depends on (but it is not limited to) the similarity, in terms of flood event and/or building characteristics, between the two countries or regions. For this reason, particular effort was made to construct depth-damage functions that were representative of the characteristics of receptors in Moldova, using local expert judgment and damage data for floods that have occurred in the past.

A basic indicator of the potential flood impacts to the environment is to measure the area of important habitats that are prone to inundation by flooding. As the flooding of an environmental protected area may be a damaging, neutral or even a beneficial event, depending on the particular characteristics of the specific area, an assessment of the vulnerability of environmental sites has been done on a site by site basis. For each site exposed to flood risk, a 'flood vulnerability factor' with a value between 0 and 1 has been evaluated by an environmental expert. Thus, the factor EN1 has been evaluated multiplying the extent of the environmental areas within the flooded area by the 'flood vulnerability factor'.

The flood risk indicator EN2 refers to the number of heritage sites flooded.

When pollution sources are flooded, it is very likely that polluted elements will be washed out by the flood water causing severe environmental problems. The last two considered flood risk indicators EN3 and EN4 take into account of such impacts. In particular, EN3 is the number of pollution sources (including petrol stations, industries, wastewater treatment plants, and sites for the disposal of waste materials) flooded and EN4 refers to the pollution due to flooding of sewage system and is calculated assuming that it is proportional to the extent of the urbanised area.

After having estimated the values of the risk indicators for each return period, each flood risk indicator is 'annualised' to estimate the long-term annual average impacts. For example, the flood risk indicator 'Annual Average Damage' is the estimated average economic damages arising from all sources of flooding which, taken over the very long-term, is likely to occur on an annual basis. Similarly, the 'Annual Average Number of People Affected' is an estimate of the number of people that may be flooded in an average year. This has been done by producing an impact–probability curve. The total area under this curve represents the annualised value of the flood risk indicator or, to put it another way, the long-term average annual value of the flood risk indicator (Fig. 2).

Finally, a unique value of 'total risk' is calculated, adding together the weighted annualised values of flood risk indicators. The final value of 'total risk' is express in €, so each weight (equal to 1 for EC1, EC2 and EC3) represents the conversion factor to express each flood risk indicators in economic terms. Even if it is recognised that these conversion factors have a degree of subjectivity, they represent a very efficient way of summarising all the different

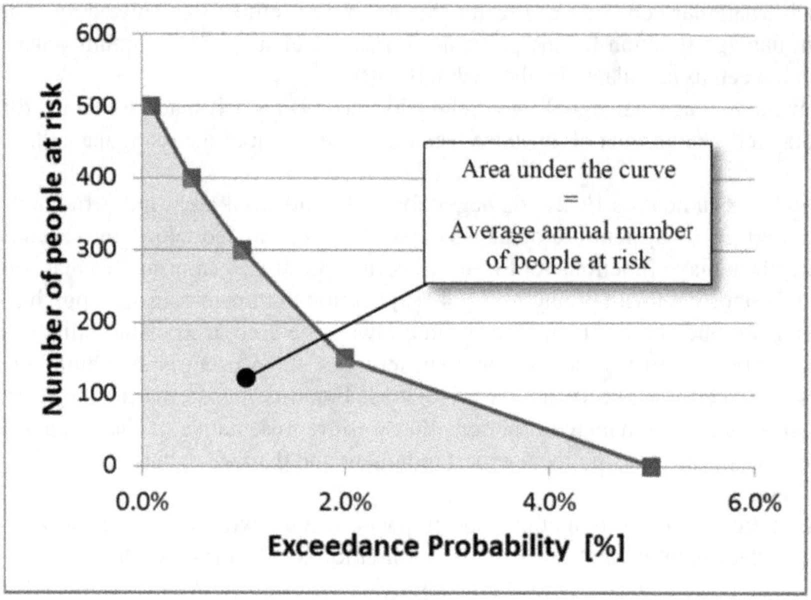

Figure 2: Calculating annual average values (annualisation).

Table 2: Levels of risk showed in the flood risk maps.

Total risk in the cell [€/year]	Level of risk
0–250	Low
250–2,500	Medium
2,500–25,000	High
>25,000	Very high

flood impacts in a unique value, the 'total risk'. This has been very useful for identifying the areas with the greatest risk and to assist the decision-making processes.

3.3 Results obtained

On the basis of the FRA conducted, flood risk maps have been prepared showing the different levels of risk (low, medium, high and very high), with respect to the value of the 'total risk' computed as described in the previous section (see Table 2). This clearly shows the areas where the flood risk is greatest and therefore the areas where flood risk management measures are most likely to be required (Fig. 3).

The total annual flood risk for the 3,400 km of studied rivers in Moldova was estimated to be €56 million. Almost half is due to direct economic damages (EN1+EN2+EN3) and the remainder is due to indirect damages including human and environmental impacts. The average annual number of people affected by flooding was estimated to be 5,200 of which 44% would be severely affected and 15% very severely affected.

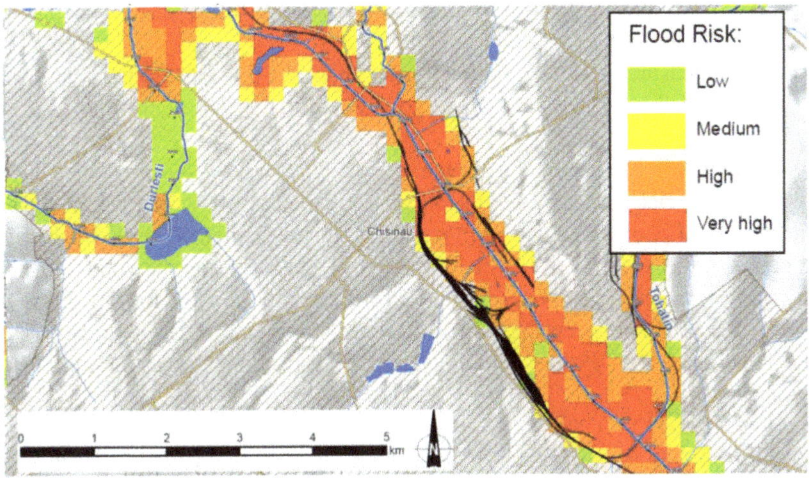

Figure 3: Example of flood risk map.

4 PRIORITISATION OF MEASURES

4.1 Objective of prioritisation

Flood management measures were identified for the areas with the greatest flood risk including both structural and non-structural measures. Flood risk is widespread on the rivers in Moldova and a total of 84 structural measures were defined to reduce flood risk for over 100 settlements. The total cost of these measures is € 325 million.

As funding is limited, it was necessary to decide which measures are better to implement first: a method for prioritisation of measures was therefore needed.

4.2 Method for prioritising the measures

A method for prioritising the measures has been discussed and agreed with local stakeholders. The method for prioritisation takes account of:

- The urgency of the measure;
- The magnitude of the risk that can be reduced with the measure;
- The benefit-cost ratio of the measure.

4.2.1 Urgency of measures

The urgency takes into account the need to develop the measure as soon as possible and includes:

- Dykes that are in poor condition and require urgent repair;
- Measures to protect urbanised areas that have been affected by flood events in the past.

The dykes that are in poor condition have been identified during a specific survey conducted as part of project activities, while urbanised areas affected by floods events in the past have been identified on the basis of data collected during the first part of the assignment.

The urgency of measures has been defined on the basis of the above criteria as shown in Table 3.

4.2.2 Magnitude of the risk that can be reduced

The magnitude of the risk that can be reduced with the measure has been calculated on the basis of the number of people that can be protected by the particular measure: the greater the number of people protected, the higher the priority of the measure, as shown in Table 4. The number of people protected has been evaluated for each measure for the 1% flood event, calculating the difference between the value of flood risk indicator HU1 'Number of people affected' (see Section 3.2) at present with the value of HU1 with the measure implemented.

4.2.3 Benefit-cost ratio

A third indicator used to assign the priority of each measure is the benefit-cost ratio of each measure: a measure with a low cost and high benefit would have a high priority. In particular, the benefit-cost ratio (*BCR*) has been calculated as:

$$BCR = (B \cdot 20) / (Ci + Cm \cdot 20) \tag{1}$$

where:

- B is the benefit of the proposed measure [€/y] and is calculated as the difference between the annualised risk at present and the annualised risk with the measure implemented;
- Ci is the cost of implementation of the measure [€];
- Cm is the maintenance cost of the measure [€/year].

Depending to the value of BCR the priority for the benefit/cost criterion is shown in Table 5.

Table 3: Priority of a measure according to its urgency.

Does the measure defend an area that has been flooded in the past?	Defences in poor condition that require urgent repair?	Urgency priority
Yes	Yes	High
Yes	No	Medium
No	Yes	Medium
No	No	Low

Table 4: Priority of a measure according to the magnitude of the risk that can be reduced with the measure.

No. of people protected with the measure	Magnitude of risk reduction priority
0–200	Low
200–1000	Medium
>1000	High

Table 5: Priority of a measure according to its benefit/cost ratio.

BCR	Benefit/cost priority
0–1	Low
1–2	Medium
>2	High

Table 6: Priority of a measure.

Total score (sum of urgency, magnitude and benefit-cost scores)	Priority of the measure
3, 4	Low
5	Medium
6, 7	High
8, 9	Very High

4.2.4 Total priority

The total priority of each measure has been assigned in the following way: for each measure and for each one of the three criteria (urgency, magnitude and benefit-cost) scores ranging from 3 to 1 have been assigned (3 for High, 2 for Medium and 1 for Low) and these are summed to give the total score for each measure. The priority of the measure has been assigned according to the value of score in Table 6.

The above methodology for prioritisation of measures has been applied both to structural and non-structural measures. In the case of non-structural measures the level (low, medium, high) of urgency, of magnitude of protected people and potential benefits that can be achieved in comparison to the costs of the measure have been based on the judgment of experts and agreed with the relevant stakeholders.

5 CONCLUSIONS

Despite the growing importance that appraising, prioritising and financing flood risk management measures is gaining in flood risk management, most of the tools needed (including damage modelling, cost-benefit analysis) can still be considered as a relatively new research area compared to other water resources fields such as hydrologic and hydraulic modelling. The methodologies that are used are not standardised, and the results are affected by large uncertainties.

This paper describes methodologies for FRA and prioritisation of measures that have been developed and applied as part of the development of a national programme of flood risk management measures in Moldova.

These methodologies proved to be valuables tools for the identification of areas with the highest flood risk, for comparing different flood risk management options and for selecting the most appropriate measures, based on an assessment of flood risks under present conditions and with the measures to reduce the flood risks in place. In particular, the final result of the FRA provides values of the parameter 'total risk' in a 200 m square grid for all high-risk rivers in the whole country that summarizes the consequences of floods with different annual exceedance probabilities on different receptors (including economy, human health, the environment and cultural heritage).

Once these relatively complex and comprehensive methodologies had been explained and agreed with the stakeholders, the simple way in which the results of the FRA and measures appraisal and prioritisation were presented facilitated the discussion with the same stakeholders and agreement of the decision-making process.

REFERENCES

[1] UNISDR, *Global Assessment Report on Disaster Risk Reduction – Revealing Risk, Redefining Development*, United Nations: Geneva, 2011.

[2] Environment Agency, *The Costs of the Summer 2007 Floods in England*, Environment Agency: Bristol, SC070039/R1, 2010.

[3] ADB & The World Bank: *Pakistan floods 2010: Preliminary damage and needs assessment*, Asian Development Bank: Islamabad, 2010.

[4] Smith, K. & Ward, R., *Floods: Physical Processes and Human Impacts,* John Wiley and Sons: Chichester, 1998.

[5] Jonkman, S.N., Bockarjova, M., Kok, M. & Bernardini, P., Integrated hydrodynamic and economic modelling of flood damage in the Netherlands. *Ecological Economics*, **66**, pp. 77–90, 2008.
http://dx.doi.org/10.1016/j.ecolecon.2007.12.022

[6] Meyer, V., Becker, N., Markantonis, V., Schwarze, R., van den Bergh, J.C.J.M., Bouwer, L.M., Bubeck, P., Ciavola, P., Genovese, E., Green, C., Hallegatte, S., Kreibich, H., Lequeux, Q., Logar, I., Papyrakis, E., Pfurtscheller, C., Poussin, J., Przyluski, V., Thieken, A.H. & Viavattene, C., Assessing the costs of natural hazards – state of the art and knowledge gaps. *Natural Hazards and Earth System Science*, **13**, pp. 1351–1373, 2013.
http://dx.doi.org/10.5194/nhess-13-1351-2013

[7] European Commission, *EU Directive on the estimation and management of flood risks* (EU/60/2007), 2007.

[8] Plate, E.J., Flood risk and flood management. *Journal of Hydrology*, **267**, pp. 2–11, 2002.
http://dx.doi.org/10.1016/S0022-1694(02)00135-X

[9] Sayers, P.B., Hall, J.W. & Meadowcroft, I.C., Towards risk-based flood hazard management in the UK. *Civil Engineering*, **150**(5), pp. 6–42, 2002.
http://dx.doi.org/10.1680/cien.2002.150.5.36

[10] European Commission, *Risk Assessment and Mapping Guidelines for Disaster Management*, Commission Staff Working Paper: Brussels, 21.12.2010, SEC (2010) 1626 final, 2010.

[11] HR Wallingford, Flood Hazard Research Centre (Middlesex University) & Risk & Policy Analysts Ltd, *Flood and Coastal Defence R&D Programme: Flood Risks to People Phase 2,* Defra & Environment Agency: London, 2006.

[12] Smith, D.I., Flood damage estimation – a review of urban stage-damage curves and loss functions. *Water SA*, **20**, pp. 231–238, 1994.

[13] Fattorelli, S. & Frank, E., A distributed technique for flood damage assessment using GIS and a 2D hydraulic model. *River Basin Management III*, pp. 433–442, 2005.

[14] Scorzini, A.R. & Frank, E., Flood damage curves: new insights from the 2010 flood in Veneto, Italy. *Journal of Flood Risk Management*, 2015.
http://dx.doi.org/10.1111/jfr3.12163

[15] Jongman, B., Kreibich, H., Apel, H., Barredo, J.I., Bates, P.D., Feyen, L., Gericke, A., Neal, J., Aerts, J.C.J.H. & Ward, P.J., Comparative flood damage model assessment: towards a European approach. *Natural Hazards Earth System Science*, **12**, pp. 3733–3752, 2012.
http://dx.doi.org/10.5194/nhess-12-3733-2012

FLOOD RISK MANAGEMENT IN NIGERIA: A REVIEW OF THE CHALLENGES AND OPPORTUNITIES

V.O. OLADOKUN[1] & D. PROVERBS[2]
[1]Department of Industrial and Production Engineering, University of Ibadan, Nigeria and a Commonwealth Academic Fellow, funded by the UK government.
[2]Faculty of Computing, Engineering and the Built Environment, Birmingham City University.

ABSTRACT
Flooding has become a major hazard in Nigeria in recent years due to a growing population, rapid urbanization and extreme weather events. This study provides a critical review and characterisation flood risk management (FRM) practices in Nigeria with a view to highlighting current weaknesses and opportunities, as well as giving recommendations for practice and for further research. Databases of academic literature, covering a wide range of FRM issues, were systematically queried and mined using suitable keywords. A structured review of the resulting literature was carried out and several past flood events and associated responses reviewed as case studies. Absence of integrated FRM systems, lack of inter agency coordination, substandard and weak infrastructures, inadequate drainage network, high urban poverty, low level literacy, cultural barriers and weak institutions characterize current FRM practices. The study recommends the adoption of an integrated approach to urban infrastructural development starting with a review of ongoing and planned infrastructural systems and projects with a view to optimizing their FRM capabilities while still meeting their intended purposes. The empowerment of more entrepreneurs into FRM solutions development and service delivery as well as the inclusion of FRM concepts and practices into the nation's educational curricula was also recommended. Nigeria also needs a multidisciplinary platform for generating effective strategic policies and efficient operational mechanisms for FRM.
Keywords blue green, flood, flood risk management, infrastructure, Nigeria, urbanization.

1 INTRODUCTION

Flooding has become a major hazard in Nigeria in recent years. It was estimated that Nigeria suffered combined losses of more than $16.9b in damaged properties, oil production, agricultural and other losses due to flood events in 2012 alone [1, 2]. Increased flood events coupled with the lack of coping capacity and high levels of vulnerability of the people have continued to put many lives and properties at risk [3]. Stakeholders are therefore increasingly concerned about the threats of flooding to communal safety and national development. While there have been various interventions in the past, there is a lack of integrated and sustainable Flood Risk Management (FRM) systems and practices in Nigeria [4]; sustainable FRM system reflects the ecological make up, infrastructural development, institutional behaviour and other techno-socio-economic characteristics of its environment [5, 6].

Meanwhile, it has been observed that the literature is sparse in many critical areas of flooding and FRM for Nigeria. For instance Egbenta *et al.* [2] observed that there is a paucity of information on the effects of flood hazard on the values of properties in Nigeria while Nkwunonwo *et al.* [7] asserted that only journalistic and non-quantitative evidence are available on some

DOI: 10.2495/SAFE-V6-N3-485-497

critical aspects of flooding in Nigeria. A recent review by Komolafe *et al.* [3] highlighted the absence of the use of state of the art flood models integrating all hydrological processes for accurate prediction and mapping of flooding and its associated risks in Nigeria.

Moreover, while Nigeria is the most populous country and has the largest economy in Africa, case studies on Nigerian flooding are scant or absent in many important global flood studies and documents. For instance in a major World Bank document [8] on guidelines for FRM in developing countries, not a single case study or reference was made to Nigeria. Therefore, it is imperative to have further research in order to have a sound and legitimate basis for effective planning and development of Nigerian FRM policies. Particularly, there is a need to identify FRM approaches and systems adopted elsewhere and how to adapt them for Nigeria. This study aimed to provide a critical review and characterisation of FRM in Nigeria with a view to highlighting current weaknesses and opportunities. This will lead towards the development of conclusions and recommendations for practice and for further research.

2 ELEMENTS OF THE NIGERIAN FLOOD SYSTEM

This section highlights some of the socio-economic, techno-environmental, ecological and other factors that characterize flooding and flood hazards in Nigeria and attempts to highlight some elements of current FRM practices.

2.1 Frequency and regularity of occurrences

Until recently major flood disasters rarely occur in Nigeria. Recorded cases of flooding in Nigeria date back to 1963 when Ogunpa River flooded Ibadan city causing loss of lives and properties with reoccurrences in 1978, 1980 and 2011 [9, 10]. The 1980 flood, which caused huge loss of lives and properties, brought some notoriety and attention to the disaster potential of the river and immediately provoked a flurry of FRM responses such as the initiation of the Ogunpa channelization project, educative radio and television jingles on FRM practices, relocation of structures and properties from flood plains [11, 12]. However the sense of urgency and priority attached to these FRM projects waned after some years of respite. For instance, the channelization project was abandoned for a long period, radio and television jingles ended, and by the late 90s, it was clear that city planning authorities had gone back to old habits of allowing structures on floodplains obstructing flood channels [13].

Agbola *et al.* [10] noted that while the rainfall of August 26, 2011 was not the highest in the recorded history of the city, the monetary value of damages to property that resulted from the event, was by far the most costly. The official estimated losses and damages from the 1980 single flood episode rainfall of 274.0 mm was put at N300 million ($1.9 million) while that of 2011 rainfall of 187.5 mm was put at N2.1 billion ($13.3 million). Some twelve anthropogenic factors were identified to have contributed to the exaggerated level of devastation that trailed the 2011 Ibadan flood [10, 13]. Also the 2011 flood event motivated the Nigerian government to approach the World Bank to finance a long term major FRM project in Ibadan city. While responses to the past flooding events have been piecemeal, mainly focusing on alleviating immediate and short-term needs such as rebuilding of destroyed assets, the Ibadan Urban Flood Management Project (IUFMP) recognizes the imperative of a long term integrated approach to FRM for the city [14]. The project with a closing date of June 2022 and objective of improving the state's capacity to manage flood risk in Ibadan is on course [15].

Meanwhile, because the meantime between floods can be irregular and deceptively long, stakeholders tend to forget the lessons of the 'last' flood disaster and become relaxed until another flood strikes. This disruption to both institutional and individual memory coincides with what obtains in the UK where FRM tends to become prominent in the immediate aftermath of a major inundation. Although the awareness of resilience measures among UK residents is high, the level of implementation is low with only 10% claiming to have implemented a full package of the measures [16]. While FRM in UK is characterized by a robust flood insurance scheme [17] as well as an advanced flood warning system, which compensate for the low level of resilience measures, Nigeria has poor flood warning systems as well as a poor flood insurance scheme [18, 19]. Hence, building resilience capability to cope with increasing climate variability remains the most viable option for FRM in Nigeria.

2.2 Nigeria flood prone areas

With increased and rapid urbanization of Nigeria, the landscape of flood prone communities and assets are on the increase. Some key causes of the increasing vulnerability of Nigerian Urban areas to flooding were summarized by Odufuwa *et al.* [19] to include inadequate drainage systems and increased replacement of natural and absorptive soil cover with impervious materials such as concrete. The authors also blamed deforestation of hillsides which leads to increase in the quantity and rate of runoff. Figure 1 depicts areas that are exposed to flooding events. The areas and sources of major flooding can be categorised as follows.

1. Coastal cities and settlements: Nigeria has over 853 km (530 mi) of coastline with extensive low-lying areas, and heavily industrialized areas prone to flooding [1]. Lagos, the main commercial hub of West Africa, Warri and Portharcout, hosting Nigeria's Petroleum infrastructures, are increasingly exposed to ocean flooding.
2. Communities and settlements along the two major Rivers, (Niger and Benue), cutting across the nation (see Fig. 1). Urban and rural communities along the River Niger and its tributaries have been shown to be most affected by flooding [20, 21].
2. Communities downstream of dammed rivers and on the banks of other major rivers. For instance, Ilorin, a North central city, has been severely flooded by River Asa that runs through the city [22].

In light of these issues, the following questions need to be answered. How many cities, towns and settlements along these flood prone zones are at risk? What is the population and demographic characteristics of the vulnerable areas? What are the major economic activities and industries along these flood prone areas and what are the risks and implications of a major flooding incident?

3 CHARACTERISTICS OF FRM SYSTEMS IN NIGERIA

Some characteristics and factors observed in the Nigerian FRM systems are discussed and benchmarked against recommended practices in the following.

3.1 Absence of integration between urban development and FRM

In most of the Nigerian urbanization experience it can be observed that there is a disconnection between FRM systems and other elements of the urbanization process. In addition to the

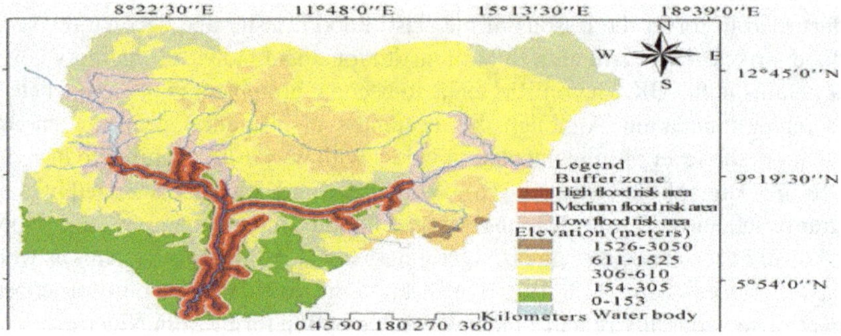

Figure 1 A map of Nigerian Flood plains [21]

technical and financial challenges faced by various government agencies, limited inter-agency coordination has militated against the success of FRM [23]. Usually, FRM systems are designed without proper integration with other developmental systems such as power infrastructures, water supply systems, transportation networks, recreational facilities and so on. An example is a situation where road construction projects can be executed by a department of public works without adequate consultation with other relevant departments such as that of water resources, environment, or agriculture. There have been several cases where road contractors damage major water pipelines serving as controlled relief outlets to dams. This in turn compromises the flood safety features of such dams and ultimately leads to the flooding of adjoining and downstream areas of the dams. There have been cases where road contractors excavate borrow pits during dry seasons without adequate consideration for flooding. Such projects end up damaging the capacity of the adjoining landscape to withstand erosion by removing the natural defence of the ecosystem against flooding [24].

An integrated and holistic approach to development whereby flood management is a high priority sub-system has been identified as the only sustainable approach to FRM [8, 25]. This approach to development will require a high level of coordination and integration among the various agencies of the Nigerian government. Inter agency coordination and integration, vis-a-vis FRM, will allow for the interrelations exiting among urbanization processes and systems to be systematically explored and exploited in a complimentary manner. For instance, in solving transportation issues, water channels built for flood control can be designed to also provide water ways to ease transportation in places like Lagos. Solving power supply demands with large numbers of wind turbines installed along the offshore line of Lagos, for instance, will reduce the landing velocity of coastal waves/storms by extracting the kinetic energy of the storms to drive power generators. This may provide a more effective and efficient protection, than the construction of crude concrete embankments and serve to benefit both energy production and FRM concerns.

Ironically, residents of many Nigerian communities ravaged by flooding also lack access to potable water. For instance Adesogan [26] reported poor water supply coverage for nearly 22% across the South West region of Nigeria while Oke *et al.* [27] noted there is a challenge of water sourcing within a major portion of this region. An integrated solution may be a rainwater harvesting system designed to meet water supply needs and also serving as an FRM scheme. Rainwater harvesting has been noted to offer some mitigation against extreme runoff events [28, 29]. Economically, Hashemi and Berndtsson [30] reported that floodwater harvesting is an inexpensive method for flood mitigation and artificial recharge of aquifers as it results in a large economic return for relatively small investment.

However, the present low level of inter agency coordination as well as the absence of solution systems integration in the Nigerian urbanization process is a hindrance to such holistic approaches. Apart from the federal government, there are 36 autonomous state governments, over 750 local authorities, hundreds of private organizations, companies, NGOs and individual property owners implementing various FRM solutions without appropriate coordination and collaboration. Often these entities put in place FRM measures that are only sub-optimal for their own immediate regions and may end up causing more damage or induce floods in other locations and regions. There have been reported cases where dams, channels, and structures built to protect a local region, community or private property causes direct runoff water into other communities downstream [31].

3.2 Focus on structural FRM measures

In tackling flood risk in Nigeria, the main focus has been on structural measures coupled with over dependence on imported expertise and technologies. The propensity to award contracts to build more structural flood defences, canals, embankments, culverts and bridges without sufficient consideration for less costly and more sustainable, non-structural solutions is evident in the budgets of the nation. Unfortunately, these structures are usually handled by foreign contractors and experts with limited understanding of the local situation resulting in the limited knowledge transfer to indigenous experts [32]. Such structures, which are usually copies of the solutions adopted in some distant countries and different socio-ecological settings without sufficient adaption for the local scenario, create other socio-technical problems. Such projects lack the right mix of soft elements like advocacy, education, stakeholders' participation, and consultation that can engender a sense of project co-production and ownership. For instance, several flood canals and drainages have turned into refuse dumps few years after commissioning [7, 10, 22, 33]).

Experience has shown that these capital intensive concrete structures and civil works rarely offer adequate and sustainable defence against the threats of flood. This coincides with major flooding in the UK and news of the failure of newly built multi-million pound flood defences in many parts of the country. It is becoming apparent that non-structural measures, institutional preparedness and coping capacity of the affected UK communities are more important in the defence against floods. Stakeholders in Nigeria are slowly recognizing the imperative of exploring non-structural FRM measures [13].

4 FLOODING HAZARD INDUCING FACTORS IN NIGERIA

There are a number of factors in Nigeria that induce and aggravate the impacts of flooding. Some of these factors are now discussed.

4.1 Substandard infrastructures induced flooding

The causal relationships that exist among various elements of the modern built environment system provide a credible platform for understanding the peculiar challenge of FRM in Nigeria. While Nigeria has made considerable investment in infrastructure projects such as road networks, bridges, and other modern facilities for the growing population, the process of planning, designing and constructing such facilities has not reflected any serious system thinking. Also due to corruption, mismanagement and incompetence, many of the facilities are substandard and collapse long before their expected life span [34].

For example, when a substandard bridge or road segment collapses during a rainy season, which is a common occurrence in Nigeria, the debris and concrete materials end up blocking or reducing the carrying capacity of the channel. This will eventually cause 'induced' flooding of the adjoining area upstream to the collapsed bridge. Also when a road segment fails due to poor and substandard civil work, commuters will ultimately divert traffic to some alternative routes thereby over stretching these other routes. For instance, heavy truck drivers are compelled to move over low capacity bridges and roads which will in turn cause more bridges to collapse and the chain continues. Invariably, substandard elements in one region of the built environment will remotely increase the risk of flooding of some other regions in the built environment. Unfortunately, in many instances, it can be a complicated process to trace back the original cause, especially if the affected infrastructures fall within different authorities.

The standards and type of materials used for many buildings in Nigeria cannot withstand mild floods. In a study by Abaje *et al.* [35]), it was observed that about 69% of the houses in a state in Northern Nigeria were built with unbaked mud and on foundations of loose sandy soil, while Anosike and Oyebade [36]; Ewa and Ukpata [37] reported that sandcrete blocks, used in over 90% of Nigerian urban houses, are below required standards. Consequently, these houses are highly exposed and vulnerable to flood damage. The activities of unqualified builders and poor culture of regulations' enforcement in the building sector have been blamed for the preponderance of weak structures [38–40]. Meanwhile, poorly built infrastructural systems, such as high voltage power lines, bridges, unprotected construction sites, and uncompleted buildings, readily collapse during flooding thereby aggravating the number of fatalities and injuries.

With the absence of reliable potable water network [26], majority of Nigerian houses depend on shallow wells and streams, mostly located close to faecal pit latrines and sewage soak away pits, for water supply. The water supply system thus has become easily and readily contaminated during flood events leading to further health and water supply issues.

4.2 Poorly planned and managed urbanization process

The growth and expansion of urban communities in Nigeria has been known to be poorly managed. Lack of proper spatial planning, poor land use management and absence of good corporate governance characterizes urban development in Nigeria [4, 35, 41]. The suburbs of big cities like Ibadan are notorious for this. Town planning officials in many of these cities have become compromised such that residents, companies, government agencies and NGOs alter the design and use of buildings without proper approvals. Potential flooding inducing alterations like dump sites, borrow pits, trenches, water dredging, and sand filling can occur without properly carried out environmental impact analysis. The use of geotechnical studies and other pre-design studies to ascertain the suitability of an area is uncommon.

This absence of a proper urban development planning process has over the years resulted in the gross distortion of ecological systems and, in the words of Odufuwa *et al.* [19], abused the flood plains especially in the low lying cities. These damaging distortions to the ecosystems include the replacement of the natural and absorptive soil cover with concrete and the silting up of drainage channels [4]. For instance, deforestation of hillsides, which has the effect of increasing the quantity and rate of runoff, and through soil erosion, may have contributed to the increase in the reported incidence of mud and landslides during floods in recent years in Nigeria [42, 43].

4.3 High prevalence of urban poverty

Poverty is a major challenge to development in Nigeria [44, 45]. Demographic data show more than 64% of the nation's population, of over 160 million, live on less than $1 a day [46]. Studies have confirmed that residents, especially women, in the lower income neighbourhoods recorded higher impacts and slower recovery after flooding Ajibade *et al.* [47]). Increased poverty levels coupled with insecurity in the Northern region in recent times have resulted in increased movement to the low-lying and coastal cities in the South like Lagos. Unfortunately, the high cost of land and housing in these cities means that most internal migrants end up living in 'affordable' slums or turn existing areas into slums. While swampy and low level areas may be cheaper to acquire per plot they are usually very costly to develop. Swamps and waterlogged lands require heavy investments in deep concrete foundations, high volume of sand filling and dredging, networks of drainage and other infrastructures required to minimize the impact of flooding.

The number and resident population of the slum areas in major Nigerian cities have continued to rise [48]. Fragile and inadequate sewage systems become overburdened and collapse, refuse and solid waste management facilities become overstretched to the extent that drainage networks get blocked and flooding becomes inevitable. So poverty is a primary driver of urban flooding, the flooding in turns creates further poverty as uninsured residents lose their properties and livelihoods to flood disasters. Hence, there is a vicious cycle of flooding, poverty and deprivation [48, 49].

Attempts in the past to relocate people from swampy slums have met stiff and violent resistance due to absence of adequate engagement of stakeholders through consultations, advocacy, and education. Such relocations should be seen to be socially equitable with fair compensation. This has not been the case in the past where relocation from flood prone areas has been used as a pretext by the ruling elites to take possession of lands close to the various beaches and water front without fair compensation. A sustainable FRM scheme must embrace social justice [50]. Involuntary resettlement from development projects should be properly managed to avoid long-term hardship and impoverishment of affected persons and communities. Unmitigated impacts often give rise to severe economic and social stress [51].

4.4 Absence of water drainage network

Inadequate and poorly maintained drainage networks have been identified as a major contributory factor to the increased frequency of urban flooding in Nigeria [41, 52]. Compared to the UK and other developed nations, Nigerian urban areas lack drainage network for waste and flood water with many depending on rivers and tributary streams flowing through them, this situation was particularly identified for Ilorin by Kolawole *et al.* [22]. For instance Abaje *et al.* [35] found that in some Northern part of the country 64.5% of the residents have no access to drainage systems.

4.5 Cultural resistance to change

Ignorance due to low level literacy, superstitious beliefs such as rivers regarded as deities to be appeased when they overflow their banks are common among many Nigerian tribes. In many cities, drainage systems have been blocked with refuse by residents who dump wastes in drainages [35].

5 OPPORTUNITIES FOR SUCCESSFUL AND SUSTAINABLE FRM SOLUTIONS

Some techno-socio-economic and geo-political factors prevailing in Nigeria offers some unique opportunities that support adoption of sustainable integrated FRM systems. These factors are identified and characterized in the following.

5.1 Emerging infrastructures

With relatively little investment in modern urban infrastructure, Nigeria has the opportunity to urbanize with Blue-Green (BG) systems. Nigeria does not face the constraint of space, which escalates the cost of retrofitting of BG infrastructures into urbanized catchments, like many developed countries [53]. Blue systems include ponds, flowing waterways, wet detention basins and wetlands existing within the drainage network. Green infrastructures refers to natural land and plant based ecological treatment systems and processes made up of assets such as open spaces, parks, recreation grounds, woodlands, gardens, green corridors etc [53]. BG FRM revolves around the concept of 'living with water', developing resilience to flooding and making space for water through BG infrastructures [54].

5.2 Stable geological zone

Nigeria is located in an environment not prone to serious natural hazards such as earthquakes, landslides, tsunamis, hurricanes, volcanic eruptions [2]. Relative to other parts of the globe, destructive geological events in Nigeria are rare and mild. Dealing with floods without the aggravated effects of these factors makes planning less complicated and allows for effective modelling.

5.3 Young population and huge land area

Relocation of residents from a flood prone area will usually be more complicated for older populations with some deep rooted attachments to 'their' communities and properties. It is usually traumatic for older people to part with their homes or be relocated involuntarily after a flood event [55]. However, with a demographic distribution dominated by young people with weak economic ties, implementing relocation programs with appropriate incentives may face less resistance in Nigeria. Moreover, given the infant state of Nigeria's infrastructural development and the nation's relatively huge land expanse outside flood plains the economic cost of relocation programs will be greatly moderated.

6 CONCLUSIONS

This study reviewed flooding and FRM in Nigeria. Some characteristics and challenges were discussed. It was noted that while the increased incidence of floods in recent times has increased levels of awareness and raised the tempo of FRM activities in Nigeria, there is a general lack of coordination and integration across several important FRM systems and processes. FRM approaches and responses have been generally geared toward sub-sector or sub-regional optimization. There is an absence of holistic or system thinking in urban system planning and development. There is a disconnection between FRM systems and other subsystems of the built environment, resulting, inadvertently, in inefficient and ineffective FRM systems. While enormous resources and investment have been put into a number of structural FRM measures, there has been a limited adaptation of these technologies to the Nigerian socio-technical environment as well the absence of the required knowledge transfer to Nigerian experts.

6.1 Recommendations

Meanwhile, there are several socio cultural features, characteristics as well as techno-economic and development opportunities which support the adoption of sustainable, integrated FRM systems in Nigeria and also favor the pursuit of Blue-Green infrastructural development approaches at relatively lower cost. These potentials can be optimally exploited if there is a framework for understanding the interactions of the urban development process and FRM system in Nigeria. The following recommendations for practice and for further research are therefore considered useful.

6.1.1 Practice and policies

1. Adoption of an integrated approach to urban infrastructural development.
2. A review of ongoing and planned infrastructural projects with a view to optimizing their FRM capabilities while still meeting their intended purposes.
3. Establishment of centre of excellence in Flood Risk Research and Capacity Development to serve as a multidisciplinary platform for generating effective strategic policies and efficient operational mechanisms for FRM in Nigeria.
4. Adoption of pragmatic steps toward developing and including suitable FRM concepts and practices into the nation's educational curricula.
5. Mobilizing and empowering more entrepreneurs into FRM solutions development and service delivery in Nigeria.

6.1.2 Further research

1. Performance evaluation of past and ongoing FRM interventions in Nigeria. This requires development of FRM performance evaluation models.
2. Development of techno-economic evaluation framework to optimize the adoption of BG technologies such as rainwater/floodwater harvesting, pumped hydro power systems, and transport canals for FRM measures.
3. Development of low cost and easy to use flood risk assessment tools at both individual property and community levels.
3. FRM knowledge/skills gap analysis across various relevant professions with a view to identifying human capacity requirements for sustainable FRM.

REFERENCES

[1] Amangabra, G.T. & Obenade, M., Flood vulnerability assessment of Niger Delta states relative to 2012 flood disaster in Nigeria. *American Journal of Environmental Protection,* **3**(3), pp. 76–83, 2015.

[2] Egbenta, I.R., Udo, G.O. & Otegbulu, A.C., Using hedonic price model to estaimate effects of flood on real property value in Lokoja Nigeria. *Ethiopian Journal of Environmental Studies and Management,* **8**(5), pp. 507–516, 2015. http://dx.doi.org/10.4314/ejesm.v8i5.4

[3] Komolafe, A.A., Adegboyega, S.A. & Akinluyi, F.O., A review of flood risk analysis in Nigeria. *American Journal of Environmental Sciences,* **11**(3), pp. 157–166, 2015. http://dx.doi.org/10.3844/ajessp.2015.157.166
[4] Adedeji, O.H., Odufuwa, B.O. & Adebayo, O.H., Building capabilities for flood disaster and hazard preparedness and risk reduction in Nigeria: need for spatial planning and management. *Journal of Sustainnable Development in Africa,* **14**(1), pp. 45–58, 2012.
[5] Storbjörk, S., Governing climate adaptation in the local arena: challenges of risk management and planning in Sweden. *Local Environmen: The International Journal of Justice and Sustainability,* **12**(5), pp. 457–469, 2007. http://dx.doi.org/10.1080/13549830701656960
[6] Tran, P., Shaw, R., Chantry, G. & Norton, J., GIS and local knowledge in disaster management: a case study of flood risk mapping in Vietnam. *Disasters,* **33**(1), pp. 152–169, 2009. http://dx.doi.org/10.1111/j.1467-7717.2008.01067.x
[7] Nkwunonwo, U.C., Whitworth, M. & Baily, B., Review article: a review & critical analysis of the efforts towards urban flood reduction in the Lagos region of Nigeria. *Natural Hazards and Earth System Science Discussion,* **3**, pp. 3897–3923, 2015. http://dx.doi.org/10.5194/nhessd-3-3897-2015
[8] Jha, A.K., Bloch, R. & Lamond, J., *Cities and Flooding: A Guide to Integrated Urban FRM in the 21st Centuries*, World bank, DC, 2012. http://dx.doi.org/10.1596/978-0-8213-8866-2
[9] Adegbola, A.A. & Jolayemi, J.K., Historical rainfall-runoff modeling of river Ogunpa, Ibadan, Nigeria. *Indian Journal of Science and Technology,* **5,** pp. 2725–2728, 2012.
[10] Agbola, B., Ajayi, O., Taiwo, O. & Wahab, B., The august 2011 flood in Ibadan, Nigeria: Anthropogenic causes and consequences. *International Journal of Disaster Risk Science*, **5**, pp. 207–217, 2012. http://dx.doi.org/10.1007/s13753-012-0021-3
[11] Sangodoyin, A.Y. & Essein, O.E., Effect of urbanization, waste disposal and hydrological factors on flooding of Ogunpa stream in Nigeria. *Discovery and Innovation,* **8**(1), pp. 11–19, 1996.
[12] Egunjobi, L., Human elements in urban planning and development: Ibadan. *Habitat International,* **10**(4), pp. 147–153, 1986. http://dx.doi.org/10.1016/0197-3975(86)90079-2
[13] Ajayi, O., Agbola, S.B., Olokesusi, B.F., Wahab, B., Taiwo, O.J., Gbadegesin, M., Taiwo, D.O., Kolawole, O., Muili, A., Adeola, M.A., Olutade, O.G., Shiji, F. & Abiola, N.A., Flood management in an urban setting: a case study of ibadan metropolis. *Special Publication of the Nigerian Association of Hydrological Sciences,* pp. 65–81, 2012.
[14] Word Bank, *Ibadan Urban Flood Management Project: Project Appraisal Document*, World Bank, 2014.
[15] Salim, R., *Nigeria - Ibadan Urban Flood Management Project : P130840 - Implementation Status Results Report: Sequence 03.* World Bank, 2015.
[16] Joseph, R., Proverbs, D. & Lamond, J., Resilient reinstatement: What can we learn from the 2007 flooding in England? *Flood Recovery, Innovation and Response IV*, **184**, pp. 175–186, 2014.
[17] Cameron, D. & Proverbs, D., Improving the uptake of flood risk adaptation measures for domestic properties in an insurance regime under transition. *Flood Recovery, Innovation and Response IV*, **184**, pp. 139–149, 2014.

[18] Adelekan, I.O., Vulnerability assessment of an urban flood in Nigeria: Abeokuta flood 2007. *Natural Hazards,* **56**(1), pp. 215–231, 2011. http://dx.doi.org/10.1007/s11069-010-9564-z

[19] Odufuwa, B.O., Adedeji, O.H., Oladesu, J.O. & Bongwa, A., Floods of fury in Nigerian cities. *Journal of Sustainable Development,* **5**(7), pp. 69–79, 2012.

[20] Amadi, L. & Ogonor, C.U.M., Climate change, environmental security and displacement in Nigeria: experience from niger delta flood disaster, 2012. *African Journal of Environment Science and Technology,* **9**(1), pp. 53–64, 2015. http://dx.doi.org/10.5897/AJEST2014.1749

[21] Nkeki, F.N., Henah, P.J. & Ojeh, V.N., Geospatial techniques for the assessment and analysis of flood risk along the Niger-Benue basin in Nigeria. *Journal of Geographic Information System,* **5**, pp. 123–135, 2013. http://dx.doi.org/10.4236/jgis.2013.52013

[22] Kolawole, O.M., Olayemi, A.B. & Ajayi, K.T., Managing flood in Nigerian cities: risk analysis and adaptation options – Ilorin city as a case study. *Archives of Applied Science Research,* **3**(1), pp. 17–24, 2011.

[23] World Bank, *Ibadan Urban Flood Management Project: Environmental and Social Management. Framework Final Report*. Oyo state Government of Nigeria, 2014.

[24] Nwachukwu, M.A. & Eburukevwe, O., Assessment of environmental hazards: Linking borrow pits, gully erosion, and road failure. *Journal of Civil Engineering and Construction Technology,* **4**(6), pp. 199–210, 2013.

[25] Baker, R., Blue space thinking. In *Flood Hazards: Impacts & Responses for the Built Environment*, eds. J. Lamond, C. Booth, F. Hammond & D. Proverbs, Taylor & Francis: New York, pp. 191–205, 2012.

[26] Adesogan, S.O., Strategies and techniques of providing adequate and affordable potable water in rural areas of Nigeria. *International Journal of Water Resources and Environment Engineering,* **6**(1), pp. 32–39, 2014. http://dx.doi.org/10.5897/IJWREE2013.0418

[27] Oke, M.O., Martins, O. & Idowu, O.A., Determination of rainfall-recharge relationship in River Ona basin using soil moisture balance & water fluctuation methods. *International Journal of Water Resource and Environment Engineering,* **6**(1), pp. 1–11, 2014. http://dx.doi.org/10.5897/IJWREE2013.0443

[28] Huang, C., Hsu, N., Wei, C. & Luo, W., Optimal spatial design of capacity and quantity of rainwater harvesting systems for urban flood mitigation. *Water,* **7**, pp. 5173–5202, 2015. http://dx.doi.org/10.3390/w7095173

[29] Magaliano, P.N., Murray, F., Baldi, G., Aurand, S., Paez, R.A., Harder, W. & Jobabagy, E.G., Rainwater harvesting in dry Chaco: regional distribution and loca balance. *Journal of Arid Environments*, **123**, pp. 93–102, 2015. http://dx.doi.org/10.1016/j.jaridenv.2015.03.012

[30] Hashemi, H. & Berndtsson, R., Water resources management through rainwater harvesting in arid area. In *7th IWA International Conference on Efficient Use and Management of Water*, Paris, 2013.

[31] Alayande, W.A., Mohammed, G., Caleb, I. & Deimode, M.I., Assessment of urban flood disaster: a case study of 2011 Ibadan floods. *Hydrology for Disaster Management: Special Publication of the Nigerian Association of Hydrological Sciences,* pp. 13–23, 2012.

[32] Ugochukwu, S.C. & Onyekwena, T., Participation of indigenous contractors in Nigerian public sector construction projects and their challenges in managing working capital. *International Journal of Civil Engineering, Construction and Estate Management,* **1**(1), pp. 1–21, 2014.

[33] Adewole, I.F., Agbola, S.B. & Kasim, O.F., Building resilience to climate change impacts after the 2011 flood disaster at the University of Ibadan, Nigeria. *Environment and Urbanization,* **27**(1), pp. 199–216, 2014. http://dx.doi.org/10.1177/0956247814547679

[34] Lanrewaju, A.F., Urbanization, housing quality and environmental degeneration in Nigeria. *Journal of Geography and Regional Planning,* **5**(16), pp. 422–429, 2012. http://dx.doi.org/10.5897/JGRP12.060

[35] Abaje, I.B., Ogoh, A.O., Amos, B.B. & Abashiya, M., Climate change, flood disaster assessment and human security in Katsina State, Nigeria. *American Journal of Human Ecology,* **4**(4), pp. 47–56, 2015.

[36] Anosike, M.N. & Oyebade, A.A., Sandcrete blocks and quality management in Nigeria building industry. *Journal of Engineering, Project, and Production Management,* **2**(1), pp. 37–46, 2012.

[37] Ewa, D.E. & Ukpata, J.O., Investigation of the compressive strengths of commercial sandcrete blocks in calabar Nigeria. *International Journal of Engineering and Technology,* **3**(4), pp. 447–482, 2013.

[38] Fagbenle, O.I. & Oluwunmi, A.O., Building failure &collapse in Nigeria: the influence of the informal sector. *Journal of Sustainable Development,* **3**(4), pp. 268–276, 2010.

[39] Mbamali, I. & Okotie, A.J., An assessment of the threats and opportunities of globalization on building practice in Nigeria. *American International Journal of Contemporary Research,* **2**(4), pp. 143–150, 2012.

[40] Ayedun, C.A., Durodola, O.D. & Akinjare, O.A., An empirical ascertainment of the causes of building failure and collapse in Nigeria. *Mediterranean Journal of Social Sciences,* **3**(1), pp. 313–322, 2011.

[41] Dalil, M., Mohammad, N.H., Yamman, U.M., Husaini, A. & Mohammed, S.L., An assessment of flood vulnerabilty on physical development along drainage channels in Minna, Niger State, Nigeria. *African Journal of Environment Science and Technology,* **9**(1), pp. 38–46, 2015. http://dx.doi.org/10.5897/AJEST2014.1815

[42] Inyang, M.P., The role of landslide education in preventing environmental hazards and promoting health in Niger Delta region of Nigeria. *International Journal of Science, Environment and Technology,* **3**(2), pp. 427–433, 2014.

[43] Ezeji, J.I., Reducing risks of contamination breaching the drinking water system through catchment sensitive farming in eastern Nigeria. *Pollution Effects and Control,* **3**(3), p. 148, 2015. http://dx.doi.org/10.4172/2375-4397.1000148

[44] Lame, S.M. & Yusoff, W.F.W., Poverty reduction in Nigeria: the role of entrepreneurship education. *Journal of Educational and Literature,* **3**(2), pp. 63–71, 2015.

[45] Ike, P.C. & Uzokwe, U.N., Estimation of poverty among rural farming households in Delta State. *Journal of Poverty, Investment and Development,* **11**, pp. 86–93, 2015.

[46] Joshua, I.A., Makama, J.G., Joshua, W.I., Audu, O. & Nmadu, A.G., Disasters in Nigeria: a public health perspective. *Journal of Community Medicine and Primary Health Care,* **26**(1), pp. 59–75, 2014.

[47] Ajibade, I., McBean, G. & Bezner-Kerr, R., Urban flooding in Lagos, Nigeria parterns of vulnerability & resilence among women. *Global Environment Change,* **23**(6), pp. 1714–1725, 2013.
http://dx.doi.org/10.1016/j.gloenvcha.2013.08.009
[48] Adelekan, I.O., Vulnerability of poor urban coastal communities to flooding in Lagos,Nigeria. *Environment and Urbanization,* **22**(2), pp. 433–450, 2010.
http://dx.doi.org/10.1177/0956247810380141
[49] Douglas, I., Alam, K., Mahgenda, M., Mcdonnell, Y., Mclean, L. & Campbell, J., Unjust waters: climate change,flooding and the urban poor in Africa. *Environment and Urbanization,* **20**(1), pp. 187–205, 2008.
http://dx.doi.org/10.1177/0956247808089156
[50] Johnson, C., Penning-Rowsell, E. & Parker, D., Natural and imposed injustices: the challenges in implementing 'fair' flood risk management policy in England. *The Geographical Journal,* **173**(4), pp. 374–390, 2007.
http://dx.doi.org/10.1111/j.1475-4959.2007.00256.x
[51] World Bank, *Resettlement Policy Framework; Nigeria,* World Bank, 2013.
[52] Ocheri, M. & Okele, E., Social impact and people's perception of flooding in Makurdi town, Nigeria. *Hydrology for Disaster Management,* pp. 97–105, 2012.
[53] Lawson, E., Thorne, C., Ahilan, S., Allen, D., Arthur, S., Everett, G., Fenner, R., Glenis, V., Guan, D., Hoang, L., Kilsby, C., Lamond, J., Mant, J., Maskrey, S., Mount, N., Sleigh, A., Smith, L. & Wright, N., Delivering and evaluating the multiple flood rik benefits in BG cities:an interdiscplinary approach. *Flood Recovery, Innovation and Response IV*, pp. 113–124, 2014.
[54] Everett, G. & Lamond, J., A conceptual framework for understanding behaviours and attitudes around 'Blue-Green' approaches to FRM. *Flood Recovery, Innovation and Response IV*, pp. 101–112, 2014.
[55] Whitlte, R. & Medd, W., Living with flood: Understanding residents' experiences of recovery. *Flood Hazards: Imapacts and Responses for the Built Environment*, pp. 281–290, 2012.

ASSESSMENT OF ENVIRONMENTAL DAMAGES IN CASE OF FLOOD IN BODVA RIVER BASIN, SLOVAKIA

MARTINA ZELEŇÁKOVÁ[1], PETER BLIŠŤAN[2], IBRAHIM ALKHALAF[1], LENKA GAŇOVÁ[3] & LENKA ZVIJÁKOVÁ[4]
[1]Institute of Environmental Engineering, Technical University of Košice, Slovakia.
[2]Institute of Geodesy, Cartography and Geographical Information Systems, Technical University of Košice, Slovakia.
[3]Slovak Water Management Enterprise, s.c., branch office Košice, Slovakia.
[4]Eurodotacie, Erste Group Bank, Košice, Slovakia.

ABSTRACT
Environmental and natural values may be affected by floods, despite the fact that floods are to some extent a natural phenomenon. Effects of floods can take many forms, physical damage, injury, death, environmental pollution, etc. A significant part of the environmental damage, the leakage of harmful substances that may be in various forms, affects ecosystems long after the flood. In the case of flood, the main damages on the environment may occur as a consequence of accidents at sources of pollution. The issue of pollution sources is a key area of environmental protection. While pollution from point sources can be disposed by suitable technology, for diffuse pollution it is essential to propose preventive measures that create conditions to prevent contamination. The purpose of this paper is to assess the impacts of the sources of pollution during floods in the catchment of Bodva River, situated in the south-east part of the Slovak Republic. We stated the classification of sources of pollution according to the degree of danger for environment and quantification of their impact on the environment by weighting method. Each source of pollution was divided into categories according to different criteria based on literature studying, consultation and experiences. The results from evaluation proved no significant environmental pollution of the studied catchment in the case of flood. Anyway in the evaluated area, the Bodva River basin, there is a need to ensure effective flood control measures not only for protection but also in terms of economic, social and environmental issues. Evidence that flood protection and prevention are justified and must be one of the most important parts of spatial planning is critical to recurrent flood damage.
Keywords: Bodva basin, flood risk, sources of pollution.

1 INTRODUCTION

The aim of flood risk management is the proposal of flood protection measures. The main objective of management as well as the entire management cycle is regulated by the Directive of the European Parliament and of the Council 2007/60/EC on the assessment and management of flood risks. The aim of this directive is to reduce and control the adverse consequences on human health, the environment, cultural heritage and economic activity associated with floods. The objective of the paper is to propose a methodology which could be used for preliminary flood risk assessment of floods.

For achieving the goals of directive 2007/60/EC [1], which is implemented in the legislative of the Slovak Republic by Act no. 7/2010 on the flood protection [2], the obligation is placed on all member states to work up a preliminary flood risk assessment, which was

DOI: 10.2495/SAFE-V6-N3-498-507

completed in December 2011; to prepare maps of flood hazard areas and maps of flood risk, which were completed in 2013; and to work up plans for management of flood risk, which were completed in 2015. Subsequently, the individual steps must then be updated every 6 years.

In the last decade, Slovakia is being increasingly affected by floods [3, 4]. Floods constantly point to the fact that the society is very vulnerable, but it has been proved that flood-related problems could be solved through planning studies and detailed projects about flood prone areas [5–8]. Floods are so dangerous for the environment mainly when sources of pollution are situated in the flooded area. A pollution that results from a human activity may cause deterioration in the status of a water body. In most cases, such pollution relates to the addition, or release, of substances into the environment. This can be the leaching of a waste product, but may also be the side-effect or by-product of some other activity, such as the leaching of nutrients from agricultural land. A pollution pressure may also be caused by an action such as a change in land use, for example sediment fluxes are modified by urbanization, forestry, and a change between winter and spring planting of crops. The most usual categorization of pollution pressures is to distinguish between diffuse and point sources (Table 1). However, the distinction between point and diffuse sources is not always clear and may again relate to spatial scale according to European Commission [9]. For example, areas of contaminated land might be considered as either diffuse or point sources of pollution.

The assessment of whether risk on a water body is significant must be based on knowledge of the pollution within the catchment area, together with some form of conceptual understanding, of water flow, chemical transfers, and biological functioning of the water body within the catchment system [9]. Finally, flood risk analysis provides a rational basis for prioritizing resources and management actions mainly in the flood vulnerable areas. Risk analysis can take many forms, from informal methods of risk ranking and risk matrices to fully quantified analysis [10].

The paper studies the impact of the sources of pollution and flooding water on the environment. Environmental impact study in Bodva river basin includes the natural condition identification as well as human activity identification which may cause environmental damages in case of floods in the river basin.

2 STUDY AREA

Bodva river basin (Fig. 1) is situated in the southwestern part of the Kosice region. The river Bodva rises in the mountains Volovske hills, on the north eastern slope of the hill Osadník (1186 m asl.). The length of its flow in Slovakia is 48.4 km. Within the territory of Hungary it flows into the Slaná River. A larger right-hand side inflowing tributary of the Bodva is the Turňa and on the left-hand side the Ida. The whole area of Kosice region belongs into the zones of Inner Western Carpathians (Fig. 2). Geological structure of the area forms the hydrogeological conditions of the sub-basin Bodva. Older Paleozoic rocks whose original character before metamorphosis was volcanic with intergranular permeability are characteristic by fissure permeability. There is a predominance of heavy loamy soils that occupy contiguous area of Košice basin. Sandy-loam soils occupy forests in the mountains Volovske hills and partially Slovak Karst MoE & SWME [11].

The orographic condition range from 168 m asl. (Hosťovce) to 1186 m asl (Osadník). Height and slope conditions affect climatic conditions, especially the size and distribution of rainfall, and thus on the overall water balance and runoff regime. Sub-basin Bodva, regarding the complex orographic ratio, ranges into several climatic zones. South and east part – the

Figure 1: Localization of Bodva river basin in Slovakia.

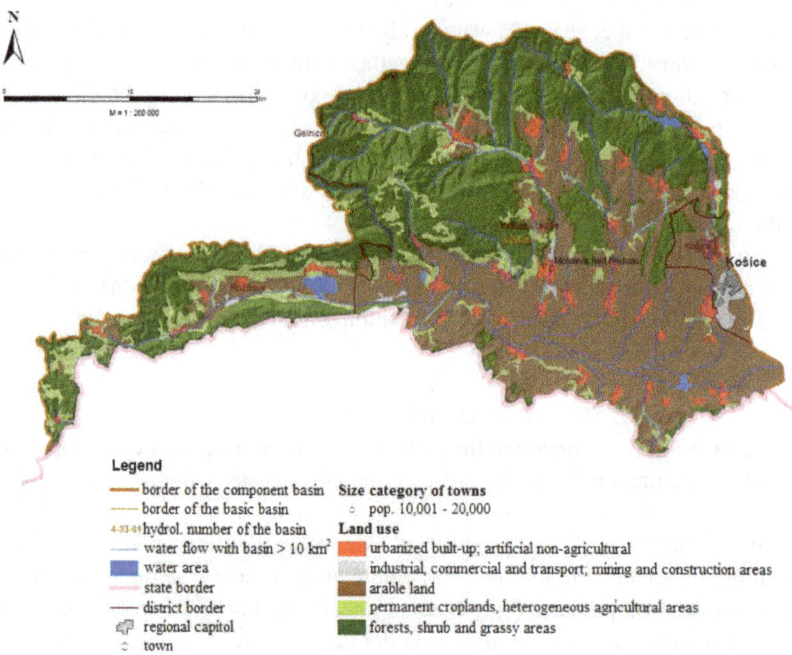

Figure 2: Characteristics of Bodva river basin in Slovakia.

largest part of the basin belongs to warm climate, which is and slightly damp with cold winters. Long-term average annual air temperatures range from 5°C to 8°C. Long-term average annual precipitations in the basin range from 600 to 1000 mm. The land use is as follows: artificial surfaces – 4.7%; agricultural areas – 48.1%; forested and semi-natural areas – 46.6%; wetland areas – 0.1%; waters – 0.5%, according to Zeleňáková [12].

The first mention of floods in Bodva river basin with serious consequences on the population which occurred in chronicles is the flood of 1813, when flooding affected the town of Moldava nad Bodvou. The next large flood is recorded in July of 1845, but from sources' descriptions its course and consequences are only known in the neighboring basins (the Hnilec and Slaná), and it probably also affected the component basin of the Bodva. Long-lasting and exceptionally profuse rains caused flooding situations in the Bodva basin also in the years 1973, 1974, 1989, 1995 and 1996. In 1997 and 1998, no water stages higher than the water stage designated for first degree of flood activity were recorded in the hydro prognosis stations of the SHMÚ in the component of the Bodva basin. In 1999 in the component basin of Bodva, 10 days occurred in which a water stage designated for degrees of flood activities was reached or surpassed in the hydroprognosis stations of the SHMÚ. And the year 2005 belonged among years with flood activities. In the Bodva basin at the water measuring station at Turňa nad Bodvou with a flow of $Q_{max} = 43.4\ m^3s^{-1}$, the level of the water rose to a water stage level of 293 cm, which corresponds to the level designated for the second degree of flood activity. And even 2006 did not pass without flood events, which occurred especially in March, April, May and June. Water stages corresponding with the degree of flood activities were achieved or surpassed in nearly all flows in the basin. In 2009, the first flood situation occurred in the final third of January, which was the warmest and at the same time when the highest total rainfall on the entire territory occurred. The next flood situation in the Bodva basin occurred at the end of 2009, in the second half of December. The year 2010 was from the viewpoint of rainfall exceptionally above average in the Bodva basin. The first flood of 2010 in the component basin of the Bodva occurred in the first half of January. In April, a second flood occurred which resulted from long-lasting and substantial rainfall which fell over the course of 3 days. These rains caused a rapid and significant rise in water levels. The floods in May and June 2010 were exceptional from the viewpoint of time and spatial distribution in the Bodva basin. In nearly all water measuring stations water stages designated as third degree flood activities were surpassed. Further above-average rainfall occurred in November. Precipitation activities continued even into the early part of December and the larger part of the precipitation was repeatedly in the form of rainfall. In the scope of a flood risk assessment in the Bodva river basin, five areas (17.2 km – the length of river sections in flood risk) were identified with the occurrence of significant flood risk, according to MoE & SWME [11], from this (Fig. 3):

- Areas with the existing potential for significant flood risk (6.7 km)
- Areas with probable potential for significant flood risk (10.5 km).

The present state of flood protection not only in the component basin of the Bodva but also throughout Slovakia is the result of long-term development. Agricultural development and the building of industry which was connected primarily with the development of towns has forced the construction of technical preventive measures for flood protection in the landscape and along watercourses. Present flood protection infrastructure in the component basin of the Bodva takes into consideration especially the alteration of watercourses, protective levees and reservoirs. No dry basins are located in the area.

In the component basin of the Bodva, there are three reservoirs, the function of which is to change the time progression and size of water flow rates or to hold water so that it can be used most usefully and not cause any damage. With the building of protective levees or anti-flood lines, the increasing capacity of the watercourse for protection of the territory from flooding

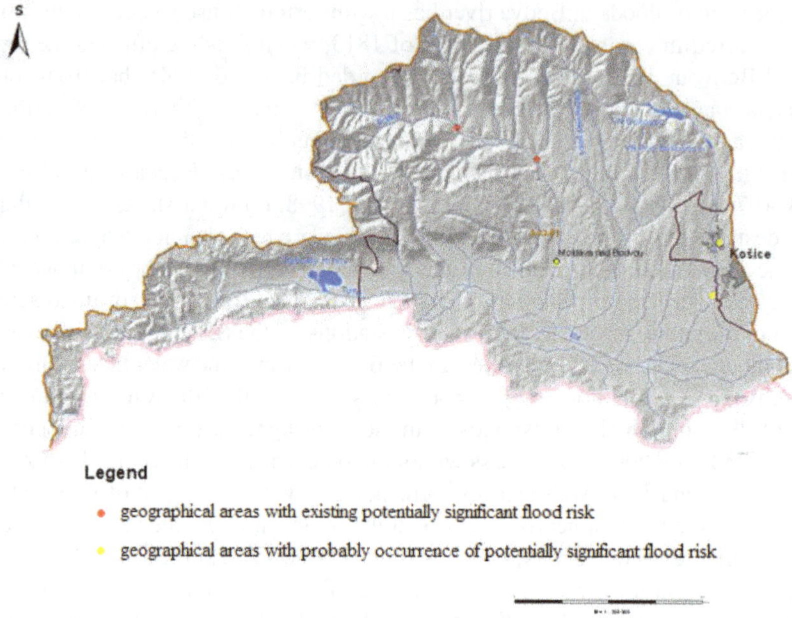

Figure 3: Geographical areas with potentially significant flood risk in Bodva river basin.

with smaller flow rates or equal to the proposed flow rate follows. The goal of modification of watercourses is to create favorable conditions for their water management use and to remove the consequences of their damaging effect. At present, several alterations of watercourses are being carried out in the component basin of the Bodva and four protective levees are being built.

A preliminary flood risk assessment from flash floods, that is the assessment of an area with this flood risk, requires a quantity of data foundations of a corresponding range.

The methods which are used in the contribution emerge from practical experience as well as from knowledge obtained from the available literature and consultations with experts dealing with the given problem in practice. The foundations of flood risk assessment are hypotheses on the impact of physio-geographical attributes of basins on the origin of floods. The conceptual framework for flood risk assessment of flash floods comes from the definition of a multidimensional conception of flood risk, and a process approach is used.

3 RESULTS AND DISCUSSION

The principle philosophy of assessing damage to the environment comes from the fact that all three environmental elements evaluated (protected biotopes, water and soil) can be damaged by the impact of a whole range of floating hazardous substances from potential sources of contamination. Attention is therefore devoted to the quality of the water, exactly to the evaluation of sources of pollution situated in flooded area, which can cause the potential environmental damages in case of floods.

Potential sources of water contamination are all activities in the catchment, all contaminated sites and landfills, industrial activities, agricultural activities, and other activities.

Point sources of pollution, that is, potentially, were also found through three dedicated databases, according to Water Research Institute [13] and Slovak Environment Agency [14]:

- KV-ENVIRO, which contains more than 13,004 potential point sources of pollution. The basis of this database is a database GeoEnviron, which contains 9177 potential point sources of pollution. These are 2279 sites, 6938 landfills and other sources of pollution.
- Register of contaminated sites (REZ), which is part of the Information System of Slovak Environment Agency, was built under the project Systematic identification of contaminated sites in the Slovak Republic. It contains 1936 sites, which are divided into three parts: likely environmental burdens (Part A) - 902 sites, environmental burdens (Part B) - 281 sites, rehabilitated and reclaimed environmental burdens (Part C) – 753 sites.
- Database: Integrated monitoring of pollution sources, which contains sources of pollution of hazardous substances, which monitor their impact on groundwater. This database currently contains more than 310 sources of pollution, mainly landfills.

Flooding of the above mentioned sources of pollution as well as diffuse sources of pollution may leak out pollutants and thus deteriorate the quality of surface water, groundwater, and soils, which can lead to environmental disasters, such as damage of habitats, fauna and flora as well as diseases and epidemics occurrence.

The starting point for assigning a category of the consequence is categorization of potential sources of contamination influencing the quality of water (Table 1).

At present, no complete register of sources of contamination exists in the Slovak Republic; therefore, for the needs of this paper one of our own designs was created. The individual sources of contamination are divided into two main categories: point and diffuse sources of contamination. Table 1 presents the divisions of the individual sources of contamination, their point classification and the weight indicating the importance of the component categories. When assigning points an inverse order is used; thus, the value 5 represents the greatest threat of contamination in the case of flooding. The individual weights of the component categories, which are standardized such that their sum is equal to one, are set in this same way. The individual points are set on the basis of a professional judgement.

The actual sources of pollution in the studied area, situated in the flooded area are marked in bold in Table 1. It is resulted as follows:

- Agricultural area (1,544,701 m^2) contains 49.83% from whole the flooded area (3,099,950 m^2) (by Q_{100}) – data from GIS (done by authors);
- Percentage of population without sewerage is 24.7%, as from 29,390 of equivalent inhabitants 46.6% is connected to sewerage systems and 28.7% has individual systems of wastewater treatment – based on data from Medzev City [15];
- Environmental burden – there is one confirmed burden in the study area 'Strojsmalt', storehouse of heavy oils and fuels, in Medzev (Fig. 4) – according to the Register of contaminated sites;

The flooding of these objects can lead to leakage and the washing away of contaminating substances and thus to worsening the quality of surface and underground water and soils, which can lead to ecological catastrophes, such as damage to biotopes, the killing of fauna and flora as well as epidemics. Very serious damage of a long-term character is related to underground aquifers, which are used as sources of drinking water. The total consequence

Table 1: Classification of sources of pollution.

Source of pollution	Characteristic	Criteria	Point classification	Weight
Point sources of pollution				
Industrial enterprises	Category of enterprise	unclassified	5	0.2
		A		0.3
		B		0.5
Sewage treatment plants	Number of equivalent inhabitants	< 2,000	4	0.14
		2,000–10,000		0.21
		10,000–100,000		0.29
		100,000 and more		0.38
Petrol stations	-	-	4	1
Diffuse sources of pollution				
Landfills	Type of landfill for	inert waste	5	0.12
		non-hazardous waste		0.29
		hazardous waste		0.59
Impoundments	-	-	3	1
Mine activities	Type	underground	2	0.4
		surface		0.6
Population in urban areas	Percentage of population without sewerage	**0–40**	**3**	**0.12**
		40–60		0.29
		60–100		0.59
Agricultural areas	Percentage of arable land in the area	0–40	**3**	0.29
		40–60		**0.59**
		60–100		0.12
Environmental burden	-	is reclaimed	**4**	0.12
		is likely		0.29
		is confirmed		**0.59**

(Table 2) is calculated as the sum of the assigned points to the individual sources of contamination located in the flooded area for the given Q_N (probability of flood occurring) multiplied by the relevant weight, and this defines the negative impact on the environment.

The resulted Flood Environmental Impact and Risk Assessment (FEIRA) is stated as the sum of the state of the environment (EIA – environmental impact assessment), predisposition for floods and sources of pollution in the area (RA – risk analysis) and present hazard in case of flood occurrence. The philosophy of the methodology is stated in Fig. 5.

In general expression environmental impact assessment is probability (P) x consequence (C) of adverse impacts of floods for the environment. Risk analysis is hazard (H) x weight (W). Hazard in environment is presented by sources of pollution in the area which present a serious risk for the environment in the case of flood. All calculations are not presented in the paper because of the length of the article. All calculations are made from presented information of Bodva river basin. The resulted FEIRA is stated as the sum of the state of the

Figure 4: Strojsmalt in Medzev situated in flooded area.

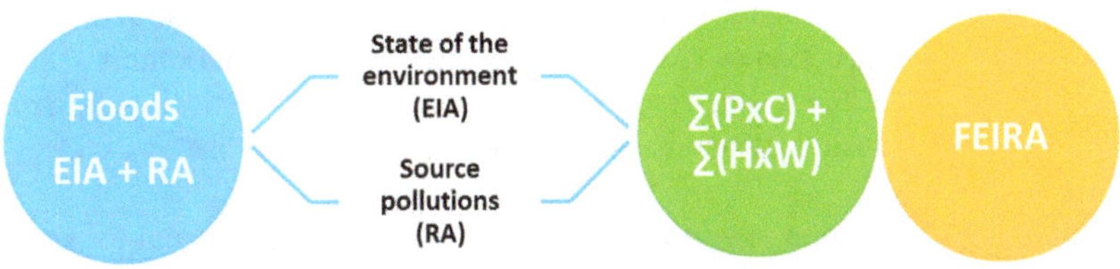

Figure 5: Methodology of FEIRA.

environment and the classification of sources of pollution by point classification. Flood risk in Bodva river basin was stated as negligible and low.

4 CONCLUSION

Basically, two phases are applied in this study to analyze flood vulnerability: firstly to identify the effective factors causing floods – the potential natural causes of flooding, and secondly to apply methods of MCA in GIS environment to evaluate the flood vulnerability of the area.

The flood vulnerable areas in the study area Bodva watershed were evaluated in four classes. Since the methods take into account some conditions of the region, the results can be as realistic only for this condition. When the characteristics would change, different results would be obtained. The subjective numbers in the weights and the values of the criteria can be changed according to the study area characteristics and experts' opinions. Our pilot study showed the possibility of using MCA and GIS for flood vulnerability assessment

On the basis of the knowledge of the current state of existing modifications we propose, in the resolved area of the town of Medzev, these possible preventive flood mitigation measures:

- Removing of sediments from the riverbed and vegetation on the bank of the water flow, thus securing the flow-rate capacity of the channel of the watercourse,
- On unmodified segments of the watercourse, carry out modification, for example, reinforce the slopes of the watercourse banks,
- Possible construction of a reservoir above the town which lowers the maximum flow rate during increased water stages.

In the Bodva river basin, according to Slovak Water Management Enterprise, state enterprise Branch Office Kosice, the total vulnerability of property is 57,306,649 euro and number of inhabitants endangered by floods is 4575. The proposed preventive measures include the following [16]: measures that slow drainage of water from the basin into waterways, increasing the retention capacity of the river basin or support the natural accumulation of water in areas of suitable territory and protecting territory against flooding by surface drainage, such as adjustments in forests, farmland modifications and adjustments to the urbanized areas,

- Measures that reduce the maximum flow of floods, such as construction, maintenance, repair and reconstruction of water works and polders
- Measures that protect the area against flooding water from a watercourse, such as the modification of water courses, construction, maintenance, repair and reconstruction of levees and flood lines along waterways
- Measures to protect the area against flooding inland waters, such as construction, maintenance, repair and reconstruction of facilities for the transfer of internal waters
- Measures to ensure the flow capacity of the watercourse bed as removing silt from the riverbed of the watercourse, and stands on the banks of the watercourse.

The proposed flood mitigation measures in Jasov (area of existing potential for significant flood risk) and Moldava nad Bodvou (area with probable potential for significant flood risk) are as follows Vágášiová [16]:

- Increasing of flow capacity by construction of riverside walls and earth dyke
- Extended riverbed, fortify river channel slopes in the local area of Budulov (part of Moldava)
- Area between the town of Moldava nad Bodvou and Jasov should be left to the natural transformation of flood waves

The total estimated costs of flood mitigation measures in Bodva river basin are 40,678,000 Euro.

ACKNOWLEDGEMENT

This paper was written thanks to support from project VEGA 1/0609/14.

REFERENCES

[1] Directive 2007/60/EC of the European Parliament and of the Council of 23 October 2007 on the assessment and management of flood risks.

[2] Act no. 7/2010 of Slovak republic of 27 October 2004 on the flood protection.
[3] Zeleňáková, M., Preliminary flood risk assessment in the Hornád watershed. In: *River Basin Management V 15,* Wessex Institute of Technology: Ramla Bay Resort, Malta. Southampton, pp. 15–24, 2009.
http://dx.doi.org/10.2495/rm090021
[4] Zeleňáková, M., Flood risk assessment and management in Slovakia. In: *River Basin Management 6,* WIT Press: Riverside, California, USA. Southampton, pp. 61–69, 2011.
http://dx.doi.org/10.2495/rm110061
[5] Hanák, T. & Korytárová, J., Risk zoning in the context of insurance: Comparison of flood, snow load, windstorm and hailstorm. *Journal of Applied Engineering Science,* **12**, pp. 137–144, 2014.
[6] Hlavčová, K., Kohnová, S., Kubeš, R., Szolgay, J. & Zvolenský, M., An empirical method for estimating future flood risk for flood warnings. In *Hydrology and Earth System Sciences,* **9**(4), pp. 431–488, 2005.
http://dx.doi.org/10.5194/hess-9-431-2005
[7] Korytárová, J., Šlezingr, M. & Uhmannová, H., Determination of potential damage to representatives of real estate property in areas afflicted by flooding. *Journal of Hydrology and Hydromechanics,* **55**, pp. 282–228, 2007.
[8] Solín, Ľ., Spatial variability in the flood vulnerability of urban areas in the headwater basins of Slovakia. *Journal of Flood Risk Management,* **5**(4), pp. 303–320, 2012.
http://dx.doi.org/10.1111/j.1753-318X.2012.01153.x
[9] European Commission: Analysis of Pressures and Impacts [online]. Luxembourg: Office for official publications of the European communities, 2003, available at: https://circabc.europa.eu/sd/a/7e01a7e0-9ccb-4f3d-8cecaeef1335c2f7/Guidance%20No%203%20-%20pressures%20and%20impacts%20-%20IMPRESS%20(WG%202.1).pdf
[10] Hall, J., Journal of flood risk management. *Journal of Flood Risk Management,* **3**(1), pp. 1–2, 2010.
http://dx.doi.org/10.1111/j.1753-318X.2009.01059.x
[11] MoE (Ministry of Environment), SWME (Slovak Water Management Enterprise), *Implementation of directive 2007/60/ES of 23rd October 2007 on flood risk assessment and management: Preliminary flood risk assessment in Bodrog river basin,* Attachments [online]. December 2011 available at: http://www.minzp.sk/files/sekcia-vod/bodrog-prilohy.pdf
[12] Zeleňáková, M., Assessment of flood vulnerability in Bodva catchment using multicriteria analysis and geographical information systems. In: *River Basin Management* WIT press: A Coruna, Spain, **8**, pp. 51–59, 2015.
http://dx.doi.org/10.2495/rm150051
[13] Water research institute (WRI), Database GeoEnviron, 2008.
[14] Slovak Environmental Agency (SEA), State list of protected areas of Slovakia, 2013, available at: http://uzemia.enviroportal.sk/
[15] Medzev City, Program of the economic and social development. *Socio-Economic Analysis,* 2013, available at: http://www.medzev.info/mesto/legislativa/phsr
[16] Vágášiová, M., *Experiences in processing of flood risk management plan in terms of Slovak Water Management Enterprise, s.e.,* Branch Office Košice. 2014, available at: http://www.urbiproof.info/wp-content/uploads/2014/09/08_Vagasiova.pdf

FLOOD DAMAGE ASSESSMENT IN A GIS – CASE STUDY FOR ANNOTTO BAY, JAMAICA

H. GLAS[1], S. VAN ACKERE[2], G. DERUYTER[1] & P. DE MAEYER[2]
[1]Department of Civil Engineering, Ghent University, Belgium.
[2]Department of Geography, Ghent University, Belgium.

ABSTRACT

Natural hazards do not only affect millions of people, but also cause material damages up to 300 billion USD per year worldwide. The SIDS (Small Island Developing States) are characterized by an extremely high vulnerability to these hazards, due to their low-lying, densely populated cities and their fragile economy. To limit the consequences of these hazards, technocratic interventions do not suffice. Therefore, new approaches that focus on flood risk management are developed. In this context, Annotto Bay, Jamaica, was chosen as a case study area to perform a flood damage assessment. In this study, a flood damage map was created for the 2001 flood caused by Tropical Storm Michelle. This map focuses on three types of damage: building, road and crop damage. The first type was calculated using the exact GPS locations of the buildings, as well as average replacement values for each building type and flood damage functions. The total building damage was then combined per land-use polygon to have an orderly visual view of the damage spread. Furthermore, the road damage was calculated, based on a road network extracted from satellite imagery. In a next step, as buildings are mostly located in proximity of roads, buffers were created around the road network, resulting in a more accurate visual view of the building damage spread. Then the crop damage was calculated based on maximum damage values for banana plantains and other crops, combined with the crop damage functions. The final result is a total damage map, visualizing the location of high risk areas with a high accuracy. Additionally, the total calculated damage was compared to the actual damage caused by the 2001 flood. This comparison shows promising results.

Keywords:damage map, flooding analysis, natural hazards, risk assessment.

1 INTRODUCTION

In the last two decades, the growing frequency and intensity of natural hazards introduced 'the era of catastrophes'[1]. The 2004 tsunami in Thailand, the 2010 earthquake in Haiti, and the 2015 earthquake in Nepal are only a few examples of natural disasters that hit these countries at social, economic, and political levels, due to devastating human and material losses. These irregular, large-scale events are not the only ones that can debilitate communities. Yearly flooding in the UK, for example, causes economic damages up to 250 million dollars [2]. To make matters worse, these damage numbers will only grow in the future, due to climate change and its associated increase in severity and frequency of hazards[3].Although natural hazards occur in developed countries as well as in developing countries, the disaster risk is disproportionately concentrated in low- and middle-income states, in particular in the small island developing states (SIDS)[4]. The extreme vulnerability is caused by the low-lying, densely populated cities and the fragile economy, characterising these regions[5].

DOI: 10.2495/SAFE-V6-N3-508-517

Flooding affects more people worldwide than any other hazard[6]. Technocratic interventions to prevent damage, such as dams, levees, and flood forecasting, have shown limited success in the past [3,7]. In order to truly minimize the socio-economic cost induced by inundations, an adequate flood risk assessment is necessary to compute the damage of future floods as accurately as possible. Since the SIDS are expected to lose 20 times more of their capital stock in disasters each year, than Europe and Central Asia [4], these new approaches that indicate high risk areas are indispensable to help decision makers to better allocate their limited resources and funds to limit the consequences of flooding, rather than to prevent flooding from occurring [7].

In this research, a damage map was created as a first step towards such a risk-based approach. Maximum damage values, or replacement values, were multiplied with a damage factor to calculate the real damage cost. This damage factor is the percentage of damage to an element at risk for a certain flood level. In order to perform this assessment, flood data and land-use values were acquired.

1.1 Case study area

Jamaica, one of the SIDS, suffers regularly from severe flooding. Over a period of four years, the estimated economic damages due to inundations were estimated at 1.5 billion USD [8]. Annotto Bay, a coastal town in the north-east of the island, has a low-lying urban area, with approximately 5,500 inhabitants. Combined with its location downhill from the Blue Mountains and four rivers traversing the city, this topography indicates an extreme vulnerability to severe inundations. Moreover, the alluvial deposits in the soil cause flash floods to happen, even after very short periods of precipitation in the mountains [9]. Therefore, Annotto Bay was chosen as a case study area for this flood damage assessment. Figure 1 shows the land use in the town, as well as the location of the main roads and buildings.

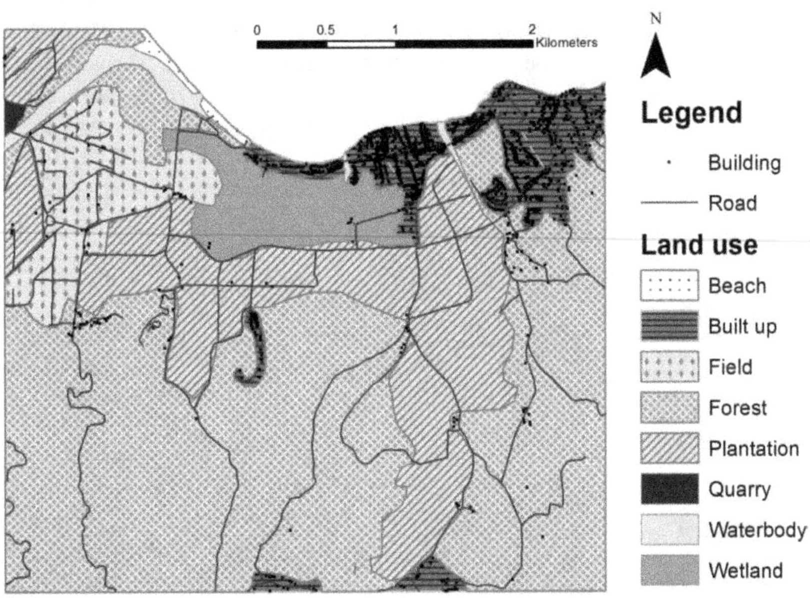

Figure 1: Annotto Bay land-use map.

2 METHODS

The approach that was used to create the final damage map for Annotto Bay is based on three types of input data, as can be seen in Fig. 2. The first type is the land-use data, or elements at risk, which can include building locations, the road network and agricultural information. The level of detail of this data determines the level of detail of the resulting damage map. The second type of input information is the economic data. This type comprises the replacement values for all land use. A replacement value is the cost to replace an element at risk after it has been completely destroyed. The last type of input data is the flood data. In this case, to create one damage map, the flood extent of the flooding caused by Tropical Storm Michelle in 2001 was used. During this storm, a large area of the town was inundated for a period of two days and many elements at risk were damaged.

By combining the replacement values with the land-use data, a maximum damage map was generated, showing the damage in the area if the flooding would destroy 100 percent of the elements at risk. This map was then combined with the flood data, using damage factors per land-use type. Consequently, the Resulting damage map shows the damage for a location, based on the element at risk and its economic value at that location for the given flood level.

This process was carried out for the three types of damage: building, road and crop damage, since the input data and calculation methods for each type differ immensely. In a final step, the separately calculated damages were combined to generate an overall damage map.

2.1 Input data

In developing countries, data availability is a major issue [11].Therefore, the data acquisition is in many cases time-consuming. However, since ODPEM[9] has performed a multi-hazard risk assessment in the area in 2013, most land-use data were already acquired. Table 1 lists the data that were used in this research. Although a road dataset was available, the road network was extracted from satellite imagery, because the original dataset was incomplete, as many dirt roads were not included. Even though the damage to these roads is minimal, they do provide access to buildings, making them an important visual and spatial element.

Since the elements at risk can be divided into three groups, buildings, roads and crops, three types of replacement values were acquired. ODPEM [9] provided average building values for different building materials. By adding the number of floors, available in the

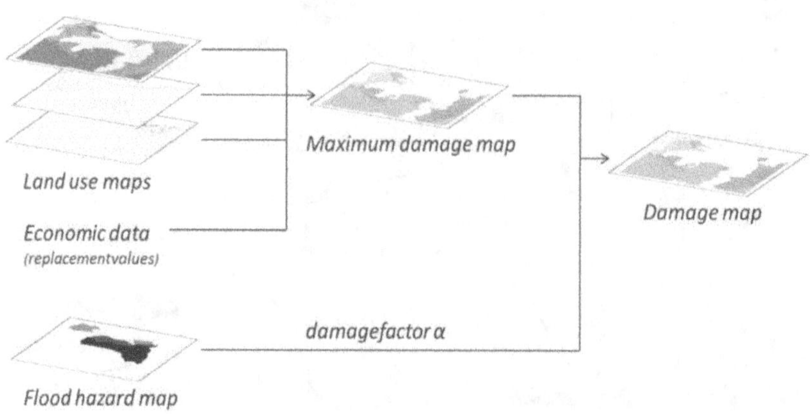

Figure 2: General methodology for Annotto Bay damage assessment [10].

building location data, the total cost of each building was determined. Average crop values were acquired from FAOSTAT [14]. To determine the last replacement value, for roads, an average cost for developing countries was used [15].

Tropical Storm Michelle induced heavy precipitation in the mountains, resulting in flash floods downstream [9]. Annotto Bay suffered from flooding up to 4 feet over a period of two days. A flood map was drawn of the maximum flood extent during these 48 hours. This map, shown in Fig. 3, was used as input flood data in the damage assessment. Since this map only shows water heights, the velocity was not taken into account in the calculations.

The flood map and the maximum damage, calculated by combining land-use data and replacement values, are merged using damage factors, extracted from depth-damage functions. The depth-damage functions for buildings and crops were acquired from Dutta *et al.* [16]. These functions were drafted for Japan, but since this island shows similar cultural, geographical and topological characteristics, they were transferred to Jamaica.

Table 1: Input land-use data.

Data	Type	Source
Land use	Polygon	NLA [12] + update based on DigitalGlobe satellite imagery [13]
Road network	Polyline	Based on DigitalGlobe satellite imagery [13]
Buildings	Point	ODPEM [9]
Critical buildings	Point	ODPEM [9]

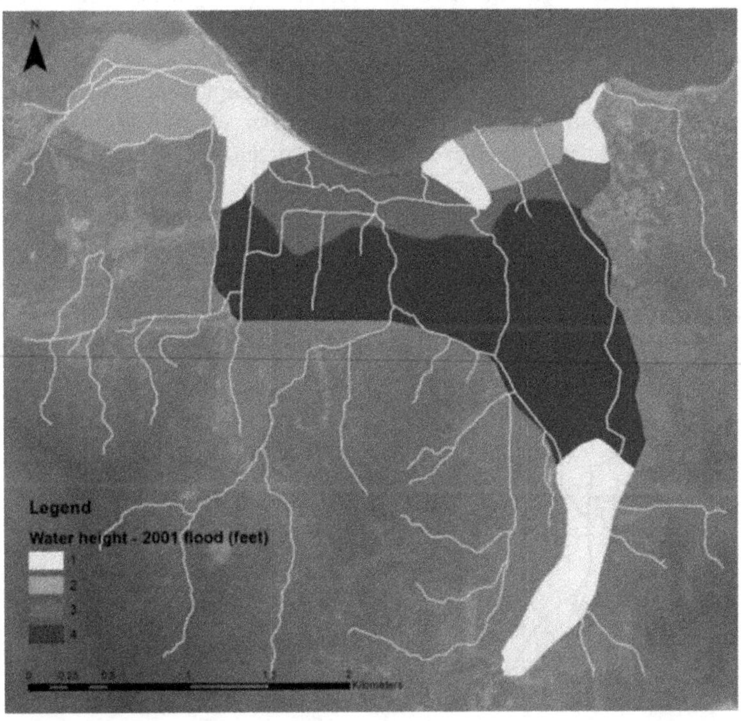

Figure 3: Flood extent of 2001 inundations caused by Tropical Storm Michelle in Annotto Bay, Jamaica [11].

2.2 Building damage

Since the exact locations, building materials and number of floors are known for all buildings in Annotto Bay (see 2.1), an adequate damage assessment could be performed. With the use of average building areas and material costs, a maximum damage value per building was calculated. Using the point damages as such in the visualisation would lead to a visual underestimation of the total damage, since the damaged points would seem very small in comparison to the entire study area. In order to generate a better visual result and facilitate the interpretation of the damage map, the building point features were aggregated per land-use polygon. All damage values of all buildings in one polygon were added up to one building damage value per land-use polygon, creating damage zones. The polygon layer was then combined with the damage factors, based on Dutta*et al.* [16] to create the building damage map.

Every building must be accessible from a road. This assumption was applied in this assessment by creating two buffers around the road network, at 25 meters and at 60 meters. Samples in the study area showed that 90 percent of all buildings are located in the 25 meter buffer and 99 percent in the 60 meter buffer. The visual result of this adaption is illustrated in Fig. 4. The spatial distribution of building damage is more accurate and high risk areas are distinctly indicated. Especially large land-use polygons without any damage except building damage, for example in the wetland areas, benefit immensely from these buffers.

2.3 Road damage

A road dataset of the area was available for this research, giving a class and corresponding width to each road [17]. Although this information allowed a more precise damage calculation, a comparison with satellite imagery showed big disparities between the available dataset and the reality. Many dirt roads were missing and the simplification of the data, due to its larger scale, caused the location of other roads to differ from reality. Additionally, damage to roads showed to be significantly smaller than building damage. Therefore, the priority was given to adding all dirt roads and losing the class information to enhance the accuracy of the building damage spread.

The road damage was calculated by multiplying the average replacement value [15] with a damage factor. This factor is based on calculations for LATIS, a Flemish flood model [7]. Since no road classes were used, an average width was assigned to each road to calculate the damage per square meter.The roads are thus visualised as polygons in the final damage map.

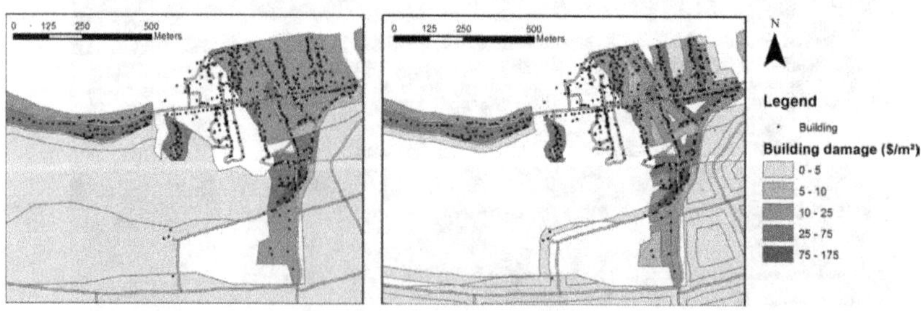

Figure 4: Visual comparison between building damage calculations without buffers (left) and with buffers (right).

2.4 Crop damage

Finally, the crop damage was calculated. Since the available land-use data distinguishes banana plantains from other crops, two different damage functions were used. For the former, a damage factor of 100% was used, since banana plants have a fragile root that cannot survive a 48 hour flood.In Dutta*et al.*[17], all crops were selected that were also found in the study area. An average of these damage factors was used for the latter land-use type.

To calculate the damage, replacement values needed to be acquired. From FAOSTAT[14], average values for the same crops as in the damage factor calculations, were gathered and combined into one value. Furthermore, the value of banana plants was acquired. Hence, these values were multiplied with the damage factors to generate the crop damage.

3 RESULTS

Before generating the final damage map for Annotto Bay, the three types of damage are visualised separately in Fig. 5. The cost of each type of damage and the damaged area is given in Table 2. It is immediately clear that the building damage cost is substantially higher than

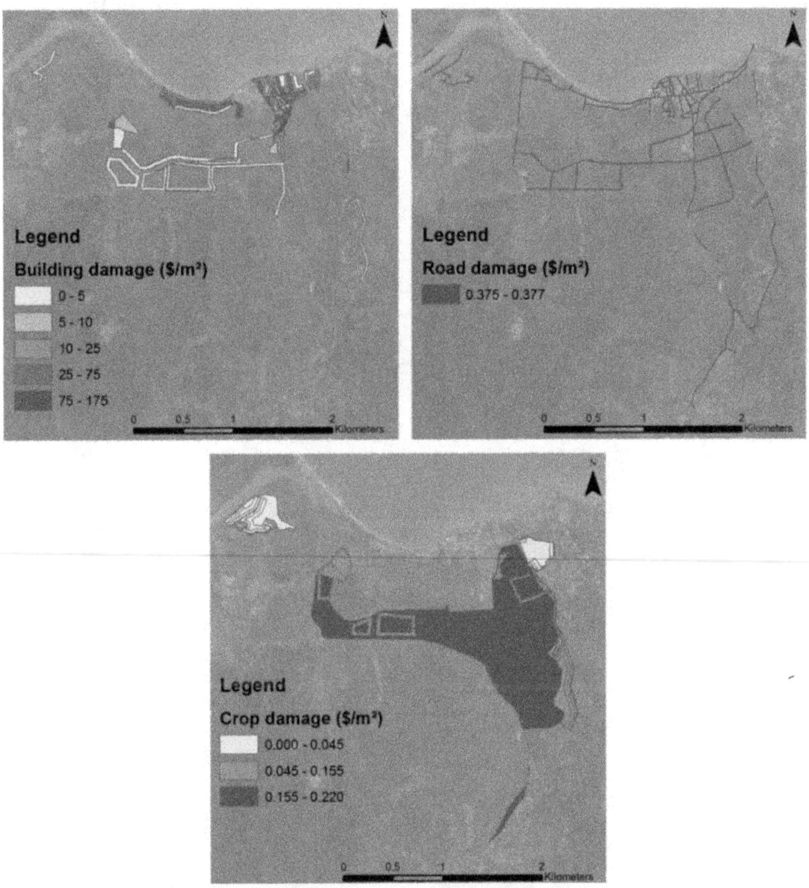

Figure 5: The damage maps for the 2001 flood in Annotto Bay, Jamaica, per damage type. (top left: building damage, top right: road damage, bottom: crop damage).

the cost of the other two types, while the crop damage has the largest spread. The total damaged area is not equal to the sum of the areas per damage type, since some polygons contain more than one type. For example, rural areas with housing will have building damage as well as crop damage.

As a result of the use of the two buffers for building locations, the total damaged building area is limited. Without the buffers, the damaged area for buildings would be 1 857 396m². However, when only looking at the average building area of the houses that were damaged, the building area of the damaged houses is only 132 903m². Hence, the buffers still cause an overestimation of the damaged building area, but the result is much closer to the reality.

In a next step, the final damage map was generated, by adding up the three separate damage types. This map is shown in Fig. 6. Due to the high damage values of buildings, the

Table 2: Overview of damage cost and damaged area per damage type.

Damage type	Damage cost ($)	Damaged area (m²)
Building damage	7 083 291	522 983
Road damage	51 210	136 284
Crop damage	371 787	2 049 042
Total damage	**7 506 288**	**2 458 702**

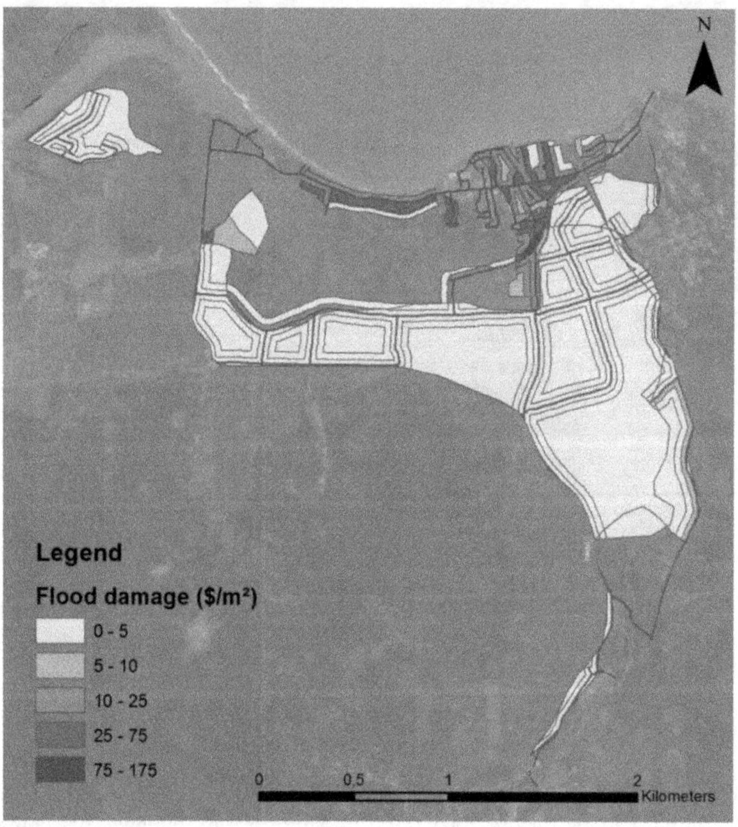

Figure 6: Final damage map for the 2001 flood in Annotto Bay, Jamaica.

distinction in crop damage has disappeared. However, it is clear that the high-risk areas are located in the low-lying urban areas and along the main road, parallel to the coastline. The wetland does not have any damage, except for the buffers along the roads, since only a few houses are located there.

With a maximum damage cost of 175 $\$/m^2$, this flooding has not completely destroyed any houses, considering that the replacement value of buildings is on average 836 $\$/m^2$.

4 DISCUSSION AND CONCLUSION

This damage assessment was executed as a first step towards a generic flood risk assessment in the SIDS. Hence, the focus was on developing a tool that can be used in other regions as well. The main goal was to visualize the high-risk areas and give an adequate estimation of the damage. Therefore, the preference was given to a road dataset based on satellite imagery of DigitalGlobe [13], rather than using the available road data with classes. This new dataset has lent the opportunity to add buffers that localise the buildings more precisely. Given that the building damage cost is the biggest share in the overall cost, this approach offered a better visual result. When this approach is used in future study areas, the buffers around the road network will help to better allocate building damage, even when the land-use data are less detailed.

By comparing the results of this flood tool with the real damage values, the accuracy of the assessment is tested. However, information about the damages done by the 2001 flood are hard to find. According to the multi-hazard risk assessment report of ODPEM[9], the number of affected houses is 749, or 46%. In this damage assessment, 799 houses, or 49%, are estimated to be affected. The small overestimation could be expected, since all houses in a certain flood level are assumed to be equally affected by the inundation. In reality, however, some houses will resist better and they will thus have less damage or no damage at all. No information was available about the damage cost to the buildings due to the flood, so this aspect could not be compared.In order to test the accuracy of the calculated damage costs, more information about the actual damages and corresponding costs, caused by the inundations, is necessary in further research.

When looking at road damage, the ODPEM report states that 3.5km of the main roads were damaged during the 2001 flood [9]. This number is impossible to compare with the results of the road damage assessment, since the road dataset that was used does not make a distinction between main roads, secondary roads and dirt roads. According to the flood tool, 22.7km of roads are affected by the inundations. Since the total road damage cost is a very small share of the total damage cost, the accuracy of this result and the importance of these calculations can be questioned, but the roads do give access to buildings and serve as evacuation routes during disasters. One could argue that the road damage cost can be neglected in urban areas, where building damage takes up a much larger part of the total damage. In rural areas, however, the road damage cost and crop damage cost are within a similar order of magnitude. Hence, further research should focus on the importance of the road damage calculations in these rural areas to determine if future flood damage analysis should only take into account the accessibility of roads, rather than the accurate damage cost.

While ODPEM noted a damaged crop area of 620 000m^2 [9], the flood tool calculated an area of 2 049 042m^2. This difference is mainly due to the crops that were taken into account in both calculations. The report only has numbers for a limited amount of crops.The tool, however, generated a result for all crops. Unfortunately, the damage cost for crops was not available. A useful comparison could thus not be made. Furthermore, the cost of crop damage is insignificant in comparison to building damage, since Annotto Bay is an urban area. A rural area, where more crop data are available, should be tested in the future to decide on the reliability of the crop damage calculations of this flood tool, visually as well as quantitative.

Although the exact damage costs could not be examined on their correctness, the visual result does give a correct view on the damage spread and the areas with the highest damage. The high-risk areas are clearly indicated and correspond to the areas that were affected most by the 2001 flood. Hence, this methodology can be used by decision makers as a qualitative tool. The quantitative results must be tested further to determine their adequacy. The focus of future research must be on study areas with sufficient information of past flood events and the corresponding damages. As many values used in the flood tool, are averages or estimations, this step is indispensable to create an accurate, quantitative flood damage assessment, usable in all SIDS.

ACKNOWLEDGEMENTS

The authors would like to thank ArpitaMandal and Sherene James-Williamson from the University of the West Indies, campus Mona, Department of Geography for their help in acquiring the necessary data.

REFERENCES

[1] Kunreuther, H.C., Michel-Kerjan, E.O., Doherty, N.A., Grace, M.F., Klein R.W. & Pauly, M.V., *At War with the Weather. Managing Large-Scale Risks in a New Era of Catastrophes*, The MIT Press, pp.xiiv–xvii, 2009. http://dx.doi.org/10.7551/mitpress/9780262012829.001.0001

[2] Penning-Rowsell, E.C., A realistic assessment of fluvial and coastal flood risk in England and Wales.*Transactions of the Institute of British Geographers,* **40**(1), pp. 44–61, 2014. http://dx.doi.org/10.1111/tran.12053

[3] Gall, M., Borden, K.A., Emrich, C.T. & Cutter, S.L., The unsustainable trend of natural hazard losses in the United States.*Sustainability,* **3**(11), pp. 2157–2181, 2011. http://dx.doi.org/10.3390/su3112157

[4] United Nations Office for Disaster Risk Reduction (UNISDR), Making development sustainable: the future of disaster risk management. *Global Assessment Report on Disaster Risk Reduction*. UNISDR, pp. 1–266, 2015.

[5] UNESCO, Changing small island developing states: A space perspective on environmental change in the Caribbean, UNESCO, pp. 1–122, 2014.

[6] International Strategy for Disaster Reduction (ISDR), *Global Assessment Report on Disaster Risk Reduction,* ISDR, pp. 5–10, 2009.

[7] Deckers, P., Kellens, W., Reyns, J., Vanneuville, W. & De Maeyer P., A GIS for flood risk management in Flanders.*Geospatial Techniques in Urban Hazard and Disaster Analysis*, Springer: The Netherlands,pp. 51–69, 2010.

[8] Office of Disaster Preparedness and Emergency Management (ODPEM), National progress report on the implementation of the hyogo framework for action (2011-2013), ODPEM, pp. 1–47, 2013.

[9] Office of Disaster Preparedness and Emergency Management (ODPEM), *Multi Hazard Risk Assessment Annotto Bay, St. Mary,* ODPEM, pp. 1–158, 2013.

[10] Glas, H., Jonckheere, M., De Maeyer, P. & Deruyter, G., Risk assessment in the Caribbean: modeling a GIS-based flood risk tool for Jamaica.*First Young Researchers Overseas' Day*, pp. 1–8, 2014.

[11] Glas, H., Jonckheere, M., De Maeyer, P. & Deruyter, G., A GIS-based flood risk tool for Jamaica: the first step towards a multi-hazard risk assessment in the Caribbean. *15th International Multidisciplinary Scientific Geoconference SGEM 2015 – Conference Proceedings Volume II,* STEF92 Technology Ltd.: Albena, Bulgaria, pp. 643–650, 2015.

[12] The National Land Agency of Jamaica (NLA), *Landuse Jamaica,* NLA, 2001.

[13] DigitalGlobe, *World Imagery (Jamaica),* 8/8/2010.

[14] FAOSTAT, *Food and Agriculture Organization of the United Nations: Statistics Division* (foastat3.fao.org/download/Q/*/E: foastat3.fao.org/download/Q/*/E), 2014.

[15] Collier, P., Kirchberger, M. & Söderbom, M., *The Cost of Road Infrastructure in Developing Countries, Centre for the Study of African Economies*, Derpartment of Economics, Oxford, 2013.

[16] Dutta, D., Herath, S. & Musiake, K., A mathematical model for flood loss estimation. *Journal of Hydrology*, **277**(2), pp. 24–49, 2003.
http://dx.doi.org/10.1016/S0022-1694(03)00084-2

[17] WRA, *Jamaica Road Network,* 2007.

FACTORS UNDERLYING THE CONCEPT OF RISK ACCEPTANCE IN THE CONTEXT OF FLOOD-PRONE LAND USE

L. ALTARAWNEH, J. MACKEE & T. GAJENDRAN
School of Architecture and Built Environment, University of Newcastle, Australia.

ABSTRACT
The determination of acceptable risk levels for planning purposes is critical for policymakers concerned with floodplain management and safety issues; including the formulation of flood-prone land use policy and risk communication strategies. However, the concept of flood risk acceptance remains vague and is not yet fully understood in terms of how it is conceived and rationalised by the individuals engaged in flood-prone land use and development. In general terms, risk acceptance involves a complex weighing up of a range of influential factors that have evolved based on three key models: revealed preferences, expressed preferences and implied preferences. By investigating these models within the broader theoretical context of flood risk-related research, this paper, in essence, categorises six typologies that describe individuals' psychophysical/cognitive states when they face the risk, respond to it and determine its acceptability contextualised within the process of flood-prone land use change. The paper then focuses on identifying the key barriers influencing the adoption of an informed, consultative approach to acceptable flood risk assessment and governance.
Keywords: barriers, consultative approach, factors, flood-prone land use change, individuals, residential satisfaction, risk acceptance, risk perception..

1 INTRODUCTION

Within the scope of both sustainable development and disaster resiliency, the power of land use planning in mitigating risks from natural hazards (including flooding hazards) while promoting a better quality of life for vulnerable communities is well documented [1–4]. For urban development in flood-prone areas to continue in a resilient way, a level of flood risk must be defined as acceptable for planning purposes by all relevant parties, such as risk regulators, local planners, developers, insurers, investors and residents [5]. In general terms, risk acceptance is the concept that underlies judgement of safety [6]. It has generally been a feature examined within a wider framework of risk perception research [7]. Research within the broader context of flood hazard management has tended to concentrate on how flood-prone land users (residential and commercial occupants, and relevant stakeholders) evaluate and perceive location-related risks and benefits, and what socio-economic and demographic circumcentres make them vulnerable to flood hazards [5, 8–14]. There is a good understanding of why people continue to locate to flood-affected environments, and how they adapt to and perceive floods and flood-prone land use policy [15–20].

Nevertheless, the concept of flood risk acceptance remains vague and not fully understood in terms of how it is conceived and rationalised by vulnerable communities and individuals in the context of flood-prone land use. Few studies have been carried out [5, 14, 21–24], and

DOI: 10.2495/SAFE-V6-N3-518-528

these have focused either on the relationship between flood risk and its socio-economic and engineering factors, or between flood risk and other risks in a multi-risk context [13]. The relationship with all relevant factors has to be now rightfully constituted and examined from the viewpoints of individuals seeking a residential location, engaged in land development and construction activity, and charged with forming and implementing public flood control policies at various levels of government. This purpose of this paper emerges from this need: to provide a distinct conceptualisation for the determinant factors of flood risk acceptance from the viewpoint of individuals in the context of flood-prone land use.

The remainder of this paper is structured as follows: the next section is devoted to a fuller review of factors underlying the concept of risk acceptance in life, from the viewpoint of individuals, and then a fuller discussion of these factors is made in the realm of flood risk. Finally, the paper concludes with an outline of some barriers influencing the adoption of an informed, consultative approach to acceptable flood risk assessment and governance.

2 RISK ACCEPTANCE: THE HOLISTIC CONCEPT

Determining what level of risk is acceptable to individuals is not straightforward. It is a relative matter, in essence, as each individual holds a unique view towards the environment they choose to occupy, with its location-related risks [25–27]. The small body of research that has attempted to examine this concept, has shown that the determination of risk acceptance involves a complex weighing up of multifarious factors that have evolved based on three key models: revealed preferences, expressed preferences and implied preferences

First, in answering the question of '*How safe is safe enough?*', Starr [28] applies a specific version of the Actor Rational Paradigm to develop what is labelled as a *revealed preference* model. This model characterises risk acceptance based on the assumption that, by trial and error, society has arrived at an '*essentially optimum*' or '*satisfactory*' outcome in which the benefits outweigh the potential risks. It considers both risks and benefits in deducing a decision rule. By applying this model to examine the relationship between risk of death and economic benefit for a number of common technologies, Starr concludes that 'the acceptability of risk appears to be crudely proportional to the third power of the benefits (real or imagined)' and that 'the public is willing to accept 'voluntary' risks roughly 1000 times greater than 'involuntary' risks.

As a consequence of the critiques around the validity of the revealed preference model, an alternative approach was developed by Fischhoff, Slovic and others [29]: their *expressed preference* model is embedded in the psychometric paradigm of risk perception. The model recommends considering individuals' decision frames – the extent of knowledge they have about the risk and their ability to utilise this knowledge optimally; the validity of past conduct and experience as an indicator of present preferences and familiarity with risk exposure; the alternatives of which they are aware and can undertake to be controllable over the risk; and the artificiality of risk source (with respect to nature, history and justifiability of risk) [30, 31]. The expressed preference model is justified by the subjectivity of safety assessment [32], which posits that even quite reasonable individuals may intuitively or heuristically judge activities with relatively low levels of risks to be unacceptably unsafe for them.

A compromise between the revealed and expressed preference approaches is the *implied preferences* approach [33], ascertained through examination of (legal) institutions that society has previously set up with regard to risk. Gough [7] argues that the proponents of this approach do not view the balance between costs, risks and benefits as represented in law as an *optimal balance*, but think of it as *society's best attempt* to date. Thus, the concept of risk acceptance transcends what is seen and evaluated by individuals, to assess what stands beyond the political, social and spatial environment of risk-related actions and decisions [6, 27]. In essence, this reveals why debates over risk acceptance are claimed to be based on the public's trust in responsible risk analysts, regulators and managers, rather than the actual estimates of a risk, especially when people lack knowledge about that risk itself.

Following from this, Raynor and Cantor suggest the adaption of the original question '*How safe is safe enough?*' to '*How fair is safe enough?*' [34]. The shift in defining safety has led to questions regarding whether the procedure by which collective consent is obtained for a course of action is acceptable to those who must bear its consequences [27]; whether the principle that will be used to apportion liabilities for an undesired consequence is acceptable to those affected; and whether the institutions that make the decisions that manage and regulate the technology are worthy of fiduciary trust ([34]:p.4). This may also evolve critical questions on the fairness of both risk and safety distributions among individuals in varying socio-economic circumstances. Fairly distributed risk is viewed as less risky and, therefore, more acceptable to individuals than unfairly distributed risks [35].

Next, a discussion is deliberated about the aforementioned factors in the realm of flood risk. This discussion will confer a deep review on the factors affecting the level of acceptable flood risk, from the viewpoint of individuals.

3 FACTORS UNDERLYING OF FLOOD RISK PERCEPTION AND ACCEPTANCE IN THE CONTEXT OF FLOOD-PRONE LAND USE

In the process of flood-prone land use change, individuals' decisions are influenced by their perception of the probabilities and consequences of actual risk [8, 9, 36]. In general terms, risk perception deals with an individual's interpretation or impression based on an understanding of a particular threat that may potentially cause loss of life or property [37]. Under this paradigm, flood risk perception is formed by the association between qualitative risk characteristics of awareness, worry and preparedness [38]. Building upon the associations between these characteristics, four different typologies can be categorised to describe individuals' psychophysical/cognitive states when they face a risk, respond to it and make decisions to determine its acceptability contextualised within the process of flood-prone land use [12, 39].

These typologies will now be discussed in relation to some explanations provided by past research that examined the persistence of people in areas subject to flood hazards ([8]: p.71 and [9]:p.135).

3.1 Unawareness

For individuals (floodplain occupants) who usually feel positive and optimistic and who thus are not unduly concerned, flood risk exposure may be deemed relatively acceptable. Risk acceptance in this case is not attributed to sufficient confidence in an individual's measures undertaken to avoid and/or reduce the risk, but mostly to their general unawareness of the presence of that risk at first place. Invisibility of flood risk (e.g., the absence of any visual clues that might have alerted individuals to the presence of a local flood risk [40]) and deficiencies in the provision of information on the risk that should be provided by those assumed have professional knowledge – —regulatory agencies, risk managers, real estate agents, insurers, financiers and so on – can adversely affect their awareness and estimates of the potential risk [18]. It follows that if they are unaware of their flood risk, they would not make fully informed decisions on the way in which they should respond to that risk and, equally importantly, the way in which they can safely invest and/or live in areas subject to floods.

3.2 Unconcern/sense of security

For individuals who usually feel safe and, therefore, would have no sense of worry, flood risk exposure may be deemed relatively acceptable. Their emotions would not change with the exposure to risk events and/or the provision of risk-related information. However, they are also insufficiently prepared as they are more willing to gamble – in the context of the infrequency of serious floods – that they themselves will not suffer losses during their persistence in areas subject to flood hazards [40]. Many such people would choose to accept the risks of investing and/or living in areas subject to floods.

3.3 Irresponsibility/risk denial

For individuals who experience a greater sense of worry and negative emotions, such as anxiety, about risks, flood risk exposure may be deemed relatively unacceptable. They show strong denial and resistance to the risk and, therefore, expect to avoid and/or reduce that risk. However, rather than being self-responsible when they are averse to being affected by any risk, they prefer to rely on others to ameliorate losses for them, for example through state compensation, insurance or other mechanisms. Some such individuals would accept the risk of investing/living in areas subject to floods.

3.4 Controllability

For individuals who characteristically believe that they have made sufficient preparations regarding the risk and thereby exhibit a strong sense of control over it, flood risk exposure may be deemed relatively unacceptable. They believe that they should be fully prepared and have undertaken, or are planning to undertake, all sufficient preventative measures to avoid/reduce the loss. What distinguishes this typology is that, because they feel positive and optimistic, they prefer to consider flood protection from its beneficial perspective – usually measured in losses avoided. It follows that, because they are usually fully aware of their flood risk, they make fully informed decisions on the way in which they can safely invest and/or live in areas subject to floods.

Nevertheless, people's persistence in areas subject to flood hazard is not explained solely with respect to their perception of location-related risks: at least two more explanations could be added to the aforementioned list, according to Fordham ([8]:p.71), He ([10]:p.12) and James *et al.* ([16]:footnote 44).

3.5 Benefit perception/ residential satisfaction

Some floodplain occupants expect to bear a loss (risk taking) but see this outcome as an acceptable cost of enjoying the derived locational benefits 'risk compensation' [41]. As urban areas grow, both geographically and demographically, flood-prone areas have long been sought as sites for urban development because of the amenities they offer [42], or their resource potential as identified by Burby [1]. A large body of empirical research on the residential choice of hazardous environments has revealed that many floodplain residents were aware of the location-related risk and tolerated it to varying degrees because of the perceived location-based benefits [10, 11, 16].

Location-based benefits, which have been viewed as essential elements of residential satisfaction on flood-prone environments choice, include [10]: natural amenities (such as scenic views, clean air and quiet surroundings, outdoor recreational opportunities, and other activities based on the location-embedded biophysical environment); neighbourhood social, economic and physical features (such as psychological bonds with a place, perceptions of privacy and safety at home, community cost of living, and other infrastructural and equipment settings, such as lighting of streets, crowding and noise level); and accessibility, preference and characteristics of housing (such as size and age of houses, interior and proximal exterior environments, and other aspects of housing, for example, building quality and disrepair).

A flow diagram is provided in Fig. 1 to illustrate the relationships between residential satisfaction, responses to dissatisfaction, the importance of location-related benefits in the trade-off process with one's perception and acceptance of location-related risks. The diagram focuses on the trade-off process within which the level of worry links risk perception to the benefits of risk taking, as follows: theoretically, a higher level of worry reduces flood-prone residents' willingness to accept flood risk, which in turn hampers their intentions to reap the benefits from flood-prone land use, and increases their flood preparedness intentions. As a consequence, this process may eventually dissuade the user intentions to invest/live in an area subject to flood. Once the intensity of the dissatisfaction exceeds the threshold of tolerance, individuals may adopt some effective responses or behavioural adjustments to relieve their dissatisfactions [10]. There is, therefore, a circulating feedback mechanism between the responses to the element of residential satisfaction.

3.6 Involuntariness

Some floodplain occupants have little or no choice about where to live and must therefore stay. This explanation is a necessary counterbalance to the dominant decision's emphasis on

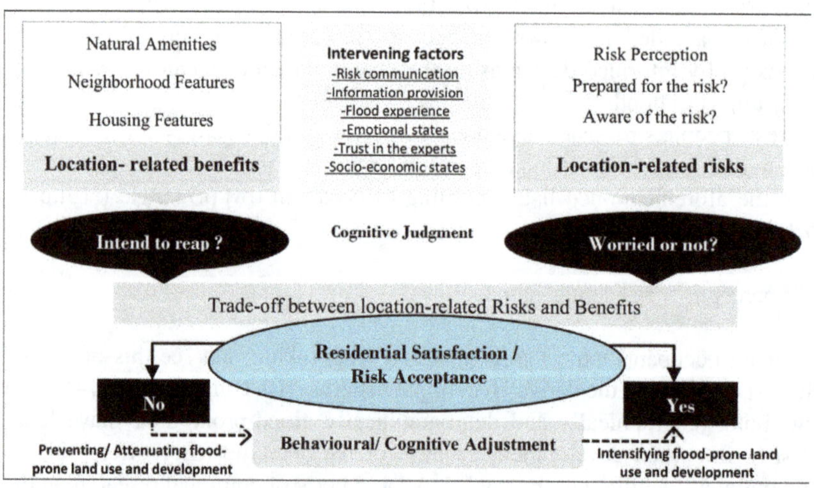

Figure 1: Trade-off between location-related risks and derived locational benefits.

voluntary individual choice as elucidated by Starr [28]. In fact, most flood types – flash floods, riverine floods and storm tides – are predominantly classified as involuntary events, 'acts of God', caused by any natural hazard [15]. On the contrary, White and Haas [43]: claim that natural hazards have a strong component of voluntary risk when they result from interactions between human activities and destructive events ([43]: p.88). Thus, if a flood threatens an urban area with existing regulatory development constraints, the risk will be classified as voluntary. This enforces voluntariness preferences in making decisions relating to the process of flood-prone land-use change [18] – that extends over time, involves numerous decision points and actors and requires the balancing of multiple objectives, future scenarios, priorities and risk reduction strategies. The process of flood-prone land use usually begins with decisions made by landowners, within constraints imposed by law and regulations. Beginning with vacant land, the alternatives for them are not to develop, develop and sell, develop and occupy, or develop and lease. If the flood-prone land is already developed, the alternatives include whether to occupy the site 'as is' or further develop the site with or without accompanying actions to reduce and mitigate of the risk.

With respect to whether people voluntarily choose to live in a flood risk area, previous analyses reveal that in some cases, voluntariness in risk taking is at times less to do with choice and more to do with lack of choice (e.g., housing affordability or changing life circumstances such as age, health, poverty and access to jobs), the inevitability of natural disasters (that the natural world is unable to be controlled), and perceived development potential and economic growth [44, 45]. Further, the lack of choice, or the involuntariness situation, is essentially based on the assumption that people in modern and urbanised societies like Australia cannot invest or live wherever they like. This assumption ignores the whole land use planning and building code system, which regulates where people can and cannot build homes, and under what conditions. It also ignores the way in which almost all new housing projects are created—by market forces and large-scale private sector land development companies that have political influence and the capacity to negotiate favourable conditions and win legal battles ([44]: p.532).

However, understanding public perception of risk can defy involuntary risk, which is determined not by the public but by controlling agents external to them. This requires policymakers to deepen their understanding of how flood risk is constituted by those who may be simultaneously 'risk takers' and 'at risk'. This understanding is of obvious importance where there is a decision to be made with respect to risk acceptance, as well as on the trustworthiness, efficacy and fairness of the associated regulatory development constraints set by experts and policymakers [8, 17, 18] – such as the 100-year flood regulatory threshold [5, 46].

In fact, the use of the 100-year flood as a regulatory constraint for the boundaries of acceptable flood risk, which informs the determination of minimum residential floor levels, has been widely criticised [47]. When it is represented in technical language, the '100-year ARI flood' (where ARI is average recurrence interval) term might be filtered differently by the general public than by those who originally produced and applied its content for residential planning purposes [48]. According to Bill and Tobin [47], this may be the initial goal behind the adoption of the 100-year flood standard in the United States: not effective communication of risk or risk policy, but efficient administration and implementation to minimise potential loss of life and property and maximise the economic benefits to be derived from floodplain development. ([47]: p.303). In light of this, Handemr [44] has indicated that such standards should be seen as the 'tolerated risk' because rarely is there genuine public debate and 'active' acceptance, as opposed to passively putting up with risk for the local population living in areas prone to flooding [44]. Here emerge the critical issues of the public's exclusion and lack of trust, and the ineffectiveness of risk communication efforts and strategies, which in turn may hamper the ability of decision makers to ensure that community and urban development occurs in a resilient manner.

To deal with such issues in a way that enhances public trust in the political environment, decision makers must accept and involve the judgement of those who actually face the complexity of making a place-based choice in consideration of their own sense of benefits, costs and land use constraints. From this perspective, 'active acceptance' is seen as a prerequisite for successful decisions enabling individuals to be active negotiators of risk [49] and therefore to ultimately settle fairly and with a strong commitment to their safety and well-being. White and Haas [43] in this respect claim that '*aggregated individual preferences towards risk must be adopted to set the threshold acceptable to the general public*' ([43]:p.89). This has important social equity implications with respect to the formulation of flood acceptance levels depending on the actual local conditions (e.g. home values and socio-economic circumstances) and the wishes of local communities. To conclude, decisions made on behalf of the community are more likely to be accepted if there has been open communication and consultation with the affected communities, so that the risk is readily understood.

However, while the involvement of individuals and agencies with regard to both technical and socio-political imperatives can lead to consensus on what can be controversial issues around flood risk perception, acceptance and floodplain management [19], engaging 'the community' can be challenging due to several barriers, such as [50–52]:

(1) Low prioritisation and acknowledgement of natural hazards compared to other immediate, day-to-day and pressing concerns such as housing affordability, unemployment and crime rate.
(2) Resource constraints that can lead to self-perpetuating short-term technical fixes rather than long-term integrated and consultative approaches to addressing problems.
(3) Lack of institutional autonomy or political will.
(4) Issues concerning stakeholder identification and knowledge levels. Thus, this is actually a dilemma over 'whose' perceptions of risk, interests in land-use etc. should be considered (whether previous, current or future residents). It is also over the awareness of the hazard, especially with the lack of both easily accessible and understandable information. Problems associated with the provision of information include impacts on land values and insurance prices, intellectual property and liability for incorrect information.
(5) Well-intentioned higher-level policies and practice (such as levee and other flood control structures) that take the risk and burden off individual property owners, communities and local authorities, discouraging risk-averse behaviour including public consultation.
(6) The uncertainty that surrounds flooding and the unique characteristics of each flood event ([53]:143).

4 CONCLUSION

Bridging the divide between those who manage disaster risk and those who bear the consequences is key to realising the promise of land-use planning to reduce disaster risk and build sustainable, hazard-resilient communities in flood-prone areas. In light of this, it is important to note that the heterogeneity of populations at risk adds to the complexity of this issue. For example, some research in this respect has revealed that more deprived populations, and those belonging to lower social classes, are the most likely to be living in zones at risk of flooding [40, 54]. Hence, we suggest that personalised risk assessment, based on qualitative risk characteristics, socio-economic circumstances, experience and feelings, is more likely to motivate risk acceptance and behavioural action than simply providing technical information about risks and consequences – supporting the findings of [55, 56].

In the absence of an integrated, holistic view that brings together all the underlying factors of individuals' willingness to accept flood risk in the context of flood-prone land use, the need for this paper is evident. Hence, the paper attempts to explicitly cover this issue by shedding more light on the articulation of six typologies that describe individuals' psycho-physical/cognitive states when they face the risk, respond to it and determine its acceptability contextualised within the process of flood-prone land use change. This would provide valuable insights into how the level of acceptable risk (based on a subjective risk assessment) can be altered by the way in which the location-related risks are perceived relative to the location-related benefits derived from flood-prone land use; by the way in which the related risk information and policy are interpreted and evaluated in terms of their efficacy, voluntariness, trustworthiness and degree of fairness; and by the way in which intervening factors including socio-economic circumstances vary to provide a holistic picture of individuals' systems.

To date, there has been little or no advancement in this area, with research highlighting such inhibitive issues as the way in which the acceptable risk levels are (actively) conceived by the public and then incorporated into acceptable risk standards and floodplain land use policy. Broadening the scope of future research is vital.

REFERENCES

[1] Burby, R.J., *Cooperating with Nature: Confronting Natural Hazards with Land-Use Planning for Sustainable Communities*, Joseph Henry Press, 1998.

[2] Godschalk, D.R., Brody, S. & Burby, R., Public participation in natural hazard mitigation policy formation: challenges for comprehensive planning. *Journal of Environmental Planning and Management*, **46**(5), pp. 733–754, 2003. http://dx.doi.org/10.1080/0964056032000138463

[3] Berke, P. & Smith, G., Hazard mitigation, planning, and disaster resiliency: Challenges and strategic choices for the 21st century. *Building Safer Communities. Risk Governance, Spatial Planning and Responses to Natural Hazards*, pp. 1–18, 2009.

[4] Burby, R.J., Beatley, T., Berke, P.R., Deyle, R.E., French, S.P., Godschalk, D.R., et al., Unleashing the power of planning to create disaster-resistant communities. *Journal of the American Planning Association*, **65**(3): pp. 247–258, 1999. http://dx.doi.org/10.1080/01944369908976055

[5] Godber, A.M., *Urban Floodplain Land-use–Acceptable Hazard? A Case Study of Flood Risk Perception on the Guragunbah Floodplain (Nerang River System), Gold Coast,* 2002.

[6] Kasperson, R.E., Acceptability of human risk. *Environmental Health Perspectives*, 52, pp. 15–20, 1983. http://dx.doi.org/10.1289/ehp.835215

[7] Gough, J.D., *A Review of the Literature Pertaining to 'Perceived' Risk and 'Acceptable' Risk and the Methods used to Estimate them,* Lincoln University & University of Canterbury. Centre for Resource Management, 1990.

[8] Fordham, M.H., *Choice and Constraint in Flood Hazard Mitigation: the Environmental Attitudes of Floodplain Residents and Engineers*, Middlesex University, 1992.
[9] Kates, R.W., *Hazard and Choice Perception in Flood Plain Management,* University of Chicago: Chicago, p. 157, 1962.
[10] He, X., *Residential Satisfaction with Home Location: Examination of the Relationship Between Location-Embedded Benefits and Risk Perception*, Texas State University-San Marcos, 2009.
[11] Vogt, M., Willis, K. & Vince, J., *Weighing Up the Risks-the Decision to Purchase Housing on a Flood Plain,* 2008.
[12] Raaijmakers, R., Krywkow, J. & van der Veen, A., Flood risk perceptions and spatial multi-criteria analysis: an exploratory research for hazard mitigation. *Natural Hazards*, **46**(3), pp. 307–322, 2008.
http://dx.doi.org/10.1007/s11069-007-9189-z
[13] Zhai, G. & Ikeda, S., Empirical analysis of Japanese flood risk acceptability within multi-risk context. *Natural Hazards and Earth System Science*, **8**(5), pp. 1049–1066, 2008.
http://dx.doi.org/10.5194/nhess-8-1049-2008
[14] Zhai, G. & Ikeda, S., Flood risk acceptability and economic value of evacuation. *Risk Analysis*, **26**(3), pp. 683–694, 2006.
http://dx.doi.org/10.1111/j.1539-6924.2006.00771.x
[15] White, G.F., *Human Adjustment to Floods,* 1945.
[16] James, L.D., Laurent, E.A. & Hill, D.W., *The Flood Plain as a Residential Choice: Resident Attitudes and Perception and Their Implications to Flood Plain Management,* Environment Resources Center, Georgia Institute of Technology: Atlanta, GA, 1971.
[17] McPherson, H.J. & Saarinen, T.F., Flood plain dwellers' perception of the flood hazard in Tucson, Arizona. *The Annals of Regional Science*, **11**(2), pp. 25–40, 1977.
http://dx.doi.org/10.1007/BF01287852
[18] Bollens, S.A., Kaiser, E.J. & Burby, R.J., Evaluating the effects of local floodplain management policies on property owner behavior. *Environmental Management*, **12**(3), pp. 311–325, 1988.
http://dx.doi.org/10.1007/BF01867522
[19] nunes Correia, F., Fordham, M., Saravia, M.A.G. & Bernardo, F., Flood hazard assessment and management: interface with the public. *Water Resources Management*, **12**(3), pp. 209–227, 1998.
http://dx.doi.org/10.1023/A:1008092302962
[20] Kates, R.W. & White,G.F., Flood hazard evaluation. Papers on Flood Problems: Department of Geography Research Paper, **70**, pp. 135–147, 1961.
[21] Vrijling, J., Probabilistic design of water defense systems in The Netherlands. *Reliability Engineering & System Safety*, **74**(3), pp. 337–344, 2001.
http://dx.doi.org/10.1016/S0951-8320(01)00082-5
[22] Vrijling, J.K., Van Gelder, P.H., Goossens, L.H.J., Voortman, H.G. & Pandey, M.D., A framework for risk criteria for critical infrastructures: fundamentals and case studies in the Netherlands. *Journal of Risk Research*, **7**(6), pp. 569–579, 2004.
http://dx.doi.org/10.1080/1366987032000081178
[23] Molino, S. & Karwaj, A., Do we understand what is an acceptable flood risk? the people of Australia have their say. In *Proceedings of the 52nd Annual Floodplain Management Authorities Conference*, 2012.

[24] Zhai, G., Public preference and willingness to pay for flood risk reduction. A better integrated management of disaster risks: toward resilient society to emerging disaster risks in mega-cities, TERRAPUB: Tokyo, pp. 1–21, 2006.
[25] Fischhoff, B., Acceptable risk: a conceptual proposal. *Risk*, **5**, p. 1, 1994.
[26] Vrijling, J., Van Hengel, W. & Houben, R., A framework for risk evaluation. *Journal of Hazardous Materials*, **43**(3), pp. 245–261, 1995.
http://dx.doi.org/10.1016/0304-3894(95)91197-V
[27] Bell, R., Glade, T. & Danscheid, M., Risks in defining acceptable risk levels. *Landslide Risk Management, Supplementary*, **400**, pp. 38–44, 2005.
[28] Starr, C., Social benefit versus technological risk. *Readings in Risk*, pp. 183–194, 1969.
[29] Slovic, P., Fischhoff, B. & Lichtenstein, S., Rating the risks. *Environment: Science and Policy for Sustainable Development*, **21**(3), pp. 14–39, 1979.
http://dx.doi.org/10.1080/00139157.1979.9933091
[30] Slovic, P., Perception of risk. *Science*, **236**(4799), pp. 280–285, 1987.
http://dx.doi.org/10.1126/science.3563507
[31] Fischhoff, B., Slovic, P. & Lichtenstein, S., Weighing the risks: risks: benefits which risks are acceptable? *Environment: Science and Policy for Sustainable Development*, **21**(4), pp. 17–38, 1979.
http://dx.doi.org/10.1080/00139157.1979.9929722
[32] Macpherson, J.A., Safety, risk acceptability, and morality. *Science and Engineering Ethics*, **14**(3), pp. 377–390, 2008.
http://dx.doi.org/10.1007/s11948-008-9058-5
[33] Rowe, W.D., Risk assessment approaches and methods. *Society, Technology and Risk Assessment*, p. 343, 1980.
[34] Rayner, S. & Cantor, R., How fair is safe enough? the cultural approach to societal technology choice1. *Risk Analysis*, **7**(1), pp. 3–9, 1987.
http://dx.doi.org/10.1111/j.1539-6924.1987.tb00963.x
[35] Covello, V. & Sandman, P.M., Risk communication: evolution and revolution. *Solutions to an Environment in Peril*, pp. 164–178, 2001.
[36] Kates, R.W., Perceptual regions and regional perception in flood plain management. *Papers in Regional Science*, **11**(1), pp. 217–227, 1963.
http://dx.doi.org/10.1111/j.1435-5597.1963.tb01899.x
[37] Bradford, R., O'Sullivan, J.J., van der Craats, I.M., Krywkow, J., Rotko, P. & Aaltonen, J., et al., Risk perception–issues for flood management in Europe. *Natural Hazards and Earth System Science*, **12**(7), pp. 2299–2309, 2012.
http://dx.doi.org/10.5194/nhess-12-2299-2012
[38] Scolobig, A., De Marchi, B. & Borga, M., The missing link between flood risk awareness and preparedness: findings from case studies in an Alpine Region. *Natural Hazards*, **63**(2), pp. 499–520, 2012.
http://dx.doi.org/10.1007/s11069-012-0161-1
[39] Meng, B., Liu, M., Yan Liufu, H. & Wang, W., Risk perceptions combining spatial multi-criteria analysis in land-use type of Huainan city. *Safety Science*, **51**(1), pp. 361–373, 2013.
http://dx.doi.org/10.1016/j.ssci.2012.08.010
[40] Burningham, K., Fielding, J. & Thrush, D., 'It'll never happen to me': understanding public awareness of local flood risk. *Disasters*, **32**(2), pp. 216–238, 2008.
http://dx.doi.org/10.1111/j.1467-7717.2007.01036.x

[41] Schmidt, M., *Investigating Risk Perception: a Short Introduction,* Vienna, Austria, 2004.
[42] Montz, B., The generation of flood hazards and disasters by urban development of floodplains. *Floods*,**1**, pp. 116–127, 2000.
[43] White, G.F. & Haas, J.E., *Assessment of Research on Natural Hazards*, Mit Press, 1975.
[44] Handmer, J., Risk creation, bearing and sharing on Australian floodplains. *International Journal of Water Resources Development*, **24**(4), pp. 527–540, 2008. http://dx.doi.org/10.1080/07900620801921439
[45] Kaiser, E.J., Burby, R.J., Bollens, S.A. & Holway, J.M., Private sector land market decision agents as targets of floodplain policy. *International Journal of Mass Emergencies and Disasters*, **5**(3), pp. 311–335, 1987.
[46] Ludy, J. & Kondolf, G.M., Flood risk perception in lands "protected" by 100-year levees. *Natural Hazards*, **61**(2), pp. 829–842, 2012. http://dx.doi.org/10.1007/s11069-011-0072-6
[47] Bell, H.M. & Tobin, G.A., Efficient and effective? The 100-year flood in the communication and perception of flood risk. *Environmental Hazards*, **7**(4), pp. 302–311, 2007. http://dx.doi.org/10.1016/j.envhaz.2007.08.004
[48] Smith, D., Problems, issues and opportunities. *Floods*, **1**, p. 254, 2000.
[49] Lupton, D., 1 Sociology and risk. *Beyond the risk society: Critical reflections on risk and human security*, p. 11, 2006.
[50] Godber, A.M., Hastings, P.A. & Childs, I.R., Local government views on addressing flood risk management on the gold coast. *The Australian Journal of Emergency Management*, **21**(3), pp. 34–40, 2006.
[51] Burby, R., Policies for sustainable land use. *Cooperating with Nature: Confronting Natural Hazards with Land-Use Planning for Sustainable Communities*, The National Academies Press: Washington, pp. 263–291, 1998.
[52] May, P.J. & Deyle, R.E., Governing land use in hazardous areas with a patchwork system. *Cooperating with nature: Confronting natural hazards with land-use planning for sustainable communities*, pp. 57–82, 1998.
[53] Deyle, R.E., French, S.P., Olshansky, R.B. & Paterson, R.G., Hazard assessment: the factual basis for planning and mitigation. *Cooperating with Nature: Confronting Natural Hazards and Land use Planning for Sustainable Communities*, pp. 119–166, 1998.
[54] Fielding, J. & Burningham, K., Environmental inequality and flood hazard. *Local Environment*, **10**(4), pp. 379–395, 2005. http://dx.doi.org/10.1080/13549830500160875
[55] Rocha, J.S. & Correia, F.N., Public perception of flood risk and flood defence policies. In *Defence from Floods and Floodplain Management*, Springer: Netherlands, pp. 457–477, 1995.
[56] Fordham, M., Participatory planning for flood mitigation: models and approaches. *Australian Journal of Emergency Management*, **13**(4), pp. 27–34, 1999.

FLOOD RISK MODELLING OF STULIANSKY STREAM IN LOPÚCHOV VILLAGE, SLOVAKIA

RASTISLAV FIJKO, MARINA ZELEŇÁKOVÁ & IBRAHIM ALKHALAF
Institute of Environmental Engineering, Technical University of Košice, Slovakia.

ABSTRACT

The paper focuses on the modelling of water flow in open channel of Stuliansky stream in the village Lopúchov, situated in the north of Slovakia. It presents: I) the proceeding of the model development in HEC-RAS; II) the creation of flooded areas in ArcGIS; III) the flood risk assessment by Cost Benefit Analysis (CBA). We will point out the modelling methodology with emphasis to collection of the data and their relevance for flood risk modelling of study area. From the developed model, it is possible to determine the maximum water level in each river profile, the extent of flooding, flow rate, or flow velocity in a particular profile, taking into account the proper Manning roughness coefficient for each profile. The benefit of this work is the definition of flooded area in the municipality Lopúchov in the eastern Slovakia. The 3D model of a given territory is created at Geographical Information System (GIS) platform and it shows the flooded part of the urban area in village Lopúchov, caused by Stuliansky stream. The CBA is essential to evaluate flood damages in the areas. Results from the flow modelling are of great importance for the field of science and practice that their interests are oriented to the proposal of flood mitigation measures in the studied territory.
Keywords: ArcGIS, flood risk modelling, HEC-RAS.

1 INTRODUCTION

Recent floods in Europe have displaced half a million people and caused more than 25 billion € in insured economic losses. Such flooding impacts are driven in part by increasing hydro-climatic extremes, but also by widespread reliance on structural measures for flood-control [1]. Structural measures to reduce negative impacts of floods include constructing new or reinforcing existing flood defenced infrastructures. There are, however, other and potentially very cost-effective ways of achieving flood protection which profit from nature's own capacity to absorb excess waters [2].

To identify flood areas, a model is required to predict spatially distributed estimates of the hydraulic variables such as flood inundation extent and depth. 1D, 2D and 3D dimensional hydraulic models are now widely used for this purpose. These models are able to compute water surface profiles rapidly in several different characterizations of the system. Recently developed tools for hydraulic modelling (e.g. Geographical Information System (GIS)) are techniques that allow one or two-dimensional representation of the computed hydraulic parameters. In hydraulic modelling, three critical factors: streamflows (flood magnitudes), topography of river channel, and Manning roughness coefficient n (friction effects) influence flow hydraulics and the resultant areal extent of the simulated flood inundation.

Reducing human casualties and damage to economic activity and the environment are key objectives shared by all EU countries. The economic analysis of how to achieve a suitable

DOI: 10.2495/SAFE-V6-N3-529-537

water resource management can be achieved by feasibility studies of water management projects. Feasibility studies of projects should be completed by using conventional methodologies of economic analysis, such as cost-benefit analysis (CBA) [3]. The CBA is one of the most widely accepted instruments since it is a rational and systematic decision-making support tool. The impact of a large number of factors should be taken into account in order to assess the financial and economic viability of projects to identify flood prone areas [4].

The paper deals with flood risk modelling of Stuliansky stream in Lopúchov village situated in the north part of Slovakia. The model of Stuliansky stream river bed was developed in HEC-RAS software. The flooded areas in the village were created in ArcGIS using digital terrain model of study area. Precisions proceeding of modelling is described in the following.

2 MATERIALS AND METHODS

The aim of this study is to present an efficient methodology for the delineation of the flood hazard mapping. For this purpose, we used HEC-RAS [5] and ArcGIS [6]. These softwares are efficient to state the inundated areas near water stream. Mutual connection of these packages offers to the user a lot of required information. These softwares are very useful for decision making in water management. Advantage of this kind of simulation is efficient (assuming that the needed data is available) visualisation of results and their use in the flood risk management.

Figure 1 illustrates the processing of flood risk modelling which we used to achieve results.

In terms of development of the model it is important to accurately focus on the area of interest in the form of input data. For model development management perspective it is important to properly define the critical points of a given territory in the form of fluvial profiles.

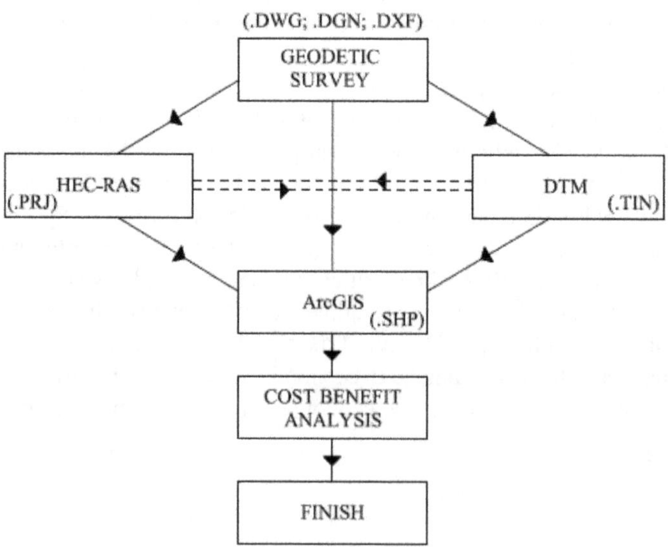

Figure 1: Flowchart of flood modelling.

2.1 Geodetic survey

We chose the area of Lopúchov village (Fig. 2) because of the variability of water flow in Stuliansky stream along its length and the uncontrolled rise of water level during the torrential rains.

The watercourse of Stuliansky stream has the character of a natural stream, river banks are damaged, and river bed is silted after erosion and also due to human intervention. Along the left side of Stuliansky stream is a local road, on the right bank are family houses with gardens [7]. For development of hydrodynamic model the geodetic survey is necessary. We used the output of geodetic survey of the area of interest – Stuliansky stream in Lopúchov village in the north of Slovakia – in various (.dwg; .dgn; .dxf) format (Fig. 3).

Geodetic survey (Fig. 3) of the area served as a base of data for development of hydrodynamic model in HEC-RAS.

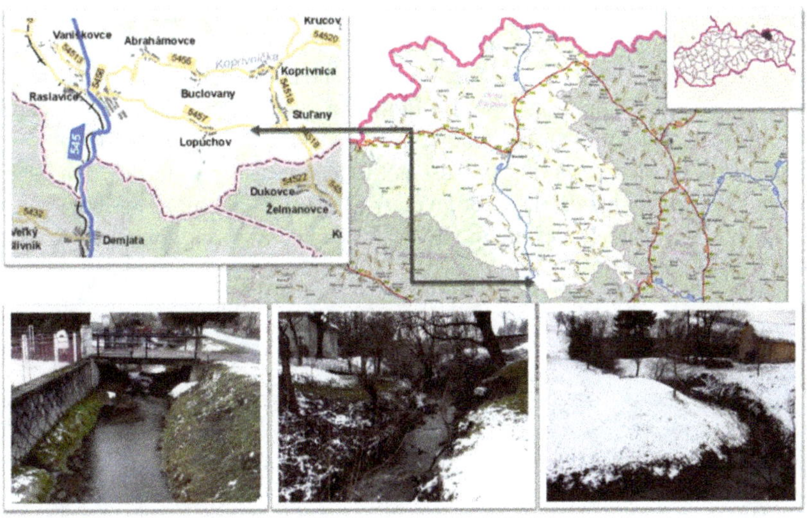

Figure 2: Location of study area.

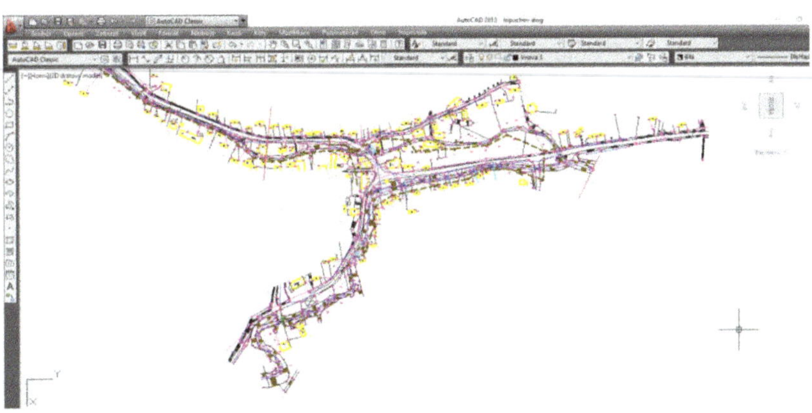

Figure 3: Geodetic survey of Stuliansky stream in Lopúchov village.

2.2 Hydrodynamic model in HEC-RAS

We determined the individual cross sections (Fig. 4) in characteristic locations – approximately each 25 m.

We stated the coefficient of roughness by visual comparison of results published by [8] (Fig. 5).

Selection of an appropriate value of Manning's n is very important for the accuracy of the modelling of the height of water level in the stream The value of Manning's n is variable and depends on a number of factors including: surface roughness; vegetation; channel irregularities; channel alignment; scour and deposition; obstructions; size and shape of the channel; stage and discharge; seasonal changes; temperature; and suspended material. In general, Manning's n values should be calibrated and values of n computed for similar stream conditions or values obtained from experimental data should be used as guides in selecting the proper n values [9].

The objects on the stream (Fig. 6) – in our case bridges – were also defined in the model as the objects influence the flow of the stream significantly.

After cross sections definition (Cross Section Data) and coefficients of roughness determination, we state boundary conditions (Steady Flow Data) as variant values of N-year flows. Such a model is ready for simulation (Steady Flow Analysis) of flood wave [10].

The partial result of the computing process is the height of the water level in each river profile. Figure 7 shows the height of water level by flow $Q_1 = 0.04\ m^3.s^{-1}$ in the whole reach of the river. The results from modelling in HEC-RAS can be exported to the prepared digital terrain model (DTM) model.

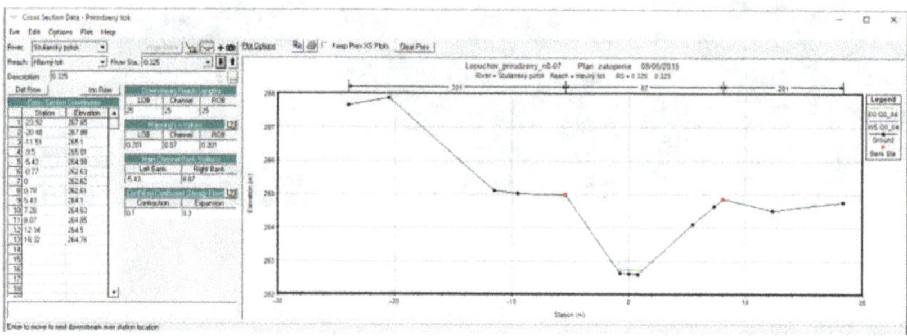

Figure 4: Example of cross section of Stuliansky stream.

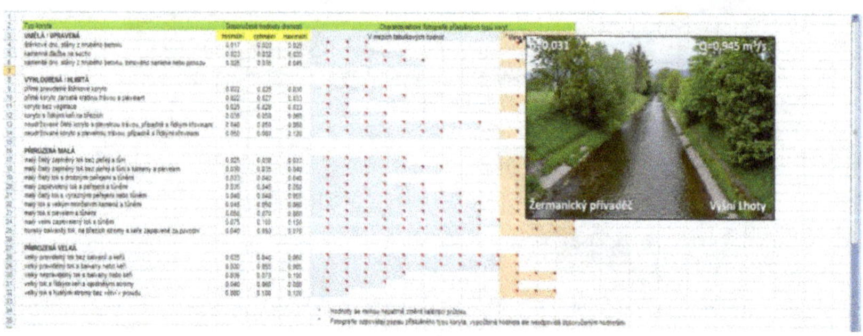

Figure 5: Coefficient of roughness for various streams.

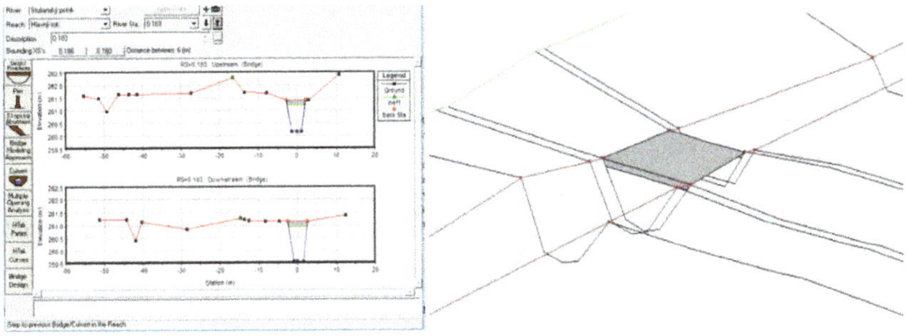

Figure 6: Defining of a bridge object.

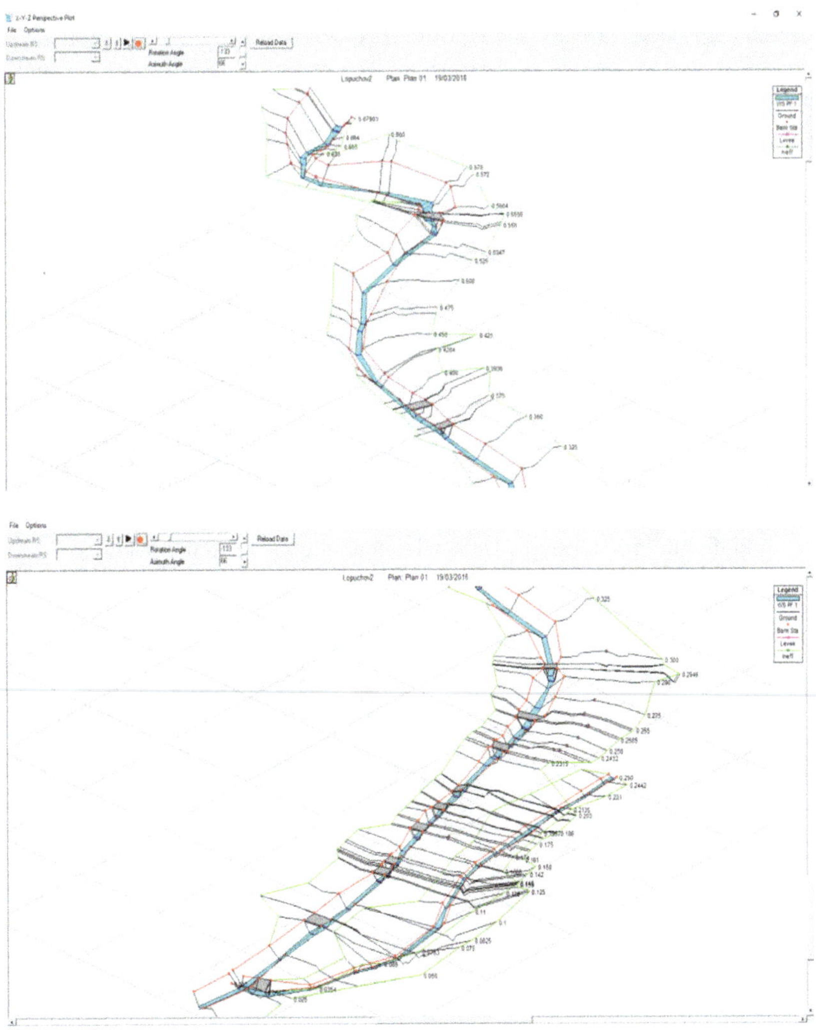

Figure 7: 1D model of Stuliansky stream in HEC-RAS (Q_1).

2.3 Digital terrain model in ArcGIS

We used the trigonometric measurement of altitudes from geodetic survey to create a DTM model. We reflected the elevation of point's data in the triangulation form and we created a ground surface using ArcGIS software (Fig. 8). Then we used the created DTM (Fig. 9) as a basis for visualizing the obtained results.

We obtained the flooded areas near Stuliansky stream in Lopúchov village by connection of developed model in HEC-RAS and DTM in ArcGIS software.

3 RESULTS

We prepared separate layers for flood risk modelling in (.shp) form from geodetic survey of Stuliansky stream and objects (houses, roads ...) in the study area. By overlaying of the layers, we created manually in ArcGIS: 'Layer Properties ⟹ Base Heights' and we used 'Floating on a custom surface'. Then we put to layers the coordinate system of the area and connected them with DTM (.tin) and model from HEC-RAS. We created Z-coordinate for 3D visualization by 'Layer Properties ⟹ Extrusion'. The result is 3D model of the study area with separate layers (Fig. 10).

Figure 11 shows the model with results of Q_{100} water using hydrodynamic model in ArcGIS. By this way we can define any flow of water as boundary conditions and to identify the total surface of inundation area.

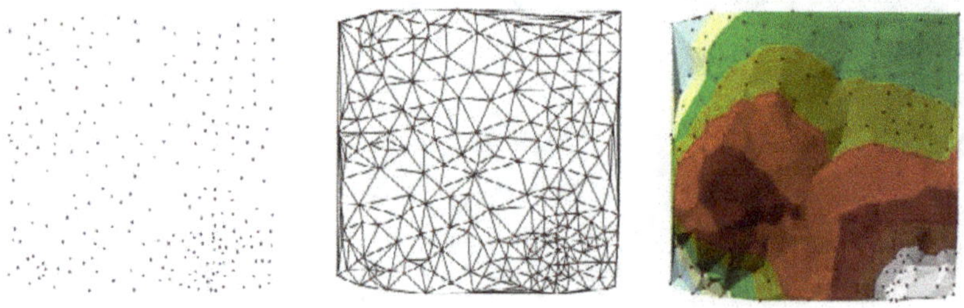

Figure 8: Process steps: Data – Triangulation – Terrain model.

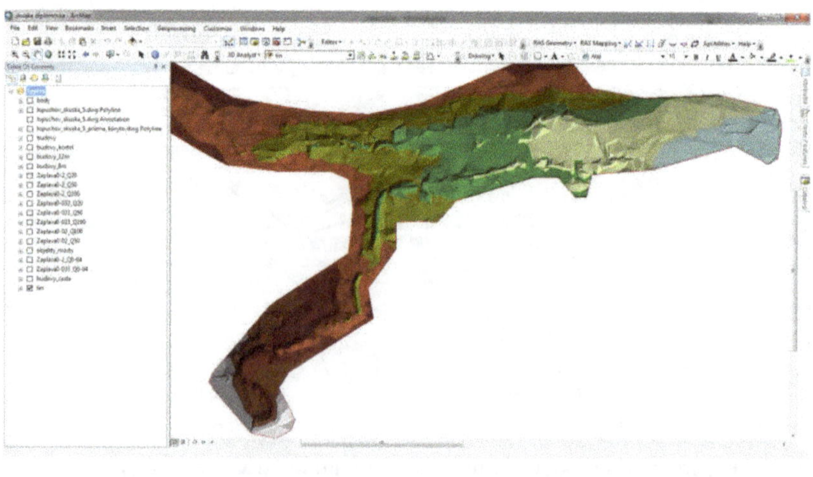

Figure 9: Digital terrain model created in ArcGIS software.

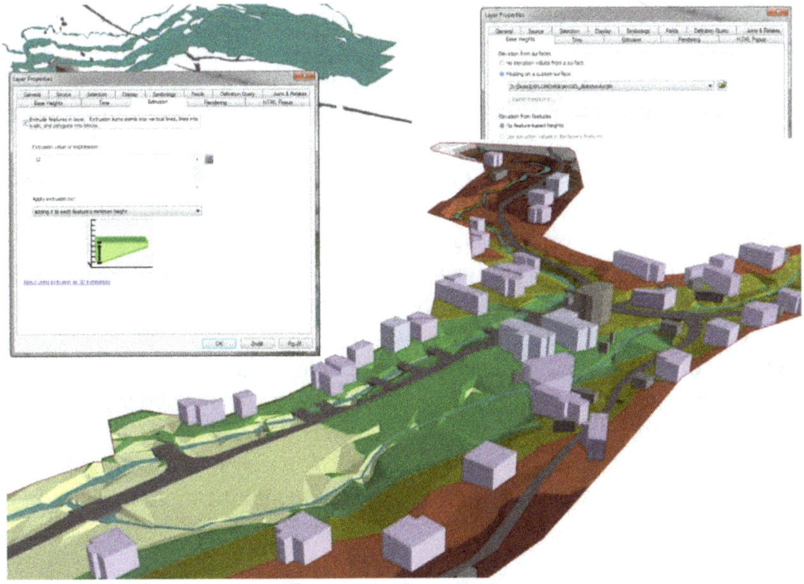

Figure 10: 3D model of the study area with separate layers.

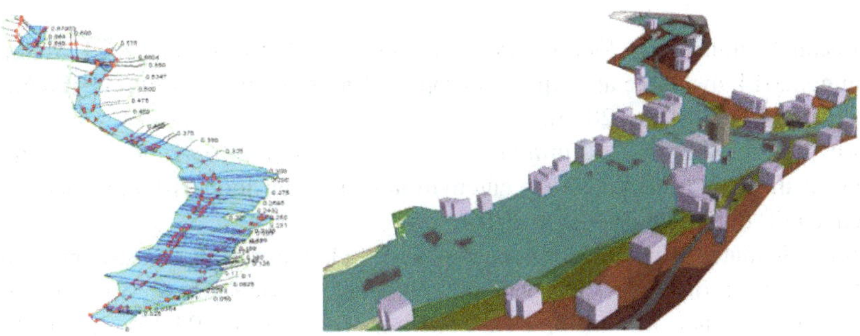

Figure 11: Flooded area by Q_{100}.

The flooded area is used as background for CBA. From the flooded area by 100-year river discharge 'Q_{100}', the potential flood damages could be identified. For the purpose of flood risk management, the costs of damages were compared with the costs of structural flood protection measures. The construction costs of stream regulation and banks fortification or protection levees have to be less than potential flood damages in the area. It is the core of CBA. We calculated potential flood damages in the study area for **977,735.00 €**. Unit cost of damaged property was calculated to be 100 €/m2 for roads and 525 €/m2 for building structures. The values were stated from UNIKA 2012: Collection of indicators of average budget value per measured unit for object. Buildings and constructions according to the structure classification [11], where the recommended minimum and maximum prices of engineering activities are determined.

4 CONCLUSION

Flood damages that arose on rivers, streams and hydraulic structures on the territory of Slovakia are huge. Under the influence of long-lasting and intensive precipitations, numerous floods arise. The water levels in many profiles on streams reached its historical maximums in the last years. Regular occurrence of floods in Slovakia occurs primarily by total land water regime, which is dominated by very low absorptive power of heavy clay soils of flysch zone and also adverse condition of forests in eastern Slovakia. Flood risk modelling is an effective tool for flood protection before flood mitigation measures implementation.

We used 1D HEC-RAS model and 2D and 3D model developed in ArcGIS to determine flood areas near Stuliansky stream in Lopúchov village. 1D mathematical model developed in HEC-RAS is very effective for the design of the proposed flood mitigation measures. It is appropriate and excellent tool for hydrodynamic modelling for water managers. However, for a visual presentation of the results to the public, it is more suitable to use results from 3D modelling in ArcGIS.

Inundation area of the village Lopúchov is quite extensive and flood protection measures (regulation of water stream) in the study area is necessary. The results of our work proved that the costs of structural flood protection measures have to be less than 977,735.00 €, which is the amount of potential flood damages in the study area. Creation of the hydrodynamic model will help in decision making of the selection of suitable flood mitigation measures in term of cost-effectiveness.

ACKNOWLEDGEMENTS

This paper was written thanks to support from project VEGA 1/0609/14.

REFERENCES

[1] European Commission, *Floods and their impacts*, A European Flood Action Programme. 2014, available at: http://ec.europa.eu/environment/water/flood_risk/impacts.htm, (accessed 02 March 2016).

[2] European Commission, *Best practices on flood prevention, protection and mitigation,* 2003., available at http://ec.europa.eu/environment/water/flood_risk/pdf/flooding_bestpractice.pdf, (accessed 23 March 2016).

[3] Molinos-Senante, M., Hernandez-Sancho, F. & Sala-Garrido, R., Cost-benefit analysis of water-reuse projects for environmental purposes: a case study for Spanish wastewater treatment plants. *Journal of Environmental Management*, **92**(12), pp. 3091–3097, 2011.

[4] Djukic, M., Javanoski, I., Munitlak, O.I., Lazic, M. & Bodroza, D., Cost-benefit analysis of an infrastructure project and a cost-reflective tariff: A case study for investment in wastewater treatment plant in Serbia. *Renewable and Sustainable Energy Reviews*, **59**(6), pp. 1419–1425, 2015.

[5] U.S. Army Corps of Engineers. (2014). *HEC-RAS River Analysis System,* HEC-RAS 4.1. Institute for Water Resources. Washington DC,, available at: http://www.hec.usace.army.mil/software/hec-ras/, (accessed 25 January 2016).

[6] ESRI. *ArcGIS Desktop. Tools for Authoring, Editing, and Analyzing Geographic Information*, available at: http://www.digitalgeo.pt/images/Produtos/31/ef02a26d-ae47-4b6d-b8a4-53133cc14a12.pdf, (accessed 25 January 2016).

[7] Zeleňáková, M., Sarka, D., Zvijáková, L. *Preventívne protipovodňové opatrenia* - Úprava *vodných tokov v obci Lopúchov*. Zámer, 2011.

[8] Smelík, L., *Katalog drsností.* Ústav vodních staveb, Fakulta stavební, VUT v Brně, available at: http://sites.google.com/site/katalogdrsnosti/6-vypocet-drsnosti (accessed 23 January 2015).

[9] Sánchez, F.J. & Román, S., *Manual introductorio a HEC-RAS. Departamento de Geología*, Universidad. Salamanca, España, 2007.
[10] Horritt, M.S. & Bates, P.D., Predicting floodplain inundation; rastered-based modelling versus the finite-element approach. *Hydrological. Processes*, **15**(5), pp. 825–842, 2001.
[11] UNIKA, Collection of indicators of average budget value per measured unit for object. *Buildings and Constructions According the Structure Classification*, Institute of Construction Economics: Bratislava, 2012.

FLOOD RISK PERCEPTION, COPING AND MANAGEMENT IN TWO VULNERABLE COMMUNITIES IN KUMASI, GHANA

D.K. AHADZIE[1], I. DINYE[1], R.D. DINYE[1] & D.G. PROVERBS[2]
[1]Kwame Nkrumah University of Science and Technology, Ghana.
[2]Birmingham City University, UK.

ABSTRACT

Flooding has become an imminent and disruptive threat in many cities around the world. In developing/emerging economies, cities are often more vulnerable to the impacts of flooding. For example, in Accra, Ghana, recently over 150 people perished in a flash flood. The high casualty rate was a consequence of the lack of a well-developed integrated flood risk management plan and exacerbated by weak institutions, slums, urban sprawl and informal land / property management. To this effect, high risk countries like Ghana are in need of urgent development of systematic mitigation measures to engender community resilience (CS). Using the combined qualitative technique of Phenomenological Studies (PS) and Focus Group Discussions (FGDs), the flood risk perception, coping and management strategies of two vulnerable communities in the second largest city of Ghana, Kumasi are examined and evaluated for evidence of a sense of community resilience through social responsibility. The findings suggest that the two urban communities are aware they are at high risk to flooding and live in a constant state of anxiety during the wet seasons. However, there is no evidence of any systematic coping and management strategies and the communities feel there is nothing they can do without government support. It is suggested that government should use the leverage of local government practices to help engender a stronger sense of community empowerment and involvement towards building resilient communities that can respond to flooding through perceptions, social responsibility and appropriate public relation models.

Keywords: community resilience, coping strategies, flood management, flood risks, Ghana, Kkumasi, social responsibility.

1 INTRODUCTION

According to the Global Facilities for Disaster Reduction and Recovery reports, sub-Sahara Africa is acknowledged to be of high degree vulnerability to flooding and its devastating effects [1, 2]. Exacerbated by climate change, population growth and increasing human activities, vulnerable communities in these countries are at more risk to urban flooding and there is the need for urgent collective action of preparedness to help mitigate the worst effects [3, 4]. Contextually, some researchers have advocated for a high sense of community resilience (CS) to help vulnerable communities respond better to flooding; through perceptions, social responsibilities and appropriate public relation models [5]. Kellens *et al.* [6] have also epitomized on CS suggesting that, as a result of increasing global socio-economic challenges facing governments, the paradigm for flood risk management is now shifting; traditional

DOI: 10.2495/SAFE-V6-N3-538-549

measures being replaced by an integrated approach that emphasises social cohesion and preparedness of communities. Wood *et al.* [7] also argues in the same tone by imputing that, the human dimension in creating resilient communities in disaster management is now fully recognized and there is the need to understand the specific contributions for integrating such knowledge, interest and values in flood risk management. The World Bank's [8] recent development of an integrated Flood Risk Management handbook is also an affirmation of the global recognition attached to the need for collective efforts by nations and the citizenry to have a holistic approach for flood risk management.

It is established that flooding is a long standing hazard in Ghana and requires the development of an integrated flood risk management plan [9]. Following Ghana's worst flood disaster in June 2015 when 150 people lost their lives in the capital, Accra, the need to expedite action on developing a robust integrated flood risk management model is now imminent. However, most sub-Saharan Africa countries such as Ghana have weak economies and are often overwhelmed by the challenges posed by competing demands for limited public funds. It is thus expected that high risk communities in such countries will be dependent on traditionally acknowledged social networks to collectively plan and take appropriate action, towards building resilient communities. The paper reports on the findings of Focus Groups Discussions (FDGs) of two high risk communities; *Sepe-Buokrom* and *Atonsu Ahinsan,* in Kumasi, Ghana to establish their preparedness to deal with floods through community initiatives. The findings have policy implications for flood risk management in Ghana in helping stakeholders to understand the level of preparedness and adaptive capacity of potential communities to flood risk management.

An overview of the flooding situation in the literature, including the vulnerability and flood risk management in African cities such as Ghana is presented, including relevant theoretical frameworks. Subsequently, the procedure for the (FGDs) and how they were executed are enumerated. The data analyses and findings are then presented followed by policy implications.

2 FLOOD HAZARDS IN DEVELOPING COUNTRIES-LITERATURE REVIEW

Developing countries are noted to be at more risk to flood hazards because of their weak socio-economic indicators [3]. Accordingly Amoako [2], citing the IPCC [10], identifies two main contributing factors with respect to African economies; firstly, the continent is the most urbanized region with an average population growth rate of 4.4% against a global average of 2.5% [11] and secondly, Africa is most vulnerable to mitigating the effects of climate change. Thus, cities and conurbations in these countries are characterised by inadequate resources, lack of preparedness and low adaptive capacities and hence described as highly vulnerable. That is, floods easily overcome communities in these cities and represent considerable risks to their sustainable development [12]. Indeed, in the Ghanaian case, flooding has been a long standing hazard and needs prompt, effective and robust risk management efforts to mitigate its worst effects.

However, in Ghana as in many other African countries, there has been a dearth of research into flood risk management despite the considerable risks of flooding. It was not until the late 2000 that the interest started manifesting and within the last five years in particular, the focus is shifting towards searching for an effective flood risk management approach for vulnerable communities, with emphasis on human dimensions [13, 14]. In Ghana some of the recent efforts are by Armah *et al.* [1]; Ahadzie and Proverbs [9]; Amoako [2]; Aboagye *et al.* [15]; Musa and Akai [14]. However, these papers focused on rural and farming communities as homogenous societies as opposed to urban settlements which are often heterogeneous and complex. Alternatively, in the sub-region, there is interest in, for example, Nigeria in also understanding how communities cope with and locally manage flood risk [13, 16–18]. Significantly, Olorunfemi and Raheem's [16] paper focusing on Ilorin, Kwara State is of much interest here because of its urban nature. Olurunfemi and Raheem [16] adopting a quantitative approach concluded that house characteristics and neighbourhood quality contribute significantly to vulnerability to rainstorm and flooding. In this study a qualitative approach is adopted towards developing insight into the coping and adaptive capacity of two Ghanaian communities described as high risk in terms of flooding.

Similarly, the global literature is also increasingly focusing on addressing integrated flood risk management from the social and community perspective; as in for example Eastern Europe [19] Western Europe [5], Asia [20] and the Americas [7]. This paper draws lessons on concepts and variables used in the increasingly global literature to shape the questions asked in the FGDs.

3 THEORETICAL FRAMEWORK AND FOCUS GROUP APPROACH

According to Jonkman and Dawson [20] recent flood events have exposed the weaknesses in engineering systems in many cities across the globe calling for the need for human operations before, during and after flooding to be examined. Here, Kellens *et al.* [6] identify various methods that could be used in understanding the perception and coping of flood risk ranging from psychometric paradigms, heuristics, and adaptive behaviours such as mitigation, preparedness and recovery and other qualitative approaches involving mental models approach. Wood *et al.* [7] also highlight key items that can be used to understand the psychological models of laypersons on topics related to flooding such as concept mapping, decision support systems, strategic environmental assessments, flood risk management diagrammatic reasoning, logical argument mapping, risk ranking, shared mental model and team mental model. As this study is focused on understanding the adaptive preparedness of communities that are at high risk of floods from their own perspective, the qualitative approach of Focus Group Discussions (FGDs) was used in gathering the data. However, lessons from the above mentioned theoretical frameworks among others were used in posing questions under four major thematic areas, namely; Flood risk, coping strategies, management and sustainability. The FGDs were utilized to collect information on the experiences of flood victims relating to the socio-economic characteristics of the victims, causes and effects of flooding, their coping mechanisms as well as their adaptation measures. Details of the specific questions have been deleted because of restrictions on the length of the paper. The FGDs also covered issues such as perceptions on flood insurance, what government could do and what the community could do to lessen the effects of flooding. The (FGDs) were conducted in *Sepe-Buokrom* and *Atonsu Ahinsan*, two communities identified as prone to repeated flooding in Kumasi [21]. In selecting the participants for the FGDs, individuals who had experienced flooding in the last 10 years were targeted for the study, including those from different socio-cultural and economic groups; including men and women, youths, artisans, food vendors, traders, farmers, and house owners among others. These participants were identified with the assistance of the government elected representative in the community locally known as "Assemblyman".

4 DESCRIPTION OF THE STUDY AREA

The study was conducted within Kumasi, the second largest city in Ghana, located in the rainforest zone of West Africa between latitude 6°30′ and 7°00′ N and longitude 1°30′ and 2°00′ W. Kumasi has a wet, semi-equatorial climate with an annual rainfall of around 1,400 mm with two distinct rainy seasons. The mean annual temperature is 25.7°C with a humidity ranging from 53% to 93% [21]. Kumasi has a population of 2,035,064 in 2010 and recorded an annual population growth rate of 5.7% between 2,000 and 2,010 [22]. In response to the population increase Owusu-Ansah [23] alludes that land use has intensified into the inner city and spilled over into the city's public parks and natural vegetation of open spaces, riparian lands and wetlands hence creating large areas with impervious surfaces and increased volumes of runoff from precipitation.

Owusu Ansah [23] further relates that since the 1980s, when a system of open drains were established as part of the major road rehabilitation programme very few improvements have been made to drainage infrastructure. Figure 1 shows the main locations in Kumasi noted to be flood prone and the study areas; *Sepe -Buokrom* and *Atonsu- Ahinsan*. *Atonsu-Ahinsan* is characterized by houses made up of cement and sand walls roofed with corrugated iron sheets and many houses are clustered together with little spaces between them; there are very few well planned roads and this gives an impression of poorly planned closed community. Choked gutters were common and the drainage system is quite poor. The economic activity in this study area is mostly small scale self-employed businesses and petty trading [21].

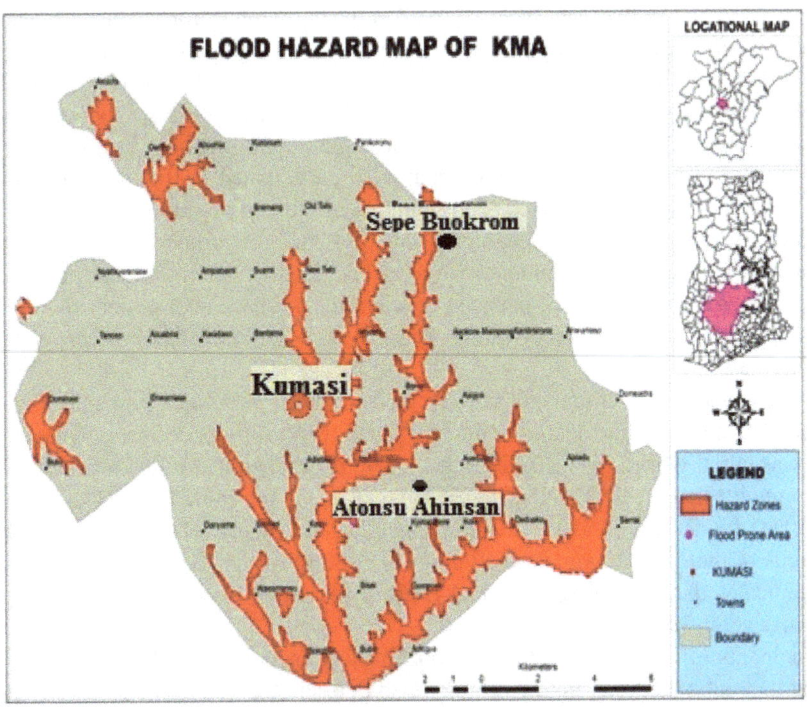

Figure 1: Map of Kumasi showing flood prone areas.

In spite of annual flooding, the *Sepe-Buokrom* area has experienced some level of urban development. In addition, unauthorized structures have further extended the settlement into floodable areas [23].

5 THE FOCUS GROUP DISCUSSIONS

The preparation towards the FGDS started in mid-February when contacts were made with the assemblymen of the various high risk communities, through a phone call to establish an audience with them on the rational for the study. Based on an initial review of literature by Owusu Ansah, (2013) in Kumasi, three main areas were identified as flood prone; *Sepe-Buokrom* and *CPC* and *Atonsu-Ahinsan* areas of Kumasi, the second largest city in Ghana. Initially, all three Assemblymen agreed to assist in identifying the potential participants, however close to the appointed times agreed, the Assemblyman for *CPC* asked for extension of time for some health reasons, however, because time was running out, the FDGs was held in the two communities, *Sepe-Buokrom* and *Atonsu- Ahinsan.*

Following the initial phone conversations with the Assemblymen of the selected communities, a visit was made to *Sepe - Buokrom* on 5/02/2016 to meet with the Assemblyman to discuss the rationale of the FDGs and the role he was expected to play in facilitating the exercise. Having reached an understanding on the modalities, the date for the FGDs was fixed for 12/02/2016. On the appointed day, three members of the research team met the group for the exercise. The research team comprised a moderator of the FGDs, a recorder and an observer of the general mood and body language of the participants. In *Sepe-Buokrom,* the FGDs began at 9.00 am and ended at 10.25 am. The participants comprised of 12 participants; 5 males and 7 females. Analysis of the general background information obtained from the participants indicates that they have all lived in the community for at least 10 years and have experienced repeated flooding in the community. The participants are mostly self-employed engaged in livelihoods ranging from artisans, food vendors, drivers and vehicle owners. The general atmosphere in the hall where the FGDs was held was a sign of cordiality and all participants were relaxed. There was no sign of intimidation as all participants expressed themselves freely. The Assemblyman also participated in the exercise both as a victim of flood and also as the elected representative of the community at the local government office. Generally, the housing characteristics of the neighbourhood are of poor quality characterised by haphazard development with some houses having their foundations exposed, suggesting signs of erosion; lack of paved streets and roads, absence of gutters and drainage facilities within the community, deep gullies in some areas and informal business and commercial activities.

As in *Sepe Buokrom*, a preliminary visit was made to meet the Assemblyman for the *Atonsu-Ahinsan* area to discuss the details of the FGDs and fix an appropriate date. Subsequently, the FGDs was initially fixed for 18/03/2016 but had to be rescheduled following bereavement in the community that week. Finally, the FGDs took place on 25/02/2016. The discussion began at 10.00 am and ended at 11.45 am lasting close to 2 hours. The FGDs comprised of 8 participants; 4 men and 4 women, including the Assemblyman. Background information gathered indicates that the participants had lived in the area for at least 10 years and had experienced several floods.

All participants of the FGDs in *Atonsu- Ahinsan* were self-employed; farmers and petty traders and they intimated that their livelihood had deteriorated due to the flooding. Similar to *Sepe-Buokrom* their housing conditions are characterised by exposed foundations of houses and cracks in buildings, lack of paved streets and roads, absence of gutters and drainage facilities within the community, deep gullies in some areas and informal business and commercial activities.

6 RESULTS

The data have been transcribed from the local dialect *Asante Twi* to English as close to verbatim as possible (the details are not included because of restrictions on the length of paper). The data were analyzed manually in the first instance individually by three researchers and later the findings emerging were compared in a collective discussion to reach a consensus on the realistic representation of the concepts emerging as listed in the quotations and Table 1 below.

With respect to risk levels in the two communities, these are the comments made by the respondents:

> *"The source of the river is at Buokrom Estate, and the river from Pakrono joins into the Buokrom Estate one. So it may not necessarily rain here but once it rains at Pankrono this whole area will be flooded. So by the time it begins to rain here the whole place will already be flooded because it is a low lying area. When it rains in the whole Kumasi it is even worse."*

Table 1: Summary of findings from focus group discussions.

Thematic Area	**Sepe Buokrom**	**Atonsu Ahinsan**	**Emerging Concepts**
Flood Risk perception	Physical observation, as soon as clouds gather, high risk area,	Physical Observation; as clouds gather, high risk area	Rain is analogous to flooding high risk area
Causes	Human activity, haphazard construction, lack of drains, dumping of refuse, fluvial flooding and flash flooding	Fluvial flooding, incomplete construction of covert, dumping of refuse into gutters	Fluvial flooding, flash flooding, weak or non-existent engineering systems, human activity
Coping strategies	Anxiety	Anxiety	Anxiety
Management	No management plan currently in place. But school buildings were considered a good gathering point during flooding.	School buildings are natural choice gathering points but not based on any community agreement.	Mostly no concise and management plan in place.
Emerging Concepts	Resilience and Social Responsibility practically non-existent. No trust in any form of insurance scheme	Resilience and social responsibility practically non-existent. No trust in any form of insurance scheme	Resilience and social responsibility practically non-existent. No trust in any form of insurance scheme

"One main area is the flooding of the Airport roundabout area. When the place floods you can't go anywhere. The whole area gets flooded from here to Dichemso" and if care is not taken you can even die. All the governments are aware of this problem but nothing has been done to alleviate the problem, if we had gutters and bridges we would not encounter these problems. All this we have come here today to say, you will take it up but nothing will be done about."

"Sometimes when it rains at Ahinsan you may think it will not flood the area, but when the water joins with that of the town then we have a problem. So for me when it begins to rain I do not sleep."

When the participants were asked to explain some indigenous signs of flooding, the general response was as follows:

"The main sign is the gathering of clouds and the possible threat of rain. In this area, whenever it threatens to rain there is a lot of apprehension because of the effects of drowning we have encountered".

On the causes of floods in the communities, this is how they responded:

"This area is an old town and when buildings were put up in the past, buildings were not properly delineated for streets and drainage systems. As a result, when it rains some houses are immersed in rain water up to the knee level. All these things bother us, whenever it rains if caution is not taken adults can even drown".

These are some revealing stories of the personal experience of some participants

"Last year my wife went to town and the vehicle she was in could not come through to our community because of the flood. I had to go out in my underwear to pick my wife. That day That day I had to carry 17 more people and also accommodate 10 more people that night".

"I was returning from town and the entire airport roundabout area had flooded, one man who was with us was carried away by the water".

"I lived in a story building (means multi-story building) and last year the entire ground floor got flooded and the tenants lost all their belongings and 8 people moved out. That day it rained from evening till dawn, the whole ground flooded".

"Some years ago I lost close to 3000 poultry which was my family's source of livelihood...Some machines got stuck in the water and there was the need to destroy my coup in order access them... There was no compensation and till date we haven't been able to work again and in spite of this we still encounter problems due to the flood".

"We stood outside for a while because there was nowhere to go. At some point we had no choice but sleep on our wet mattress...we purchased a rubber (polythene sheet) to cover it in order to sleep with the water still draining out in the process".

These are some feelings expressed regarding willingness to participate in any kind of flood insurance scheme:

"We have heard of flood insurance, in the past I had four cars but when there was a problem I got tossed around so in the end I sold the cars. As a result, I believe even if I purchase any insurance, I will not be able to reap the benefits unless I'm dead...When they visit you, their packages seem very attractive but when you require assistance they can even tell you they can't find your file".

"Here in Ghana the insurance companies do not abide by the rules unlike other countries, so no one really trusts them"

"If a community initiative to collect funds to assist flood victims is introduced, it will not work because some community members will not participate fully. The programme will work if only a community forum is organized. Some education will be required about the insurance."

"It would be good but I think it shouldn't be done, because of the tendency of those in charge to spend the funds."

Some flood coping strategies of the respondents were as follows:

"For individuals living in low lying areas, whenever flooding occurs they stand on their beds with their belongings. So when it is not raining we are all relaxed but the moment we see signs of rain there is a lot of anxiety."

"I am a widow and I am alone with my children... so when it begins to rain I let my children sleep first and then I keep vigil because my wall is already broken as a result of the flooding".

In relation to community participation in lessening the effects of flooding the participants expressed that:

"We have self-help projects that the community should involve themselves. But in this community if you speak to people about something that doesn't directly affect them they are unwilling to participate. There are some people who will not even come and share their condolences so how much more will they contribute to a programme like this".

"Personally I feel that the victims of flooding should form a group and forget about the entire community because they cannot understand our problems".

"In the past, the traditional leaders had more control of the communities, but right now the jurisdiction is in the hand of the local government who charges us for the use of every utility. Now we have to pay for everything including toilet facilities and dumping refuse. So now we don't want to participate in any community activities because we feel it's the duty of the local government"

These findings suggest that in the *Sepe-Buokrom* area, the main cause of flooding is the absence of gutters and drains within the community. Coupled with this, is the problem of residents dumping refuse in the gutters of the main streets of the communities and the rivers. The most vulnerable groups are residents in the low lying grounds in the *Moshie Zongo* area who are notably migrants. Among the effects of flooding are the loss of lives, livelihoods and property, the inability to access social amenities and infrastructure and the inability of individuals to move from one place to the other during rains and subsequent flooding. Similarly, in the *Atonsu Ahinsan* area flooding is mainly as a result of the gathering of rain water in an uncompleted culvert and choked gutters as a result of the dumping of refuse. The major effect of flooding in this area is mostly loss of livelihoods and property making it difficult for victims of flood disasters to put their lives back in order after the floods has subsided. In both communities there was a strong request for government to provide services to prevent flooding and or otherwise curb flood disasters. Apart from monthly communal labour activities to clear gutters as well as the filling of gullies with big stones, there are no community or individual initiatives to lessen the effects of flooding. In spite of the frequent flooding of the area, there is no flood risk management plan in place which is exhibited by the absence of a designated gathering point or contact person during flooding. There was the general consensus that an introduction of flood insurance is a good initiative, however, both communities were highly sceptical based on the general perception of the poor image of the insurance industry in Ghana.

7 DISCUSSION

This study has established that the participants used for the FGDs from the two communities *Sepe Buokrom* and *Atonsu-Ahinsan* are fully aware that their communities are high risk flood areas and that both communities are vulnerable. Flooding has emerged as the most devastating natural hazard and developing countries such as Ghana are at most risk demanding the development of robust mitigation measures [9]. Against the background of recent flooding, Ghana now requires having a holistic approach towards flood control and management, especially as the vulnerability of engineering and economic system and also communities are being exposed. An effective flood risk management approach has to include the connection between construction infrastructure, economic system and the role of human factors [20]. However, the FGDs have revealed that some vulnerable communities in Ghana which could be described as high risk are generally in a state of constant trepidation and looking for government support. Given that flooding is often widespread, uncontrollable and devastating, it is indeed normal for communities to look up to Governments who often have the leverage to provide the kind of assistance needed. For instance, Wood et al [7] has noted that, even in developed economies such as the US where there is relatively high awareness and involvement of communities, many communities still expect the government to take the full burden of flood risk management, including those effecting their personal loss and liability. However, it is also now known that societal drivers in terms of social responsibility and willingness to prepare for an event and knowing how to prepare is also crucial in building CS to flood hazards [5, 6]. The emerging trend is that, flood risk management is shifting from the traditional objective approach to an integrated approach with the focus on social aspects and preparedness of communities incorporated into community knowledge [6].

In Ghana, the National Disaster Management Organisation (NADMO) is the government agency responsible for the management of floods as well as other emergencies in Ghana. NADMO performs specific functions which are all aimed at ensuring that in times of emergency, the government is ready to support with relief efforts. The evidence on the ground is that the NADMO has been slow in reacting to the emerging trends and concentrate too much on providing relief items (which are limited anyway) to the victims. Flood hazard is expected to exacerbate due to increases in human activity and climate change, and consequently flood vulnerability will be expected to increase. It is therefore high time that the NADMO of Ghana re-align by expanding their activities to help build resilience communities by tapping into the concept of social responsibility.

Traditionally, many African societies are said to be closely knit as a result of the extended family system. The literature in Ghana has demonstrated this perhaps is still workable in the rural communities [4] However, in the urban centres it is not exactly clear whether this assertion holds true. The FDGs has revealed that there is not enough collective action in the communities and that it will require more active intervention at all levels to engender this sense of social responsibility so that Ghanaian communities would begin to embrace the concept as a core societal obligation towards mitigating flood hazards.

8 CONCLUSION

This paper used FDGs to establish whether there is any sense of CS through social responsibility in *Sepe-Buokrom* and *Atonsu-Ahinsan,* two high-risk communities to flooding in Kumasi, Ghana. The findings revealed that even though the two communities are not oblivious of the fact that, their communities are high risk, they lack any systematic effort in taking the initiative to contribute in developing their adaptive capacity to help manage floods. Emerging issues suggest that it will need a lot of deep understanding of community perceptions, actions and behaviours for them to be convinced in living to their social responsibility [7]. The decentralization concept in Ghana allows for the election of local representatives of community members called Assemblymen and their Unit committee members. The FDGs has revealed that, the Assemblymen and the committee members have not been able to provide the necessary leadership to the respective communities regarding their social responsibility. As a matter of policy, it would be important for the Assemblymen to be empowered to meet the community regularly and provide regular plans to the NADMO and the local government on efforts towards building an effective integrated flood risk management system for their communities.

REFERENCES

[1] Armah, F.A., Yawson, D.O., Yengoh, G.T., Odoi, J.O. & Afrifa, E.K., Impact of floods on livelihoods and vulnerability of natural resource dependent communities in Northern Ghana. *Water*, **2**(2), pp. 120–113, 2010. http://dx.doi.org/10.3390/w2020120

[2] Amoako, C., Emerging issues in urban flooding in African cities-the case of accra, Ghana. In *35th AFSAAP Annual Conference Proceedings*, 2012.

[3] Jha, K.A., Bloch, R. & Lamond, J., *Cities and Flooding: A Guide to Integrated Urban Flood Risk Management for the 21st Century*, The World Bank: Washington DC, 2011.

[4] Fabiyi, O.O. & Oloukoi, J., Indigenous knowledge system and local adaptation strategies to flooding in coastal rural communities of Nigeria. *Journal of Indigenous Social Development,* **2**(1), pp. 1–19, 2013.

[5] Mullins, A. & Soetanto, R., *International. Journal of Safety and Security Engineering,* **1**(2), pp. 115–125, 2011. http://dx.doi.org/10.2495/SAFE-V1-N2-115-125

[6] Kellens, W., Terpstra, T. & De Maeyer, P., Perception and communication of flood risks: a systematic review of empirical research. *Risk Analysis*, **33**(1), pp. 24–49, 2013. http://dx.doi.org/10.1111/j.1539-6924.2012.01844.x

[7] Wood, M., Kovacs, D., Bostrom, A., Bridges, T. & Linkov, I., Flood risk management: US Army Corps of Engineers and layperson perceptions. *Risk analysis*, **32**(8), pp. 1349–1368, 2012.
http://dx.doi.org/10.1111/j.1539-6924.2012.01832.x

[8] World Bank, Cities and Flooding: A Guide to Integrated Urban Flood Risk Management for the 21st Century, 2015.

[9] Ahadzie, D.K. & Proverbs, D.G., Emerging issues in the management of floods in Ghana. *International Journal Safety Security Engineering*, **1**(1), pp. 182–192, 2011.
http://dx.doi.org/10.2495/SAFE-V1-N2-182-192

[10] Intergovernmental Panel on Climate Change, IPCC, Summary for Policymakers. In *Managing the Risk of Extreme Events and Disasters to Advance Climate Change Adaptation, Special Report of Working Groups 1 and 11 of the Intergovernmental Panel on Climate Change*, eds. C.B. Field, V. Barros, T.F. Stocker, D. Qin, D.J. Dokken, K.L. Ebi, M.D. Mastrandrea, K.J. Mach, G.K. Plattner, S.K. Allen, M. Tignor & P.M. Midgley, Cambridge University Press: Cambridge, UK, and New York, NY, USA, pp. 1–19, 2012.

[11] UN Habitat 2011, Global Report on Human Settlements 2011: Cities and Climate Change, United Nations Human Settlements Programme (UN Habitat), Earthscan: London UNISDR, 2012.

[12] Perdikaris, J., Gharabaghi, B. & McBean, E., A methodology for undertaking vulnerability assessments of flood susceptible communities. *International Journal of Safety and Security Engineering,* **1**(2), pp. 126–146, 2011.
http://dx.doi.org/10.2495/SAFE-V1-N2-126-146

[13] Olugunorisa, T.E., Strategies for mitigation of flood risk in the Niger Delta, Nigeria. *Journal of Applied Sciences and Environmental Management*, **13**, (2), pp. 17–22, 2009.

[14] Musah, B.A.N. & Akai, C.Y., Effects of flood disasters on livelihood coping mechanism in Tolon/Kumbumgu District of Northern Region of Ghana. *International Journal of Agricultural Policy and Research*, **2**(1), pp. 33–40, 2014.

[15] Aboagye, D.D., Dari, T. & Koomson, J., Risk perception and disaster management in the savannah region of Ghana. *International Journal of Humanities and Social Science,* **3**(3), pp. 85–95, 2013.

[16] Olorunfemi, F.B. & Raheem, U.A., Floods and rainstorms impacts, responses and coping among households in Ilorin, Kwara Stat. *Journal of Educational and Social Research*, **3**(4), pp. 135–144, 2013.
http://dx.doi.org/10.5901/jesr.2013.v3n4p135

[17] Thompson, P. & Tod, I., Mitigating flood losses in the active floodplains of Bangladesh. *Disaster Prevention and Management: An International Journal*, **7**(2), pp. 113–123, 1998.
http://dx.doi.org/10.1108/09653569810216333

[18] Anikpo, M., Ogbanga, M.M., Ifeanacho, M., Flood disaster, adaptation and coping strategies in selected communities of Ahoada East and Ahoada West local government areas, rivers state. *IIARD International Journal of Environmental Research,* **1**(3), pp. 1–12, 2015.

[19] Duží, B., Vikhrov, D., Kelman, I., Stojanov, R. & Juřička, D., Household measures for river flood risk reduction in the Czech Republic. *Journal of Flood Risk Management*, pp. 1–14, 2015., DOI:10.1111/jfr3.12132.

[20] Jonkman, S.N. & Dawson, R.J., Issues and challenges in flood risk management—editorial for the special issue on flood risk management. *Water*, **4**(4), pp. 785–792, 2012. http://dx.doi.org/10.3390/w4040785

[21] Tay, S.C.K., Danuor, S.K., Morse, A., Caminade, C., Badu, K. & Abruquah, H.H., Entomological survey of malaria vectors within the kumasi metro area-a study of three communities: Emena, Atonsu and Akropong. *Journal of Environmental Science and Engineering B1(2012),* pp. 144–154, 2015.

[22] Ghana Statistical Services, Population and Housing Census, 2010.

[23] Owusu -Ansah, J.K., The influences of land use and sanitation infrastructure on flooding in Kumasi, Ghana. *Geojournal of Spatially Integrated Social Science and Humanities*, **80**(1), 2015.
http://dx.doi.org/10.1007/s10708-015-9636-4

[24] Government of Ghana. Local Government Act 462:1993.

COST-EFFECTIVE MITIGATION STRATEGIES FOR RESIDENTIAL BUILDINGS IN AUSTRALIAN FLOOD PLAINS

T. MAQSOOD, M. WEHNER, K. DALE & M. EDWARDS
Geoscience Australia and Bushfire & Natural Hazards Cooperative Research Centre, Australia.

ABSTRACT

In recent years, floods have impacted many Australian communities. The floods have resulted in significant logistics for emergency management and considerable costs to all levels of government and property owners to undertake damage repair and enable community recovery.

These impacts are fundamentally the result of inappropriate development on floodplains and a legacy of high risk building stock in flood-prone areas. The Australian Bushfire and Natural Hazards Collaborative Research Centre project entitled "Cost-effective mitigation strategy development for flood-prone buildings" aims to address this issue and is targeted at assessing mitigation strategies to reduce the vulnerability of existing residential building stock in Australian floodplain area.

This paper presents the outcomes of this ongoing project. Key tasks of the project include: (1) a classification of residential building stock, (2) a review of flood mitigation strategies, (3) vulnerability assessment of typical building types with and without mitigation, and (4) benefit cost analyses of all retrofit options for a range of severity/likelihood of flood hazard covering a selection of catchment types. The work will provide information on the optimal retrofit strategies in the context of Australian construction costs and catchment characteristics.

The research will also entail experimental testing of preferred materials to ascertain their resilience to flood water exposure. The outcome of this research will be an evidence base to inform decisions about mitigating the risk posed by buildings on floodplains. The information will be targeted to all levels of government, insurance industry and private property owners.

Keywords: Australia, cost-effectiveness, mitigation, vulnerability.

1 INTRODUCTION

Globally, floods cause tremendous damage with loss of life and property. An analysis of global statistics conducted by Jonkman [1] showed that floods (including coastal flooding) caused 175,000 fatalities and affected more than 2.2 billion people between 1975 and 2002.

In Australia, floods cause more damage on an average annual cost basis than any other natural hazard [2]. The fundamental causes of this level of damage and the key factor contributing to flood risk, in general, is the presence of vulnerable buildings constructed within floodplain areas due to ineffective land-use planning. Recent events in Australia (2011, 2013 and 2015) have highlighted the vulnerability of housing to flooding (see Fig. 1).

While there is now a construction standard published by the Australian Building Code Board [3] for new construction in some types of flood-prone areas, there is a large proportion of the existing building stock that has been built in flood-prone areas across Australia [2].

The Bushfire and Natural Hazards Collaborative Research Centre (BNHCRC) project entitled "Cost-effective mitigation strategy development for flood-prone buildings" is examining

DOI: 10.2495/SAFE-V6-N3-550-559

Figure 1: Examples of building damage after 2011 Queensland flooding.

the opportunities for reducing the vulnerability of Australian residential buildings to flood [4]. It addresses the need for an evidence base to inform decision making on the mitigation of the flood risk posed by the most vulnerable Australian building types. This project investigates methods for the upgrading of existing residential building stock in floodplains to increase their resilience in future flood events.

2 PROJECT ACTIVITIES

Information on the vulnerability of buildings and factors affecting vulnerability is fundamental to evaluating mitigation strategies to reduce future losses. Therefore, this BNHCRC project is systematically developing information about residential building types in Australia, their vulnerability and possible mitigation measures to reduce their vulnerability.

To date, a building classification schema has been developed to categorise Australian residential buildings into a finite set of typical building types [5]. A literature review of flood mitigation strategies applied internationally has also been conducted. Each mitigation strategy is being evaluated and will be costed through the engagement of professional quantity surveyors. Cost benefit analyses will be conducted to determine optimum retrofit strategies for selected building types applicable to a range of catchment behaviours.

2.1 Development of building schema

This research requires the context of a building vulnerability classification, or schema. The classes identified within the schema have to represent the variety of housing within the nation's residential building stock and, more specifically, the variation in vulnerability across the nation's building stock. Furthermore, the schema must identify specific classes for which the project will develop mitigation strategies.

In this research a literature review has been conducted which reviewed building schemas developed nationally and internationally for a range of uses within different projects. The reviewed schemas are from the USA [6], Germany [7], Philippines [8], New Zealand [9], Australia [10] and UNISDR Global Assessment Report [11].

After the literature review, a new schema was proposed that represents a fundamental shift from describing the complete building as an entity to one that focuses on sub-components. The schema divides each building into its major components (i.e. foundation, ground floor, upper floors (if any) and roof) enabling the vulnerability of each of these components to be assessed separately (Fig. 2). Each storey type is then classified using the following six attributes.

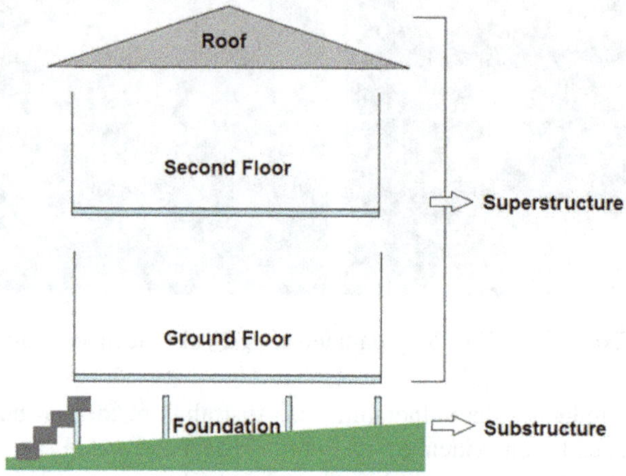

Figure 2: Building structure divided into main components.

- Construction period (pre-1960 or post-1960)
- Fit-out quality (standard or low)
- Storey height (3.0 m or 2.7 m or 2.4 m)
- Bottom floor system (slab-on-grade or raised timber or raised chipboard)
- Internal wall material (masonry or plasterboard or timber)
- External wall material (brick veneer or weatherboard or masonry)

With the exclusion of combinations that are invalid in an Australian context, the schema defines 60 discrete vulnerability classes based on the above- mentioned attributes. Furthermore, the schema proposed six roof types based on material and pitch of the roof.

This new approach facilitates the development of vulnerability models for taller buildings, buildings with basements, buildings with mixed usages and those with different construction materials used at different floor levels.

2.2 Literature review of mitigation strategies

A literature review has been conducted to identify mitigation strategies used in several countries and for various severities of flooding. The review has considered literature available through peer-reviewed journals, international conferences, research reports and guideline documents.

Bouwer *et al.* [12] classified the different types of retrofit or mitigation measures into nine basic categories in which a distinction was made between mitigation measures that focus on hazard reduction and those that focus on vulnerability reduction. The use of insurance to recover from a disaster and to provide incentives for mitigation works was studied by Kunreuther [13] and Crichton [14]. The use of spatial zoning and land-use changes was presented by Burby *et al.* [15] and Poussin *et al.* [16]. Another widely used broader classification of mitigation measures is based on whether the strategies utilise engineering and administrative methods to reduce flood risk or modify the flood characteristics and human occupancy of the floodplain. These are broadly divided into structural and non-structural approaches [17] or hard and soft measures [18]. Both approaches have their own benefits and limitations.

A summary of mitigation strategies that have been applied in Australia and internationally to minimise vulnerability and future losses of residential buildings is given below.

2.2.1 Elevation

Elevation of a structure is one of the most common mitigation strategies which aims to raise the lowest floor of a building above the expected level of flooding (Fig. 3). This can be achieved, for example, by extending the walls of an existing structure and raising the floor level; by constructing a new floor above the existing one; or through raising the whole structure on new foundations.

The technical considerations that need to be taken into account in raising buildings are structure type, construction material, foundation type, building size, flood characteristics and other hazards [19].

Generally, the least expensive and easiest building to elevate is a low-set single storey timber frame structure [20]. The procedure becomes complicated and expensive when other factors are included such as a slab-on-grade construction, walls of masonry or concrete or a multi-storey building [21]. Currently elevation of a structure is one of the strategies which result in incentives from the insurance industry in the form of reductions in annual premiums for flood insurance [22]. An analysis conducted by FEMA [23] showed that house owners can break even on their investment in adopting this mitigation strategy in little over 5 years due to reductions in their flood insurance premiums.

2.2.2 Relocation

Relocation of a building is the most dependable technique in mitigation of flood risk. However, it is generally the most expensive as well [21]. Relocation involves moving a structure to a location that is less prone to flooding and normally involves placing the structure on a wheeled vehicle and then transporting it to a new location and setting it on a new foundation [19]. The bulk of relocation cost includes cost of new land, transportation and reconnection of services. This strategy is much easier and cost effective for low-set timber frame structures. The relocation of slab-on-grade structures is more complicated and expensive. In this case, there are two approaches to relocating, that is, detaching the structure from the slab or moving the structure with the slab attached [24].

Relocation is most appropriate in areas where flood conditions are severe such as a high likelihood of deep flooding, or where there is high flow velocity with short warning time and a significant quantity of debris.

The technical considerations for relocation include structure type, its size and condition. Light weight timber structures are easy to transport compared to heavy masonry and concrete

(A) House before elevation (B) House after elevation

Figure 3: Example of elevating floor levels.

buildings. Similarly, relocation of single storey compact size structures is far easier than for large multi-storey structures. Further, the structure should be in good condition and able to withstand the stresses imposed when the structure is lifted and relocated [25].

2.2.3 Dry floodproofing

In dry floodproofing, the portion of a structure that is below the expected flood level is sealed to make it substantially impermeable to floodwaters. Such an outcome is achieved by using sealants which include wall coatings, waterproofing compounds, impervious sheeting over doors and windows and a supplementary leaf of masonry [19]. The expected duration of flooding is critical when deciding which sealant systems to use because seepage can increase with time making flood proofing ineffective [21]. Preventing sewer backflow by using backwater valves is also important in making dry floodproofing effective [25, 26]. Sump pumps are also required to drain out the water which may leak through small openings or exterior walls [27].

The flood characteristics which can affect the success of dry floodproofing are flood depths, flow velocity, flood duration, flood borne debris and length of warning time [25].

Dry floodproofing is generally not recommended in flood depths exceeding one metre based on tests carried out by the US Army Corps of Engineers as the stability of the building becomes an issue over this threshold depth [26, 28]. Dry floodproofing may also be inappropriate for light timber frame structures and structures which are not in good condition and may not be able to withstand the forces exerted by the floodwater [19].

2.2.4 Wet floodproofing

In this measure, the building is modified and floodwater is allowed to enter into the building to equalise the hydrostatic pressure on the interior and exterior of the building and thus reduces the chance of building failure [21, 25]. With this technique, as all the building components below the flood level are wetted, all construction material and fit-outs should be water-resistant and/or can be easily cleaned following a flood extent [29, 30]. Wet floodproofing involves raising utilities including heating, ventilation, air conditioning and electrical systems. It also include raising important contents above the expected flood level, installing flood openings to equalise the hydrostatic pressure exerted by floodwaters and installing pumps to remove floodwater if the building has a basement [31].

Wet floodproofing may not be suitable in floods with duration of more than a day as longer exposure to flood water leads to damage to structural components of the building and also results in the growth of algae and mould [25]. In a study conducted by Kreibich *et al.* [26], wet floodproofing through adapted use (changing ground floor usage), adapted interior finishes (use of flood resistant materials) and installing utilities at higher levels has resulted in reductions of building mean damage by 46%, 53% and 36%, respectively, in Germany during the 2002 floods.

2.2.5 Flood barrier

Flood barriers considered in this research are those built around a single building and are normally placed some distance away from it to avoid any structural modifications to the building. There are two kinds of barriers: permanent and temporary. An example of a permanent barrier is a floodwall which is quite effective because it requires little maintenance and can be easily constructed and inspected. There are also several types of temporary flood barriers available in the market which can be moved, stored and reused. There are a number of considerations with regard to the use of these barriers such as the need for prior warning and enough time to be set up to be effective [32]. They also require periodic inspection and maintenance to address any repairs required. Further, access to the building could be difficult [25].

A number of vendors make temporary flood barriers that can be assembled relatively easily, moved into place and anchored. Example flood barriers include sand bags, PVC tubes, box wall, flexible barriers, box barrier and metal fence [33]. In a study in Germany, the mean damage ratio was reduced by 29% for the cases where water barriers were available during the 2002 floods [26].

2.3 Development of Australian specific retrofit options and costing

A limited number of typical residential storey types have been selected for the balance of the research which are based on the schema proposed earlier in this paper. Fig. 4 presents the six floor types for which retrofit strategies will be investigated. Key characteristics of these floor

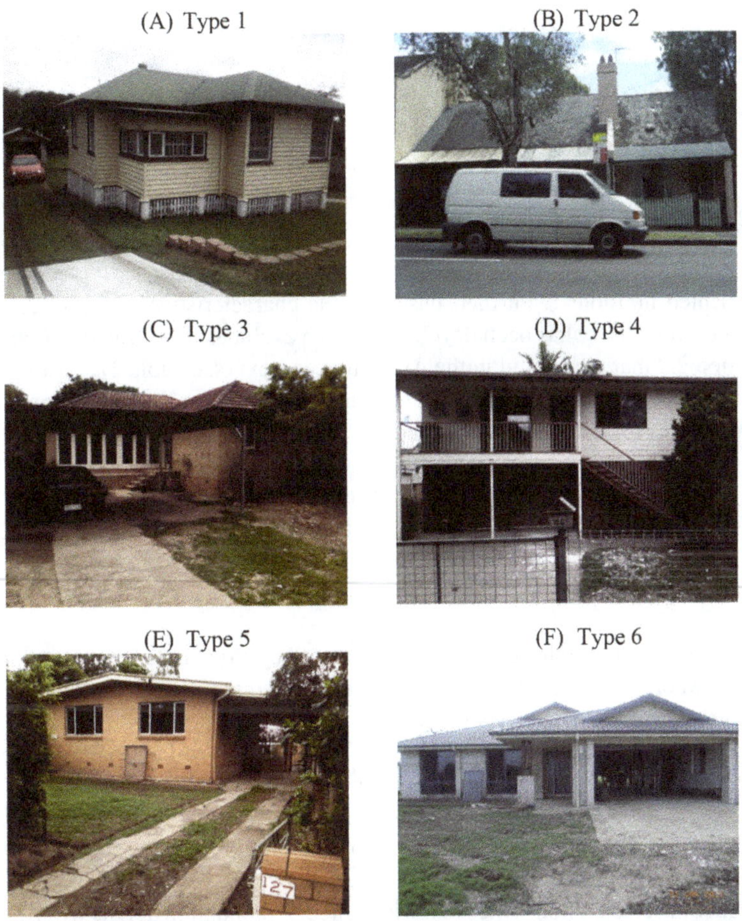

Figure 4: Selected typical residential floor types.

Table 1: Characteristics of selected storey types.

Type	Const. period	Fit-out quality	Storey height	Floor type	Internal wall material	External wall material
1	Pre-1960	Standard	3.0 m	Raised Timber	Timber	Weatherboard
2	Pre-1960	Standard	2.7 m	Slab-on-grade	Masonry	Solid masonry
3	Pre-1960	Low	2.7 m	Raised Timber	Timber	Cavity masonry
4	Post-1960	Standard	2.4 m	Raised Timber	Plasterboard	Weatherboard
5	Post-1960	Low	2.4 m	Raised Timber	Plasterboard	Brick veneer
6	Post-1960	Standard	2.7 m	Slab-on-grade	Plasterboard	Brick veneer

Table 2: Floodproofing matrix for selected Australian floor types.

Type	Flood mitigation strategies				
	Elevation	Relocation	Dry floodproofing	Wet floodproofing	Flood barriers
1			N/A		
2	N/A	N/A			
3		N/A	N/A		
4		N/A	N/A		
5		N/A	N/A		
6	N/A	N/A			

types are presented in Table 1. Further, based on the characteristics of the selected storey types a floodproofing matrix has been developed which excludes the mitigation options mentioned in Section 2.2 that are invalid in the Australian context (see Table 2). Costing modules are being developed for quantity surveying specialists to estimate the mitigation cost of all appropriate strategies for these six floor types.

3 FUTURE ACTIVITIES

A brief overview of the future activities of the project is given below.

3.1 Experimental testing of selected building materials

In this project, the strength implications of immersion of key structural elements will be examined to ascertain where deterioration due to wetting and subsequent drying needs to be addressed as part of repair strategies. An analysis will be conducted to identify research gaps in building material susceptibility to flood water in Australia. This research will also entail experimental testing of preferred material types to ascertain their resilience to flood water exposure.

3.2 Vulnerability assessment for current and retrofitted building types

The vulnerability of selected floor types to a wide range of inundation depths will be assessed. It will also be supplemented by both a significant body of flood vulnerability research by Geoscience Australia and a body of damage survey activity in Australia.

3.3 Benefit versus cost analysis

Retrofit options entail an investment that will realise a benefit over future years through reduced average annualised loss. Decisions to invest in reducing building vulnerability, either through asset owner initiatives or incentives provided by government or the insurance industry, will depend upon the benefit versus cost of the retrofit. In this research, all retrofit options will be assessed through a consideration of a range of severity and likelihood of flood hazard covering a selection of catchment types.

3.4 Outcomes

The result will be an evidence base to inform decision making by government and property owners on mitigation of flood risk by providing information on the cost-effectiveness of different mitigation strategies and optimal solutions for different cases of building and catchment types. The work will provide information on the optimal retrofit types and design levels in the context of Australian construction costs and catchment behaviours.

4 SUMMARY

The economic losses due to floods have been increasing in recent decades due to vulnerable construction types and because of rapid urban development in floodplains. The increase in loss emphasises the need to improve flood risk management and to reduce future flood losses. These need to be built upon a sound analysis of flood hazard, typical floor types, potential losses and the effectiveness of different mitigation measures [29].

Flood risk management not only includes the measures taken by government but also includes mitigation measures adopted by private property owners to reduce the potential losses. These measures include elevating structures above the expected flood level, relocating the structure outside the floodplain, dry floodproofing to make the structure water tight, wet floodproofing by using water-resistant materials and installing storey barriers to keep water away from the building.

These efforts have a significant potential to reduce flood damage to buildings and contents particularly in low to moderate flood levels. Selection and implementation of any of these strategies would require comprehensive analyses of the characteristics of flood, local building standards and a cost benefit analysis to evaluate the optimum strategy.

This BNHCRC project aims to conduct a comprehensive analysis of mitigation options and evaluate each of them through cost benefit analysis for use in Australian conditions. The result would be a clear understanding of cost and benefits involved in implementing any of these mitigation measures. This evidence base will facilitate and encourage governments and property owners to make informed and optimal decisions to reduce flood risk.

ACKNOWLEDGEMENTS

The support of the Commonwealth of Australia through the Cooperative Research Centre program is acknowledged.

This paper is published with the permission of the CEO, Geoscience Australia.

REFERENCES

[1] Jonkman, S., Global perspectives of loss of human life caused by floods. *Natural Hazards*, **34**, pp. 151–175, 2005.
http://dx.doi.org/10.1007/s11069-004-8891-3

[2] HNFMSC, *Reducing Vulnerability of Buildings to Flood Damage: Guidance on Building in Flood Prone Areas*, Hawkesbury-Nepean Floodplain Management Steering Committee: Parramatta, 2006.
[3] Australian Building Code Board (ABCB), *Construction of Buildings in Flood Hazard Areas*, Standard Version 2012.2, Canberra, Australia, 2012.
[4] BNHCRC, Cost-effective mitigation strategy development for flood prone buildings, bushfire and natural hazards cooperative research centre, available at http://www.bnhcrc.com.au/research/resilient-people-infrastructure-and-institutions/243
[5] Maqsood, T., Wehner, M., Dale, K. & Edwards, M., A schema to categorise residential building in Australian floodplains. *Proceeding of the Floodplain Management Association National Conference*, Brisbane, Australia, 2015.
[6] FEMA, Multi-hazard loss estimation methodology, flood model, *HAZUS. Technical Manual. Federal Emergency Management Agency*, Department of Homeland Security, Washington DC, USA, 2007.
[7] Schwarz, J. & Maiwald, H., Damage and loss prediction model based on the vulnerability of building types. *Proceeding of the 4th International Symposium on Flood Defence*, Toronto: Canada, 2008.
[8] Pacheco, B., Hernandez, H., Castro, P., Tingatinga, E., Suiza, R., Tan, L., Lonalong, R., Vreon, M., Aquino, H., Macuha, R., Mata, W., Villalba, I., Pascua, M., Ignacio, U., Germar, F., Dino, J., Reyes, G., Tirao, J. & Zarco, M., *Development of Vulnerability Curves of Key Building Types in the Greater Metro Manila Area, Philippines*, Institute of Civil Engineering, University of the Philippines Diliman, 2013.
[9] NIWA, Riskscape User Manual Version 0.2.30, available at https://riskscape.niwa.co.nz/
[10] Wehner, M., Maqsood, T., Corby, N., Edwards, M. & Middelmann-Fernandes, M., *Augmented vulnerability models for inundation*, Technical report submitted to DCCEE, Geoscience Australia, Canberra, Australia, 2012.
[11] Maqsood, S., Wehner, M., Ryu, H., Edwards, M., Dale, K. & Miller, V., *GAR15 Vulnerability Functions: Reporting on the UNISDR/GA SE Asian Regional Workshop on Structural Vulnerability Models for the GAR Global Risk Assessment, 11–14 November, 2013,* Geoscience Australia, Canberra, Australia, Record 2014/38. Geoscience Australia: Canberra, Australia, 2014.
http://dx.doi.org/10.11636/Record.2014.038
[12] Bouwer, L., Poussin, J., Papyrakis, E., Daniel, V., Pfurtscheller, C., Thieken, A. & Aerts, J., Methodology report on costs of mitigation. CONHAZ project report, Report number WP04_2, 2011.
[13] Kunreuther, H., Disaster mitigation and insurance: learning from Katrina. *The Annals of the American Academy*, **604**, pp. 208–227, 2006.
http://dx.doi.org/10.1177/0002716205285685
[14] Crichton, D., Role of insurance in reducing flood risk. *The Geneva Papers*, **33**, pp. 117–132, 2008.
http://dx.doi.org/10.1057/palgrave.gpp.2510151
[15] Burby, R., Deyle, R., Godschalk, D. & Olshansky, R., Creating hazard resilient communities through land-use planning. *Natural Hazards Review*, **1**, pp. 99–106, 2000.
http://dx.doi.org/10.1061/(ASCE)1527-6988(2000)1:2(99)
[16] Poussin, J., Bubeck, P., Aerts, J. & Ward, P., Potential of semi-structural and non-structural adaptation strategies to reduce future flood risk: case study for the Meuse. *Natural Hazards and Earth Systems Science*, **12**, pp. 3455–3471, 2012.

http://dx.doi.org/10.5194/nhess-12-3455-2012
[17] Brody, S., Kang, J. & Bernhardt, S., Identifying factors influencing flood mitigation at the local level in Texas and Florida: the role of organizational capacity. *Natural Hazards*, **52**, pp. 167–184, 2010.
http://dx.doi.org/10.1007/s11069-009-9364-5
[18] Productivity Commission, Natural disaster funding arrangements. Productivity commission inquiry report no. 74, Canberra, Australia, ISBN 978-1-74037-524-5, 2014.
[19] FEMA, *Engineering Principles and Practices for Retrofitting Flood-Prone Residential Structures*, FEMA P-259, 3rd., Federal Emergency Management Agency, USA, 2012.
[20] USACE, *Flood proofing; techniques, programs and references*. National Flood Proofing Committee. US Army Corps of Engineers. Washington D.C., USA, 2000.
[21] USACE, *Flood proofing; how to evaluate your options*. National Flood Proofing Committee. US Army Corps of Engineers. Washington D.C., USA, 1993.
[22] Bartzis, N., Flood insurance pricing. *Proc. of the Floodplain Management Association National Conference*, Tweed Heads, Australia, 2013.
[23] FEMA, *Natural hazards and sustainability for residential buildings*. FEMA P-798. Federal Emergency Management Agency, USA, 2010.
[24] USACE, *Raising and moving the slab-on-grade house with slab attached*. National Flood Proofing Committee. US Army Corps of Engineers. Washington D.C., USA, 1990.
[25] FEMA, *Selecting appropriate mitigation measures for floodprone structures*. FEMA 551. Federal Emergency Management Agency, USA, 2007.
[26] Kreibich, H., Thieken, A., Petrow, T., Meuller, M. & Merz, B. Flood loss reduction of private households due to building precautionary measures – lessons learned from the Elbe flood in August 2002. *Natural Hazards and Earth Systems Science*, 5, pp. 117–126, 2005.
http://dx.doi.org/10.5194/nhess-5-117-2005
[27] FEMA, *Floodproofing non-residential buildings*. FEMA P-936. Federal Emergency Management Agency, USA, 2013.
[28] USACE, Flood proofing tests; tests of materials and systems for flood proofing structures. *Floodplain Management Services Programs. National Flood Proofing Committee*. US Army Corps of Engineers: Washington D.C., USA, 1988.
[29] Kreibich, H. & Thieken, A., Assessment of damage caused by high groundwater inundation. *Water Resources Research*, **44**, W09409.
http://dx.doi.org/10.1029/2007WR006621
[30] FEMA, Flood damage-resistant materials requirements for buildings located in special flood hazard areas. NFIP Technical Bulletin 2. Federal Emergency Management Agency, USA, 2008.
[31] FEMA, *Protecting Building Utilities From Flood Damage; Principles and Practices for the Design and Construction of Flood Resistant Building Utility Systems*, FEMA P-348, 1st edn., Federal Emergency Management Agency, USA, 1999.
[32] Kreibich, H., Christenberger, S. & Schwarze, R., Economic motivation of households to undertake private precautionary measures against floods, *Natural Hazards and Earth Systems Science*, **11**, pp. 309–321, 2011.
http://dx.doi.org/10.5194/nhess-11-309-2011
[33] Bluemont, Flood prevention, available at http://www.bluemont.com.au/flood-prevention

DEVELOPMENT OF A 3D DYNAMIC FLOOD WEB GIS VISUALISATION TOOL

S. VAN ACKERE[1], H. GLAS[2], J. BEULLENS[1], G. DERUYTER[2], A. DE WULF[1] & P. DE MAEYER[1]
[1]Department of Geography, Ghent University, Krijgslaan 281 (building S8), Ghent, Belgium.
[2]Department of Civil Engineering, Ghent University, V.Vaerwyckweg 1, Ghent, Belgium.

ABSTRACT

Low elevation coastal areas are vulnerable to the effects of sea level rise and to an increase in the frequency and severity of storm surge events due to climate change.Coastal urban areas are at risk because coastal flooding causes extensive damage to energy and transportation infrastructure, disruptions to the delivery of services, devastating tolls on the public's health and,occasionally, significant loss of life. Although scientists widely stress the compelling need to mitigate and adapt to climate change, public awareness lags behind. Because WebGIS maps (web-based geographic information systems) quickly convey strong messages, condense complex information, engage people on issues of environmental change, and motivate personal actions, this paper focuses on searching the ideal flood assessment WebGIS method to encourage people to mitigate and adapt to climate change. Surveys demonstrated that 3D visualisations have an enormous added value because they are more vivid and therefore more understandable and make it easier to imagine the consequences of a flood than2D visualisations. In this research, the WebGIS will be created using Ol3-Cesium and openlayers to visualise a flood event by dynamic layers in a 2D/3D environment.

Keywords:coastal flooding, flooding analysis,landscape visualisation, WebGIS visualisation.

1 INTRODUCTION

More than 634 million people, or approximately 8.7% of the world's population, live in low elevation coastal areas below 10 m [1,2]. A recent study published in the journal *Nature* reports that sea levels may rise considerably faster than expected in the coming decades because the Antarctic ice sheet is less stable than previous thought [3]. Although floodings are commonly referred to as *natural disasters*, they are directly attributable to the actions of human beings[4].

The Intergovernmental Panel on Climate Change (IPCC) concluded that low elevation coastal areas are vulnerable to the effects of sea level rise and an increased frequency and severity of storm surge events due to climate change[6]. Coastal urban areas are at risk because coastal flooding causes extensive damage to energy and transportation infrastructure, disruptions to the delivery of services, a devastating toll on the public's health and, occasionally, significant loss of life. At low elevation coastal areas, the wind contribution to storm surges is usually dominant compared to other forcing mechanisms as short-waves, atmospheric pressure, rainfall and river flow because the wind effect is inversely proportional to the water depth. Nevertheless, loss of human lives and property damages caused by coastal flooding are frequently larger compared to those directly related to wind [7].

DOI: 10.2495/SAFE-V6-N3-560-569

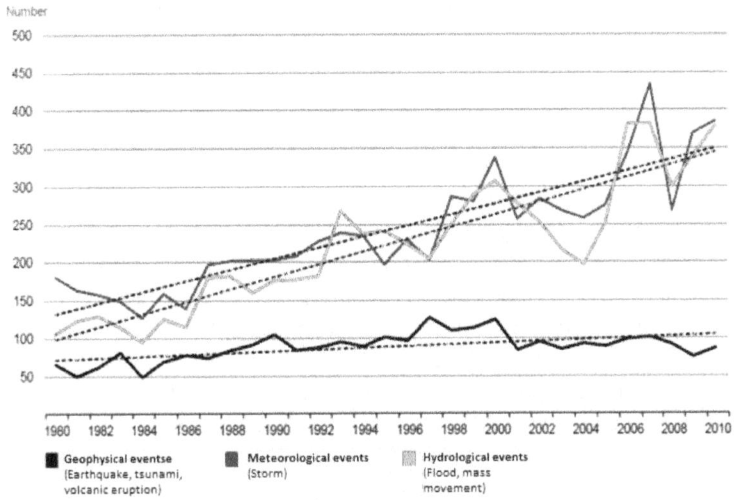

Figure 1: Natural catastrophes, worldwide (1980–2012)[5].

2 WEB GIS FLOOD VISUALISATION

Since the mid-1960s, numerical models have been used to investigate storm surges[8], which resulted in studies on mitigating the impacts of floods using artificial physical flood control (via the construction of dams, dykes,levees, diversion channels and related structures)[9]. To date, researchers have shown an increased interest in computational power and the advent of parallel computing, which allowed very accurate simulations of coastal floodings[7]. Zoran-Vojinovic emphasizes in his book 'Flood risk: the holistic perspective': "*Climate change (or climate extremes), population growth and urbanisation processes are certainly bringing some of the greatest challenges of our time which in turn reinvigorate the significance of adaptation,resilience and transformation (i.e. transcendence)*[4]." Although scientists widely stress the compelling need to mitigate and adapt to climate change, public awareness lags behind[10]. Therefore, there is a compelling need to inform people in low elevation coastal areas of the potential hazards with tools that quickly convey strong and clear messages of this sword of Damocles. In many cases,WebGIS applications are often used, as it quickly conveys strong messages, condenses complex information, engages people in environmental change issues, and motivates personal actions[10].In this paper, the researchers are searching for the ideal webGIS application method to visualise potential flood hazard events to engage people to adapt to climate change.

2.1 Evaluating water level visualisation tools

Reviewing the academic literature, considerable attention has been devoted to creating visualisation tools of sea level rise. Good examples of WebGIS of sea level rise visualisation tools are Sea Level Rise and Coastal Flooding Impacts Viewer[11], New Jersey Flood Mapper [12], Lakes Entrance Visualisation[13], Flood map Brussels[14],Explore Sahul Time [15], Interactive Sea Level Rise Web Map [16], Sea Level Rise Visualisation for AL, MS and FL[17], Surging Seas[18], What Could Disappear[19], Impacts of Sea Level Rise on the California Coast[20].These interactive WebGIS tools allow the user to visualise the sea level rise at a certain location, which is interesting to study when searching for the ideal WebGIS application method of flood events.

Among the many WebGIS tools, there are several tools that use incorrect methods to calculate and visualise information of sea level rise. Therefore, the most common mistake is that these tools neglect natural and artificial coastal defence infrastructures. This finding is clearly observed when the sea level is visualised for a sea level rise of +0 m (see Fig. 2). Therefore, each area with an elevation lower than current sea level is visualised as areas that are flooded. This incorrect calculation can easily be corrected by only displaying areas which are physically connected with the sea. For example, when sea level rise is visualised using PostGIS, it is possible to take into account only water level polygons that are directly connected to the sea and rivers (using the function ST_ClusterIntersecting[21]). Hereby, the representation of a certain sea level rise will be correct. In addition to this incorrect visualisation, this interactive map has the misleading capture "data provided from NASA"[22], which refers to the used land elevation data and not to future sea level rise data.

It is of importance that sea level rise is well-visualised such that the inhabitants are well-informed, allowing them to assess the seriousness of the case, although many interactive maps have the opposite effect.

2.2 2D visualisation versus 3D visualisation

As already mentioned, lack of public awareness can be remedied with the use of better understandable and user friendly interactive WebGIS mapping tools. Surveys such as those conducted by Leskens (2015) showed that 3Dvisualisations have an enormous added value: *"The open question about what the added value of the 3Dvisualisation was with respect to the 2Dvisualisation was answered with the following statements:*

- *It makes it more likely to imagine the consequences of the flood.*
- *It enhances prediction of what a flood would mean for an area and helps to better empathize with the situation.*
- *It is more realistic and detailed, and it is easier to interpret what a flood means for the area.*

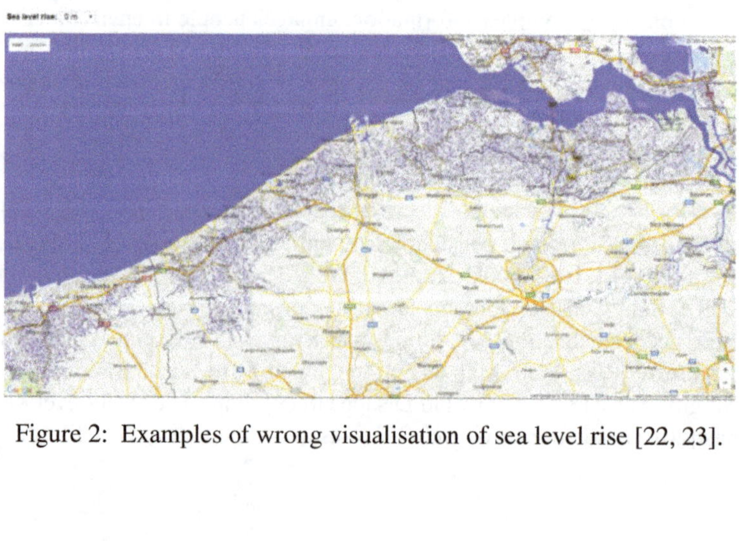

Figure 2: Examples of wrong visualisation of sea level rise [22, 23].

- *It is more vivid and therefore more understandable.*
- *Less interpretation is required to estimate the consequences of the flood.*
- *It helps the user to better imagine how serious the flood could be.*
- *It better shows the consequences for the environment.*"[24]

3Dvisualisationhas been successfully employed for the assessment of damage to infrastructures from fire [25] and earthquakes [26]. However, despite their potential benefits for flood risk management [27],only limited efforts have been made in this area. Several studies [24, 28, 29]have employed 3D models in flood risk management. However, the aim of these studies was merely the static 3Dvisualisation of flood hazard which was not embedded in a WebGIS tool.

A 3D map is a tremendous added value for the interpretation of a flood and for visualising reality. Nevertheless, 3D maps cannot take over all functionalities and perfect representations which is inherent to 2Dvisualisation. For example, it's easier to assess on a 2D map where the maximal contour of the flood occurs, in comparison with the estimation of the position of the maximal flood contour on a 3D map. The addition of a 2Dvisualisation together with a 3D visualisation (e.g. using a split-screen) is therefore a necessity.

2.3 Dynamic versus static visualisation

Many WebGIS applications make use of a single layer to visualise a sea level situation or a snapshot in time of a specific flood condition (for example, a flood condition where maximal damage to infrastructures occurs). Unfortunately, too much information gets lost by this one-sided visualisation. Therefore, it is impossible to visualise the intensity or velocity of a flooding, or the rising velocity of the water on a specific location. Furthermore, it is difficult to assess how long it takes before a specific area is flooded. This information is desperately needed for evacuation and transport possibility assessments and damage to infrastructure estimation.

Therefore, there is a need for a complete turnaround and a tremendous progress in this part of visualisation methods of flooding. For visualising changes occurring (e.g.approaching hurricanes and altering wind velocity) a Web Map Service (WMS) is developed, called Time-enabled WMS layer. Time-enabled WMS layers are already used in other research areas, exhibiting the advantages of visualising altering and shifting events [30]. With this time-enabled WMS layer, it is possible to embed the output of such flood modelling software as TUFLOW, Flood Modeller, MIKE FLOOD, TELEMAC, and OpenFoam in a WebGIS environment.

Figure 3: 3D visualisation of flooding [40].

2.4 Flood damage visualisation tools

Numerous flood damage determination and assessment tools visualise the relationship between flood depth and the maximal damage to infrastructures (e.g.[31]), whereas the intensity, horizontal velocity and rising velocity of flood events are major decisive factors to estimate the damage to infrastructures.

In addition to flood depth, damage to infrastructures has to be calculated on the basis of dynamic flood event characteristics (intensity, horizontal velocity and rising velocity). Analogous to 2.3, it's possible to visualise the damage to infrastructures in sync with the visualised dynamic flood situation. Therefore, it's possible on basis of theWebGISvisualisation tool to assess how long it takes before a certain infrastructure becomes dangerous to, for instance, enter, traverse, or pass by. Logically, the user of the WebGIS tool should also have the possibility to request the maximal damage that occurs in a specific flood event, to provide sufficient information to reach the maximal usefulness of the tool.

2.5 Intern flood model visualisation and damage assessment

In the past, the vast majority of researchers have not considered visualising flood infiltration inside buildings on a WebGIS application. The most likely cause of this is that a large portion of these buildings'information has to be available before it is possible to calculate the

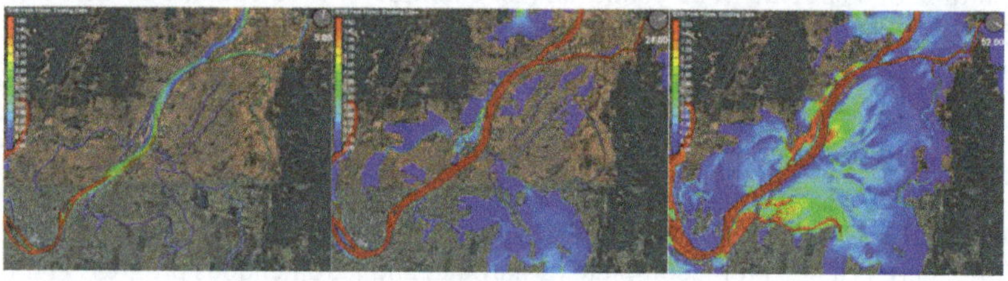

Figure 4: TUFLOW flood model simulation [32].

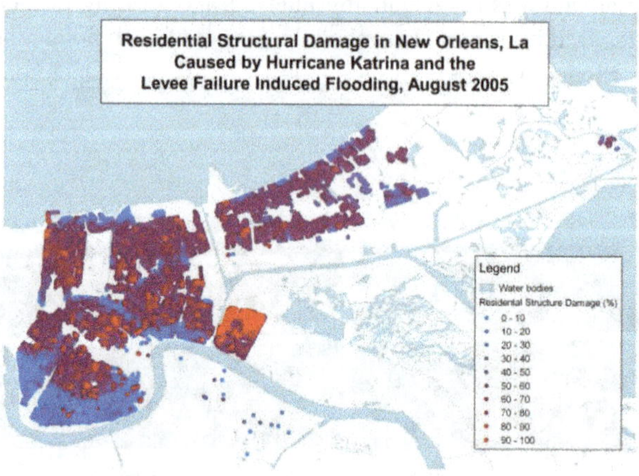

Figure 5: Example of a web-based flood damage map [33].

infiltration of floods and intern flood paths. When a certain flood event is modelled for a fairly large area (for example a coastal town), it is impossible to question all inhabitants to collect the necessary data of all the buildings in the area (e.g.permeability of brick walls, location of air bricks, location of pipes and cables, position of doors and windows, estimation of inadequate seals between windows, doors and frames). Subsequently, building information models (BIM) and geographical object-relational database (e.g.PostGIS) are difficult to link to each other because they have been developed separately, and they have pursued different standards and technologies. A BIM is a digital model of an existing and/or planned structure made up of objects attached with information. Thus, in addition to the geometry and position of, for example, a wall, such a model also adds such information as the construction materials (masonry or reinforced concrete), use, cost, and course of pipe work. Therefore, only a small number of researchers were able to model intern flood patterns with a geographical object-relational database in combination with a BIM [34–36]. Unfortunately, to date, there are no WebGIS tools created which visualise these intern flood patterns in buildings.

3 DEVELOPMENT OF A 3D FLOOD VISUALISATION TOOL

From the foregoing (see 2), it is clear that more is required to create a useful WebGIS tool for flood hazard visualisation than onlyvisualising static 2D layers. For instance, to make it easier to imagine the consequences of a flood and making a map more realistic and detailed, an additional 3Dvisualisation of the landscape (for example in the form of a split screen) is needed. To make a realistic 3D flood visualisation tool in this research,Ol3-CesiumOpenlayers is used [37, 38]. Ol3-Cesium is written in Javascript, uses WebGL. WebGL and is an open standard for visualising3D-computer graphics in a webpage without using special plug-ins for web browsers.

This Ol3-Cesium web library adds 3Dvisualisation to an OpenLayers3 map by automatically creating a virtual globe side-by-side the existing OpenLayers3 map. When manipulating the OpenLayers3 map (zooming, panning, etc.) the synchronised Cesium virtual globe follows automatically. The globe camera and the map view (centre, resolution, rotation) are bidirectionally synchronized: interacting with one updating the other. Layers are synchronized from the openlayers3map. Additionally, ol3 vector layers render on the Ol3-cesium screen using the same style and is displayed in the 3D environment.

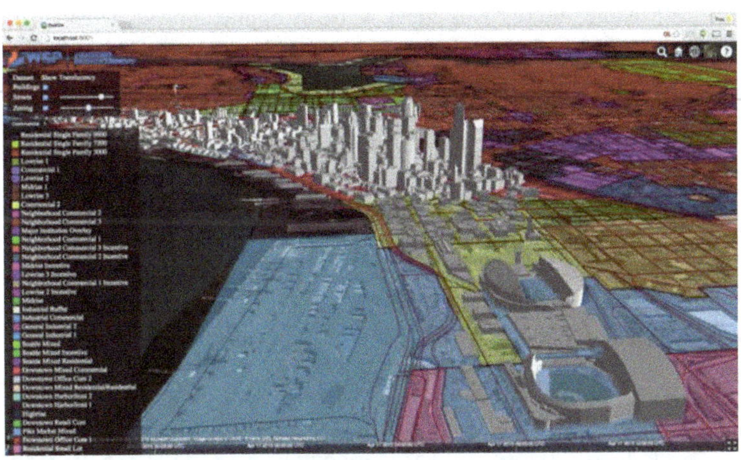

Figure 6: Example Ol3-Cesium web-based map [37].

In November of 2014, the OpenWebGIS group published a WebGIS tool where Atlantic hurricane tracks shown from 1851–2004 using time-enabled layers in the Ol3-Cesium environment[30]. Unfortunately, the user was forced to identify an interval before its visualized layer was available instead of showing the successive layers using a play button. Research has yet to prove that it is possible to visualisetime-enabled WMS layer in the 3DOl3-cesium (see 2.3) environment whereby dynamic flood events can be visualised.

4 CONCLUSIONS

Although scientists widely stress the compelling need to mitigate and adapt to climate change, public awareness lags behind that of scientists. Because WebGIS maps quickly convey strong messages, condense complex information, engage people on issues of environmental change, and motivate personal actions, this paper focusses on searching the ideal flood assessment WebGISmethod to encourage people to mitigate and adapt to climate change. In comparison with other WebGIS tools, many WebGIS applications make use of a single layer to visualise a sea level situation or a snapshot in time of a specific flood condition. Unfortunately, too much information is lost by this one-sided visualisation.

Surveys such as that conducted by Leskens (2015) showed that 3D visualisations have an enormous added value.These findings suggest that 3D condenses more complex information, and conveys strong messages quicker than2D. Nevertheless, 3D maps cannot take over all functionalities and perfect representations, which is inherent to 2D visualisation. For example, it is easier to assess on a 2D map where the maximal contour of the flood occurs in comparison with the estimation of the position of the maximal flood contour on a 3D map. The addition of a 2D visualisation together with a 3D visualisation (for example by using a split-screen) is therefore a necessity.

Analogous to the dynamic visualisation of a flood event, it is possible to visualise the damage to infrastructures in sync with the visualised dynamic flood situation. Therefore, it is possible to assess how much time is required before a certain infrastructure is dangerous to, for instance, enter, traverse, or pass by on the basis of the WebGIS visualisation tool.

In this research, the initial steps have been taken to create a WebGIS tool that visualises property damages caused by coastal flooding and flood events themselves in a 2D/3Dol3-cesium environment. Dynamic time-enabled WMS layers will visualise flood events, whereby the user can imagine how serious the flood will be. This tool will be elaborated during the CREST-project [39] in the future. This ideal webGIS application will visualise potential flood hazard events to engage people to adapt to climate change in the future.

ACKNOWLEDGEMENT

This paper is written in the framework of the research project CREST (Climate REsilient-coaST) [39]. This project aims to increase resilience of coastal flooding caused by extreme weather events.

REFERENCES

[1] Lichter, M., Vafeidis, A.T., Nicholls, R.J.&Kaiser, G.,Exploring data-related uncertainties in analyses of land area and population in the "Low-Elevation Coastal Zone" (LECZ). *Journal of Coastal Research,* **27**, pp. 757–68, 2010.

[2] McGranahan, G., Balk, D.&Anderson, B.,The rising tide: assessing the risks of climate change and human settlements in low elevation coastal zones. *Environment&Urbanization,* **19**, pp. 17–37, 2007. doi:10.1177/0956247807076960. http://dx.doi.org/10.1177/0956247807076960

[3] Deconto, R.M.&Pollard, D.,Contribution of Antarctica to past and future sea-level rise. *Nature*, **531**, pp. 591–597, 2016. doi:10.1038/nature17145. http://dx.doi.org/10.1038/nature17145
[4] Vojinovic, Z.,*Flood Risk: The Holistic Perspective*. vol. 1. London IWA Publishing: London, 2015.
[5] Re, M.,Natural disaster - Annual statistics 2015,available at: http://www.munichre.com/en/reinsurance/business/non-life/natcatservice/annual-statistics/index.html (accessed 15 March 2016).
[6] Nicholls, R.J., Wong, P.P., Burkett, V.R., Codignotto, J.O., Hay, J.E., McLean, R.F., SR and CDW. Coastal systems and low-lying areas. *Climate Change 2007: Impacts, Adaptation and Vulnerability. Contribution of Working Group II to the Fourth Assessment Report of the Intergovernmental Panel on Climate Change*, Cambridge, UK, 2007.
[7] Bertin, X., Li, K., Roland, A., Zhang, Y.J., Breilh, J.F.&Chaumillon,E.,A modeling-based analysis of the flooding associated with Xynthia, central Bay of Biscay.*Coastal Engineering*.**94**, pp. 80–89, 2014, doi:10.1016/j.coastaleng.2014.08.013. http://dx.doi.org/10.1016/j.coastaleng.2014.08.013
[8] Jelesnianski, C.P.,A numerical calculation of storm tides induced by a tropical storm impinging on a continental shelf. *Monthly Weather Review,***93**, pp. 343–358, 1965. doi:10.1175/1520-0493(1993).
[9] Sayers, P., Yuanyuan, L., Galloway, G., Penning-Rowsell, E., Fuxin, S., Kang, W., et al. *Flood Risk Management: A Strategic Approach*, UNESCO: Paris, 2013.
[10] Sheppard, S.R.J.,Landscape visualisation and climate change: the potential for influencing perceptions and behaviour. *Environmental Science&Policy,***8**, pp. 637–654, 2005. doi:10.1016/j.envsci.2005.08.002. http://dx.doi.org/10.1016/j.envsci.2005.08.002
[11] NOAA Coastal Services Center. *Sea Level Rise and Coastal Flooding Impacts* 2015,available at: https://coast.noaa.gov/slr/ (accessed March 29, 2016).
[12] JCNERR, CRSSA, NOAA CSC SN, Rutgers University. *Sea Level Rise and Coastal Flooding Impacts* 2015,available at: http://54.243.129.238/SLR.html# (accessed 29 March 2016).
[13] Monash University. *Lakes Entrance Flood-Risk Visualisation* n.d,available at: http://sahultime.monash.edu.au/LakesEntrance/ (accessed 29 March 2016).
[14] Leefmilieu Brussel. *Overstroming* n.d,available at: http://geoportal.ibgebim.be/webgis/Overstroming_kaart.phtml (accessed 11 April 2016).
[15] Monash University. *Explore SahulTime* n.d,available at: http://sahultime.monash.edu.au/explore.html (accessed 29 March 2016).
[16] SBEP & MOTE Marine Laboratory. *Interactive Sea Level Rise Web Map* n.d. available at: http://geology.com/sea-level-rise/florida.shtml (accessed 29 March 2016).
[17] NOAA—MS-AL Sea Grant U. *Sea Level Rise Visualization for AL, MS, & FL* n.d. available at: http://gom.usgs.gov/slr/slr.aspx (accessed March 29, 2016).
[18] Climate Central. *Surging Seas: Risk Zone Map,* n.davailable at: http://ss2.climatecentral.org/#12/40.7298/-74.0070?show=satellite&projections=0-RCP85-SLR&level=5&unit=feet&pois=hide (accessed 29 March 2016).
[19] NY Times. *What Could Disappear* n.d,available at: http://www.nytimes.com/interactive/2012/11/24/opinion/sunday/what-could-disappear.html?_r=0 (accessed 29 March 2016).
[20] Pacific Institute. *Pacific Institute: Sea-Level Rise* 2009,available at: http://www2.pacinst.org/reports/sea_level_rise/gmap.html (accessed 29 March 2016).

[21] PostGIS community. *ST_ClusterIntersecting* n.d. available at: http://postgis.net/docs/ST_ClusterIntersecting.html (accessed 29 March 2016).

[22] Tingle, A.,*Flood Maps* n.d,available at: http://flood.firetree.net/?ll=48.3416,14.6777&z=13&m=7 (accessed 29 March 2016).

[23] Burle, S.,*Flood Map: Water Level Elevation Map* n.d. available at: http://www.flood-map.net/?ll=51.256222,4.120335&z=9&e=0 (accessed 29 March 2016).

[24] Leskens, J.G., Kehl, C., Tutenel, T., Kol, T., de Haan, G.,Stelling, G., et al. An interactive simulation and visualization tool for flood analysis usable for practitioners. *Mitigation-and Adaptation Strategiesfor Global Change*, 2015. doi:10.1007/s11027-015-9651-2. http://dx.doi.org/10.1007/s11027-015-9651-2

[25] Ruppel, U.&Schatz, K.,Designing a BIM-based serious game for fire safety evacuation simulations.*Advanced Engineering Informatics*,**25**, pp. 600–611, 2011. doi:10.1016/j.aei.2011.08.001.
http://dx.doi.org/10.1016/j.aei.2011.08.001

[26] Christodoulou, S.E., Vamvatsikos, D.&Georgiou, C.,A BIM-based framework for forecasting and visualizing seismic damage, cost and time to repair. *eWork and ebusinessin Architecture Engineeringand ConstructionECPPM 2010*,CRC Press: Boca Raton, 2010:33–38.

[27] Isikdag, U.,A SWOT analysis on the implementation of building information models within the geospatial environment. *Urban and Regional Data Management: UDMS Annual,*Taylor & Francis: London, pp. 15–30, 2009.

[28] Kemec, S., Zlatanova, S.&Duzgun, H.S.,A framework for defining a 3d model in support of risk management. *Geographic Informationand Cartographyfor Risk and Crisis Management*,pp. 69–82,2010, doi:10.1007/978-3-642-03442-8.
http://dx.doi.org/10.1007/978-3-642-03442-8

[29] Mioc, D., Anton, F., Nickerson, B., Santos, M., Adda, P., Tienaah, T., Ahmad, A., Mezouagh, M., MacGillivray AM and PT. Flood progression modelling and impact analysis. Gi4DM (Geoinformation Disaster Manag. Conf., 2011.

[30] OpenWebGis. *Atlantic Hurricane Tracks with OpenWebGIS* n.d. available at: https://www.youtube.com/watch?v=y0UAV9jYsXc (accessed April 11, 2016).

[31] Pistrika, A.,Flood damage estimation based on flood simulation scenarios and a GIS platform. *European Water*,**30**, pp. 3–11, 2010.

[32] TUFLOW n.d. available at: http://www.tuflow.com/ (accessed 4 April 2016).

[33] Disastermap. *Residential structural damage in New Orleans* n.d. available at: http://disastermap.net/wordpress/wp-content/uploads/2015/08/Residential-Damage-Points.jpeg (accessed 12 April 2016).

[34] Amirebrahimi, S., Rajabifard, A., Mendis, P.&Ngo, T.,A framework for a microscale flood damage assessment and visualization for a building using BIM–GIS integration. *International Journalof DigitalEarth*,1–24, 2015. doi:10.1080/17538947.2015.1034201.
http://dx.doi.org/10.1080/17538947.2015.1034201

[35] Amirebrahimi, S., Rajabifard, A., Mendis, P.&Ngo, T.,A data model for integrating GIS and BIM for assessment and 3D visualisation of flood damage to building.*CEUR Workshop Proceeding*,**1323**, pp. 78–89, 2015.

[36] Varduhn,V. & Mundani R.P.&Rank, E., Multi-resolution models: recent progress in coupling 3d geometry to environmental numerical simulation. In: 3D Geoinformation Science,eds. M. Breunig,M. Al-Doori, E. Butwilowski, P.V. Kuper, J. Benner &K.H. Haefele, Springer International Publishing,pp. 55–69, 2015. doi:10.1007/978-3-319-12181-9.
http://dx.doi.org/10.1007/978-3-319-12181-9

[37] Cozzi, P., *3D Buildings in Cesium* n.d,available at: http://cesiumjs.org/2015/04/27/3D-Buildings-in-Cesium/ (accessed 8 April, 2016).
[38] Cozzi, P.,*Ol3-Cesium by openlayers* n.d,available at: http://openlayers.org/ol3-cesium/ (accessed 8 April 2016).
[39] Flanders Innovation & Entrepreneurship. *Climate Resilience Coast - CREST* 2015,available at: http://www.crestproject.be/en (accessed 12 April 2016).
[40] DeVito K. *Assessing Flood Risk with 3D GIS - A NEMAC study of Ft. Lauderdale* 2015. https://www.linkedin.com/pulse/assessing-flood-risk-3d-gis-nemac-study-ft-kevin-de-vito.

DEVELOPMENT OF A STORM RUNOFF MODEL CONSIDERING PROCESS OF INDIVIDUAL BUILDING INUNDATION

H. AMAGUCHI & A. KAWAMURA
Deptment. of Civil and Environmental Engineering, Metropolitan University, Japan.

ABSTRACT
A numerical simulation model for rainfall-runoff and flood inundation model considering process of individual building inundation was developed. The set-up of this model is based on so-called "urban landscape GIS delineation" that faithfully describes the complicated urban land use features in detail. The flow between single spatial elements is based on established hydraulic and hydrological models with equations that describe all aspects of storm runoff generation in an urban environment. The model was set up and applied for the upper Kanda catchment in Tokyo Metropolis, Japan. The runoff response to a storm event was simulated. It was demonstrated how the model can be used to evaluate the process of building inundation in the urban hydrological system.
Keywords: sewer network flow, storm runoff model, urban hydrology, urban land use GIS.

1 INTRODUCTION

Heavy rainfall and poor drainage facilities cause inundations in urban catchments in Japan [1]. The inundations are complex because of the problem of river overflow along with poor drainage system along rivers. Moreover, such areas are at high risk of flood damage because of the concentration of houses. As a result, not only the improvement of rainwater drainage facilities, but also assessment and review of inundation risks through inundation simulations are considered important for flood control [2, 3]. Inundation depth is the most important parameter, and it is used for calculating the cost of damage for assessing flood control projects [4].

Simultaneous flow analysis of rainwater is generally applied in inundation models. Examples of such models include regular mesh-shaped grids, unstructured grids and street network models [5–7]. Generally, in these analytical grids, buildings are assumed to be resistant to inundation, and inundation depth is not calculated individually [8, 9] this is because the geographical distribution details of buildings within a ward were not available [10]. In most cases, the inundation depth of the buildings is considered the same as the calculated maximum water depth of the analytical grid [4, 10]. For the precise evaluation of building damage, object-oriented land-use information of buildings should be evaluated, since the analytical grid enables the calculation of the water depth of buildings. Only few published studies explicitly explain the approaches for the studying individual building inundation. The lack of such studies may be attributed to the demand for detailed, local input data and the large effort required per unit area to obtain such information [4].

DOI: 10.2495/SAFE-V6-N3-570-581

In this study, a numerical simulation model for rainfall-runoff and flood inundation model considering inundation process of individual building is developed here, which can divide the computational region into individual land use feature. Especially this model aims to treat the structure and distribution of individual buildings suitably and calculate the water depth of the buildings in a direct way.

2 MODELLING OF BUILDING INUNDATION PROCESS

2.1 Urban landscape GIS delineation

The hydrologic characteristic of an urban surface depends on its land use. The types of surfaces present ranges from the relatively impervious character of streets, parking lots and roofs, to the more pervious character of gardens, bare soil and parks. The geometric composition of these different types of surfaces that forms a block is usually complicated. This complicated and inhomogeneous nature of urban catchments makes it very difficult to model the runoff process with accuracy. However, by the use of what will be denoted "urban landscape GIS delineation," which is described in detail below, the complex configuration of urban catchments can be faithfully reproduced in a runoff model [11]. The model is thus designed to be a tool which makes it possible to simulate the flooding process in an urban catchment in a comprehensive, detailed and accurate way.

Figure 1a shows a schematic of the rainfall-runoff process as represented in the model. When rainfall begins, water falling on a land use element inside a block or a road forms pools, where water falling on a river adds to the river discharge. Rainfall excess from blocks flows out directly or indirectly through different types of surfaces and finally out into the nearest road. When a manhole exists inside the road, water flows through it to the rainwater sewer pipe conduit. When no manhole exists, water flows down the road to an adjacent road element. In a manhole, the water level is obtained considering the inflow from the road together with the upstream inflow from connected pipe conduits. In a pipe conduit, the water flow is obtained considering the water levels in the manholes located upstream and downstream, respectively. When the water level in a manhole exceeds ground level, water flows out and inundates the associated road. The inundated water either flows to adjacent road elements until a manhole that has not reached full inflow capacity is found. It may also flow into and

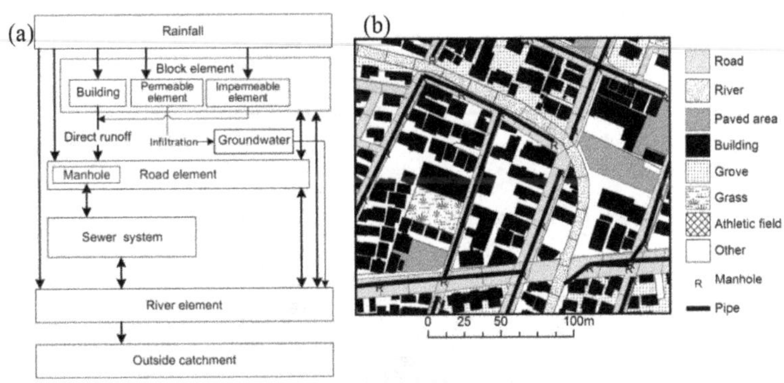

Figure 1: Schematic of the rainfall-runoff process (a) and (b) Map representing the different spatial elements considered in urban landscape GIS delineation.

flood a block, if the water level in the road is higher than that in the block. The water in the sewer pipe conduits eventually reaches the river channel, which finally drains in the catchment outlet. Infiltrated water from pervious land use elements inside blocks finally drains out into the river as long-term groundwater runoff, which is however at present not considered in the model.

2.2 Model equations

2.2.1 Direct runoff

Figure 2 shows the schematic of urban landscape GIS delineation and the assumed flow processes. In land use segments inside block elements, direct runoff from a non-building segment to an adjacent block segment is assumed. The direct runoff on a building segment is as follows. Although rainfall on buildings usually runs off to the sewer system through a gutter, it is rare to have information on the connection between a building and the sewer system. In model, the kinematic wave model [12] is applied to calculate the flow from a building segment to the nearest road or river segment eqns (1) and (2). Since the actual shapes of individual buildings are sometimes complicated, the shape (flow distance x_b and flow width B) is supposed to be square with the same size as the actual segment.

$$\frac{\partial h_b}{\partial t} + \frac{\partial q_b}{\partial x_b} = r_e \tag{1}$$

$$q_b = \alpha h_b{}^{\beta} (\alpha = \sqrt{\sin\theta} / \mathrm{N_b},\ \beta = 5/3) \tag{2}$$

where q_b denotes the discharge per unit width from the building (m²/s), h_b the flow depth (m) and x_b the distance along the building (m). In (2), α and β are constants related to slope and overland roughness, θ the slope of the building (rad) and N_b is the equivalent Manning's roughness coefficient (s/m$^{1/3}$). Rainfall intensities in excess of the final infiltration capacity become effective rainfall eqn (3).

$$r_e(t) = r(t) - I_i \tag{3}$$

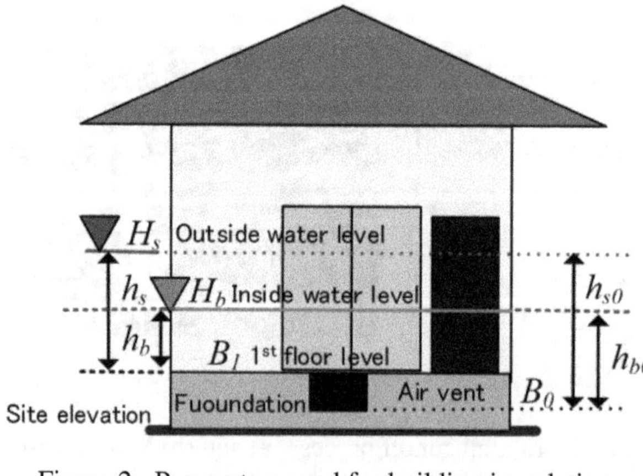

Figure 2: Parameters used for building inundation.

2.2.2 Overland flow

To calculate the discharge in overland component, one-dimensional unsteady flow without convective acceleration is assumed according to eqn (4). Water level changes in the overland segments are computed by considering the effective rainfall and outflow/inflow from manhole elements in addition to the outflow/inflow from overland segments. After calculating water storage, the water depth is obtained by eqn (5)

$$\frac{\partial Q_s}{\partial t} + gA_s \frac{\partial H_s}{\partial x_s} + \frac{gn_s^2 Q_s |v_s|}{R_s^{4/3}} = 0 \tag{4}$$

$$\frac{dh_s}{dt} = \frac{\Sigma Q_s + \Sigma Bq_b + \Sigma Q_{div} + r_e Area_s}{Area_s} \tag{5}$$

where Q_s is the overland discharge (m^3/s), A_s the overland flow cross sectional area (m^2), H_s ($= z_s + h_s$) the overland water level (m), z_s the overland elevation(m), h_s the overland water depth (m), x_s the longitudinal distance along overland segment (m), n_s the overland Manning's roughness coefficient($s/m^{1/3}$), v_s the velocity of overland flow (m/s), R_s the overland hydraulic radius (m), B the width of building segment (m), Q_{div} the discharge to/from manhole from/to overland (m^3/s) and $Area_s$ the overland area (m^2) (excluding the building area in block segments).The numerical analysis technique used for the overland flow, as well as for the calculations of sewer pipe flow and river flow described below, is the unsteady flow equation by the explicit finite difference method.

2.2.3 Building inundation

In order to calculate the building inundation depth, buildings in the block element are used for the analysis of the overland flow and the flow to the buildings is considered based on the water level around the buildings eqn (6). Figure 3 shows the parameters used in the model.

$$Q_{b0} = \begin{cases} \mu_1 L\alpha_0 h_{s0}\sqrt{2gh_{s0}} & (h_{b0}/h_{s0} \le 2/3) \\ \mu_2 L\alpha_0 h_{b0}\sqrt{2g(h_{s0} - h_{b0})} & (h_{b0}/h_{s0} > 2/3) \end{cases} \tag{6}$$

where Q_b is the flow from the surrounding features to building (m^3/s), and are flow coefficients, is the opening ratio below the 1st floor, is the opening ratio above the 1st floor, and L is the outer perimeter of the building (m). In addition, h_b is the indoor inundation height above the 1st floor ($= H_b - B_1$), h_s is the outdoor water height above floor ($= H_b - B_1$). The same height applies to the water level to the underfloor air vacant, and h_{b0} is the indoor inundation height ($= H_s - B_0$), h_{s0} is the outdoor inundation height ($= H_s - B_0$). The overflow equation is used for the discharge from/to a building. The inundation depth in the building is calculated by eqn (7).

$$\frac{dh_b}{dt} = \frac{\left(\sum Q_{b0} + \sum Q_b\right)}{0.9A_b} \tag{7}$$

where, A_b denotes the building roof area which is obtained from the urban landscape GIS. In this study, we assume the floor area is approximately 90% of the roof area.

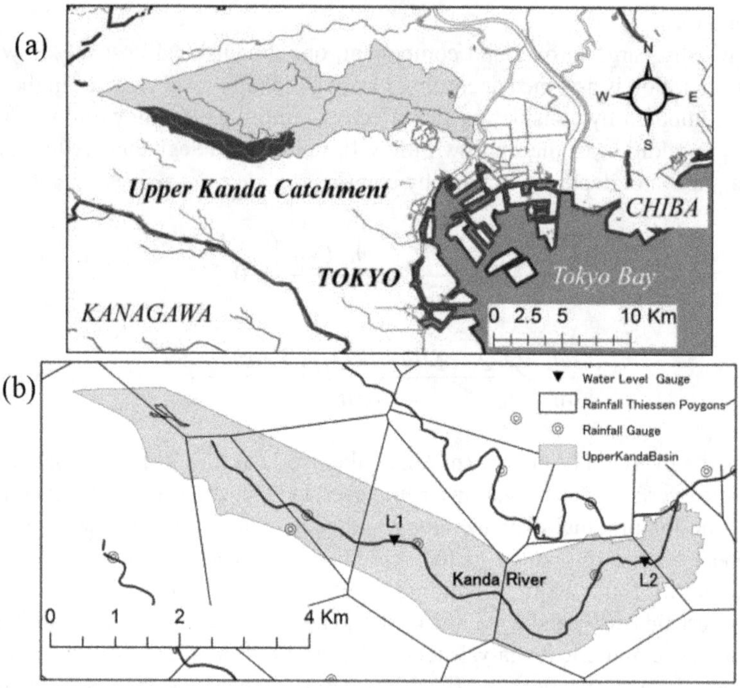

Figure 3: Location of the upper Kanda catchment in Tokyo and (a) and overview of the upper Kanda catchment selected for model application (b).

2.2.4 Sewer pipe flow

The runoff in a sewer is usually in the state of free surface flow. However, in the case of a storm, both free surface flow and surcharged flow occur, and their regions vary temporally and spatially. In order to describe these processes in detail, a particular model for sewer pipe flow with surcharge is required and here dynamic waves are used to describe both free surface flow and surcharged flow. For the surcharged flow, the technique based on the assumption of a hypothetical slot is applied [13]. The equation of motion without convective acceleration and continuity of the slot model is applied as

$$\frac{\partial Q_p}{\partial t} + gA_p\left(\frac{\partial H_m}{\partial x_p}\right) + \frac{gn_p^{\,2}Q_p\left|v_p\right|}{R_p^{\,4/3}} = 0 \tag{8}$$

where Q_p is the pipe discharge (m^3/s), A_p the pipe flow cross sectional area (m^2), H_m the manhole water level (m), x_p the distance along pipe (m), n_p the pipe Manning's roughness coefficient (s/m$^{1/3}$), v_p the velocity of pipe flow (m/s), R_p the pipe hydraulic radius (m).

In a manhole, the storage quantity and water level are calculated by eqn (9) and eqn (10) respectively. The storage is based on the size of the manhole and its connection according to

$$\frac{dS_m}{dt} = \sum Q_{div} + \sum Q_m \tag{9}$$

$$H_m = f_{mh}(S_m) \tag{10}$$

where S_m is storage quantity in a manhole and connected pipe (m^3), Q_m the flow from/to the connected pipe (m^3/s) and f_{mh} a function relating storage to water level in the manhole (m).

When the water level in a manhole exceeds ground water level, water flows out and inundates the associated overland segment. The discharge between a manhole and a overland segment is evaluated according to

$$Q_{div} = \begin{cases} \dfrac{f_{mv}(H_m) - f_{mv}(H_s)}{dt} & \text{(outflow from manhole)} \\ \mu Area_m \sqrt{g\Delta h_s} & \text{(inflow to manhole)} \end{cases} \tag{11}$$

where f_{mv} is a function relating manhole water level to volume ,μ a coefficient and $Area_m$ the manhole area.

2.2.5 River flow

The river channel flow considering the inflow from the rainfall and sewer pipes, as well as the side inflow from overland segments, is calculated by the equations of motion and continuity as

$$\frac{\partial Q_r}{\partial t} + \frac{\partial (Q_r^2 / A_r)}{\partial x_r} + gA_r \frac{dH_r}{dx_r} + g\frac{n_r^2 Q_r |v_r|}{R_r^{4/3}} = 0 \tag{12}$$

$$\frac{\partial A_r}{\partial t} + \frac{\partial Q_r}{\partial x_r} = q_r \tag{13}$$

where v_r is the velocity of the river flow (m/s), x_r the distance along river channel (m), H_r the river water level (m), n_r river Manning's roughness coefficient ($s/m^{1/3}$), R_r the river channel hydraulic radius (m), A_r the river flow cross sectional area (m^2) and q_r the cumulative discharge from sewer system and overland segments to river segment in addition to rainfall (m^3/s).

3 APPLICATION TO THE UPPER KANDA CATCHMENT

3.1 Study area

The study area selected for the model application is an urban catchment, which is located in the Kanda catchment Tokyo Metropolis, Japan, as shown in Fig. 4a. The study catchment will be termed "upper Kanda catchment" and Fig. 4b shows this catchment in some detail. The boundary of the study catchment is specified based on two conditions, the topography and the extension of the sewer pipe network. The upper Kanda catchment area is ~11 km² and the length of river inside it is ~10 km. It is essentially a residential area with some minor parks, groves, fields, etc. There are several water level gauges and rainfall gauges in the catchment. Concerning the land use, ~65% of the surface is impervious. Rainfall from the upper part of the Kanda river reaches the study catchment only by gravity flow through the combined sewer system. There is a main sanitary sewer along the Kanda River. During intense storms, this sewer soon flows full and discharges downstream of the catchment, which makes the catchments vulnerable to flooding. Concerning rainfall and runoff observation, there are seven rainfall gauges around catchment, where observations are made with a 1-min time resolution

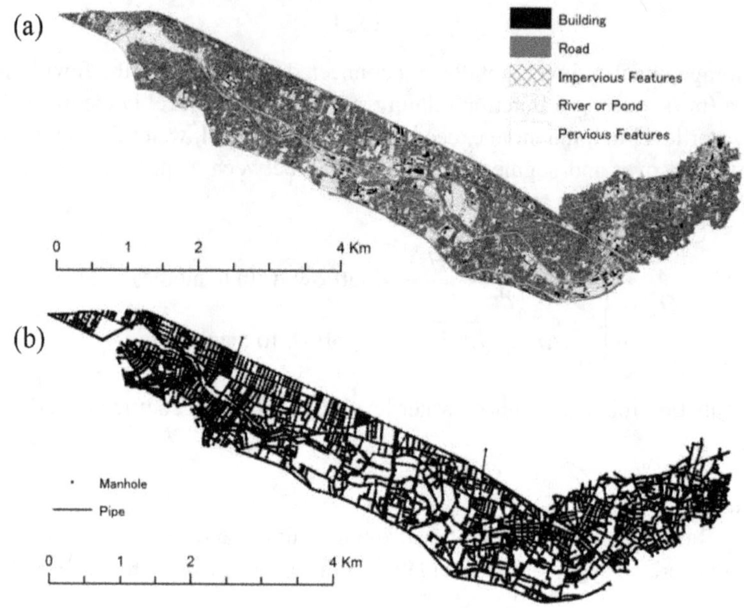

Figure 4: Final map of overland component (a) and sewer element (b) in the upper Kanda catchment.

by Tokyo Metropolitan Government. Fig. 3 shows the final maps of overland component (a) and sewer element with road, block and river segments (b) in the entire study catchment. It should be emphasized that more than 180 000 homogeneous elements (land use, road, river, manhole and pipe) were used to completely specify the urban catchment.

3.2 Application condition

Table 1 gives the parameters of the model for the building inundation process. The ground level of each building was calculated as an averages obtained from a 5 m DEM with 0.1 m precision in elevation, with 0.1 m added to that value to set the elevation of the building. The floor height and water intrusion rate of a building can have significant impact on the risk of building inundation. Therefore, we classified buildings into residential houses and non-residential buildings, such as commercial facilities and factories, in order to approximate differences in building accessibility. For residential houses, the Japanese building standards stipulate that the height of the foundation should be at least 45 cm above the ground surface elevation, and underfloor ventilation holes of 300 cm^2 in cross-sectional area should be installed at least every 5 m along the foundation [14]. The height of the foundation is assumed to be 50 cm in our model, while that of the ventilation holes is assumed to be 10 cm from the top of the foundation with a width of 30 cm every 5 m in foundation length. As a result, the opening ratio below the 1st floor is 6/100 (= 0.3/5.0). For factories, commercial facilities and other non-residential use buildings, the height of the 1st floor is assumed to be 25 cm. The foundation height of buildings with basement rooms is assumed to be 3 m in depth below the site elevation.

Table 2 gives the parameters required by the model. Only standard values

Table 1: Parameter of the building inundation model.

	Wooden	Non-wooden	Commercial use
B_0	+40 cm	-	-
B_1	+50 cm	+50 cm	+25 cm
a_0	6/100	-	-
a_1	6/100	2/100	2/100
Numbers of elements	2,449	28,708	2,893
Total Area (m^2)	188,919	2,601,434	273,530

Table 2: List of required input data and parameters of the model.

Parameter (Unit)		Value
Initial infiltration capacity (mm/hr)	Green space	30
	Others	10
Final infiltration capacity (mm/hr)	Green space	20
	Others	5.0
Building roughness coefficient n_b ($s/m^{1/3}$)		0.035
Overland roughness coefficient n ($s/m^{1/3}$)	Between roads	0.043
	Other	0.067
Channel roughness coefficient n ($s/m^{1/3}$)		0.0225
Pipe roughness coefficient n ($s/m^{1/3}$)		00.013

were used in this study, and no attempt was made to adjust the parameter values is made but standard values are used [5, 15–17].

3.3 Simulation of actual storm events

To test the applicability of the model for storm runoff analysis, simulations were performed for two historical storm events. The first represents a small scale event in October 2004, during which the storm water runoff in the sewer system was mainly free surface flow and no inundation occurred. The second represents a flooding event that followed heavy rainfall caused by typhoon no. 14 (Nagi) in September 2005. Although there was no significant inundation damage in the upper Kanda catchment, due to the use of an underground retention basin for the river, we evaluated the building inundation model using inundation characteristics assumed in the absence of the underground retention basin.

In the simulations, the initial water levels and discharge in the river channel were set at steady-state conditions, after an adequate model warm-up period for the calculations to attain stability, during which river flow was calculated using the first water level observation in each event. The initial flow in the sewer system, overland component was set to zero. The water level at the catchment outlet was estimated from the calculated discharge using the outlet rating curve.

3.4 Results

The validity of the model was judged using the water levels in river channels. Observed and simulated water levels at gauge locations L1 and L2, together with the observed rainfall during the first storm event, are shown in Fig. 5a and b. The calculated and observed discharges are in good overall agreement at these two locations, although no calibration or tuning of model parameters was performed. Also the runoff peak is generally well reproduced, although it is underestimated by 4 m^3/s at gauge L2. Overall, the accuracy and trends in the simulated data were very similar to the observed discharge data.

In order to also evaluate the model's ability to reproduce inundated conditions, a second simulation was performed using the major storm event that occurred on 2005-09-04. The maximum observed 1-h rainfall volume was 112 mm within the Kanda catchment.

Figure 5c and d show observed and calculated river water levels at L1 and L2. The observed water level at L1 is lowered intermittently because of the inflow to the retention basin. Overall, the simulated levels, including the peak water level agree with the observed levels, although the initial rise time of the simulated water level is 20 minutes earlier than that of the observed water levels. There was no difference between the model and a conventional model without considering the process of individual building inundation in terms of simulated river water levels.

In the simulation, to evaluate the area affected by building inundation, the distribution of the maximum inundation depth in the catchment was determined. A sample map of the maximum inundation depth near the river channel is shown in Fig. 6. The symbol inside the building denotes whether it is intended for residential or commercial use. In the proposed model, which employs the explicit process of building inundation, rainwater is assumed to flow into residential houses through ventilation holes in the foundation, which are set 40 cm higher than the site elevation. We found that residential buildings are inundated when the

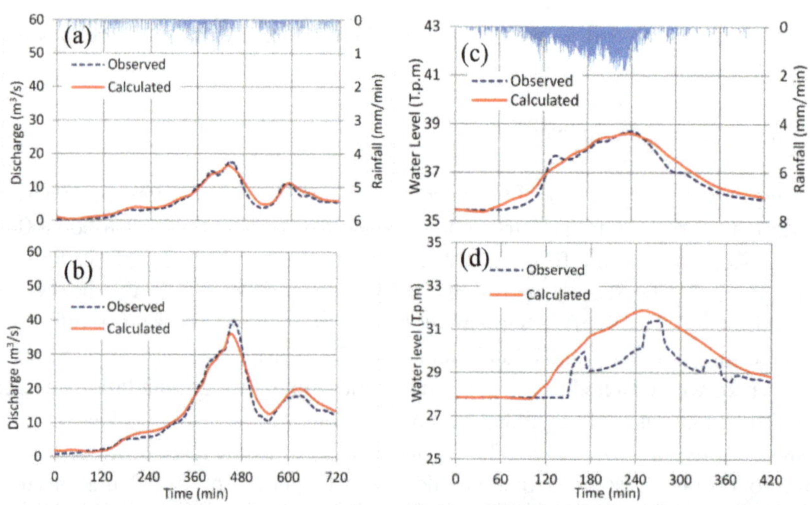

Figure 5: Observed and simulated river discharge in gauges L1 (a) and L2 (b) during event 2004/10/9 and Observed and simulated river water level in gauges L1 (c) and L2 (d) during event 2005/9/4.

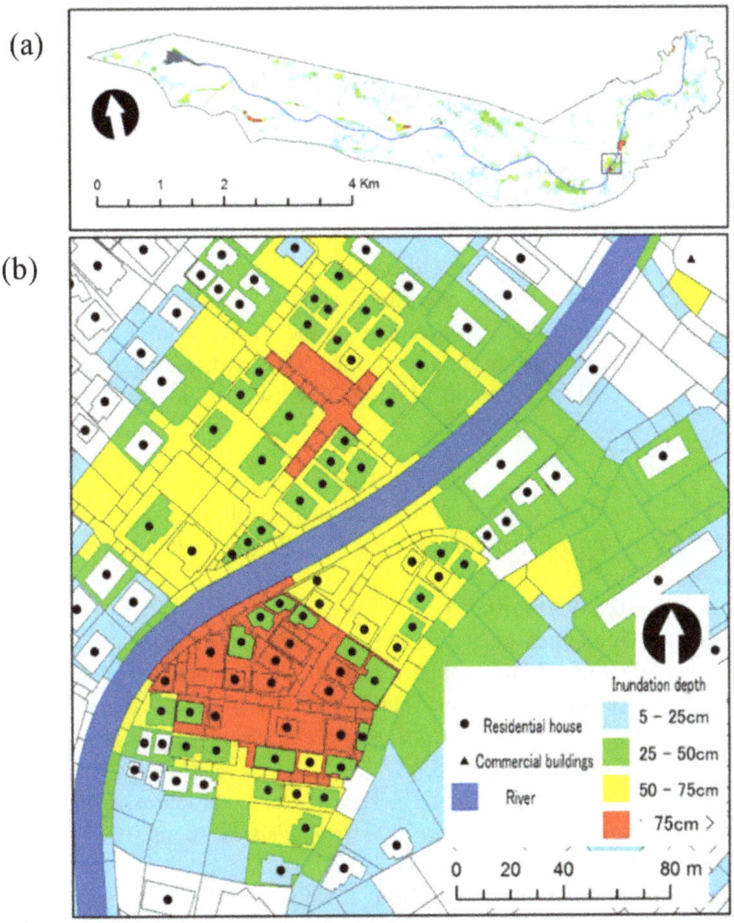

Figure 6: Distribution of maximum inundation depths (a) Overall vies, (b) Expanded view.

maximum inundation depth of their surroundings is in the range of 25–50 cm. As the inundation depth increases from 50 to 75 cm and exceeds 75 cm, the inundation depth of the buildings also increases gradually. This means that in this simulation, the space under the floor of residential houses does not get flooded unless the water level around these buildings exceeds 40 cm. In conventional modelling approaches which are usually based on regular mesh-shaped grids, unstructured grids and street network data [5, 6, 10], the inundation depth of the building is usually set the same as that of the analytical cell, and hence, there is a tendency to overestimate the inundation risk of buildings as compared to the proposed model.

4 SUMMARY AND DISCUSSION

In this study, a numerical simulation model for rainfall-runoff and flood inundation was developed; this model considers the inundation process of individual building. The model was built and evaluated for the upper Kanda catchment, Tokyo Metropolis, Japan. In the model, the parameters for the foundation height and floor height of each building and the opening ratio below and above the 1st floor were set for individual buildings based on their structure and purpose. The model was applied to simulate one major storm event. It could reproduce differences in the inundation depth of individual buildings reflecting the additional input parameters. The use of the model to evaluate the inundate condition for specific properties of the building was also demonstrated.

Overall, the results show that the suggested approach, which is based on a detailed reproduction of all relevant elements in an urban catchment, can simulate the process of individual building inundation. Although in principle, the approach is relatively straight-forward, it has been developed and applied practically until now because of limitations in GIS data availability. We believe the methodology has a wide range of applications with regard to improving flood protection facilities and installing new runoff control facilities. Another potential use of the model is detailed urban impact assessment of the higher rainfall intensities that are expected to occur more frequently in the near future. The condition at specific critical locations can be easily extracted from the model output and used for flood risk assessment.

REFERENCES

[1] Kawata, Y., Urban flood control policy for sustainability under global warming in Japan. *First International Conferences on Flood Recovery, Innovation and Response*, pp. 273–282, 2008.
http://dx.doi.org/10.2495/friar080261

[2] Andjelkovic, I., Guidelines on non-structural measures in urban flood management. *International Hydrological Programme*, technical document in Hydrology, No. 50, 2001.

[3] Genovese, E., A methodological approach to land use-based flood damage assessment in urban areas: prague case study. *Mission of the Institute for Environment & Sustainability*, European Commission, 2006.

[4] Merz, B., Kreibich, H., Schwarze, R. & Thieken, A., Review article assessment of economic flood damage. *Natural Hazards and Earth System Sciences*, 10, pp. 1697–1724, 2010.
http://dx.doi.org/10.5194/nhess-10-1697-2010

[5] Inoue, K., Kawaike, K. & Toda, K., Refuge analysis under storm surge based on the street network model. *Proceeding of the International Symposium on Flood Cont*, Beijing, China, pp. 507–514, 1999.

[6] Kawaike, K., Inoue, K. & Toda, K., Inundation flow modelling in urban area based on the unstructured meshes. *Hydraulic Engineering Software VIII*, eds. C.A. Brebbia & W.R. Blain, WIT Presss, 2000.

[7] Dey, A.K. & Kamioka, S., An integrated modeling approach to predict flooding on urban basin. *Water Science & Technology*, **55**(4), pp. 19–29, 2007.
http://dx.doi.org/10.2166/wst.2007.091

[8] Schubert, J.E. & Sanders, B.F., Building treatments for urban flood inundation models and implications for predictive skill modelling efficiency. *Advances in Water Resources*, **41**, pp. 49–64, 2012.
http://dx.doi.org/10.1016/j.advwatres.2012.02.012

[9] Chang, T., Wang, C. & Chen, A.S., A noble approach to model dynamic flow interaction between storm sewer system and overland surface for different land covers in urban area. *Journal of Hydrology*, **524**, pp. 662–679, 2015.
http://dx.doi.org/10.1016/j.jhydrol.2015.03.014

[10] Dutta, D., Herath, S. & Musiake, K., A mathematical model for flood loss estimation. *Journal of Hydrology*, **277**, pp. 24–49, 2003.
http://dx.doi.org/10.1016/S0022-1694(03)00084-2

[11] Amaguchi, H., Kawamura, A., Ollson, J. & Takasaki, T., Development and testing of a distributed urban storm runoff event model with a vector-based catchment delineation. *Journal of Hydrology*, **420–421**, pp. 205–215, 2012. http://dx.doi.org/10.1016/j.jhydrol.2011.12.003

[12] Singh, M.B., *Kinematic Wave Modelling in Water Resources*, John Wiley & Sons: USA, 1996.

[13] Preissmann, A. & Cunge, J.A., Calcu des Intumescences sur Machinas Electroniques. *Proceeding 9th Congress of IAHR*, Dubrovnik, Yugoslavia, pp. 656–664, 1961.

[14] Japanese Ministry of Internal Affairs and Communications, order for enforcement of the building standards act, available at http://law.e-gov.go.jp/htmldata/S25/S25SE338.html

[15] Ando, Y., Takahashi, Y., Izumi, K. & Kanno, K., Urban flood modeling considering infiltration various land uses. In *Urban Drainage Modelling*, eds. C. Maksimović & M. Radojković, pp. 229–238, Pergamon press: Oxford, England, 1986.

[16] Yen, B.C., Hydraulic resistance in open channel. In *Channel Flow Resistance: Centennial of Manning's Formula*, ed. B.C. Yen, pp. 1–135, Water Resour. Pub.: Colorado. USA, 1991.

[17] Mays, L.W., *Water Resources Engineering*, John Wiley and Sons: USA, 2001.

VULNERABILITIES OF COASTAL COMMUNITIES RESULTING FROM CLIMATE CHANGE: A CASE STUDY OF SAN MATEO, BELIZE

A. BELL, R. DUGGLEBY III & A. KINCH
Florida State University.

ABSTRACT

This paper focuses on the migration patterns of residents of the community of San Mateo, San Pedro, Belize. Recent research conducted in the community of San Mateo by Florida State University faculty and students revealed numerous vacant lots along the coast and in areas where water has risen and not been displaced. In total, the vacant lots accounted for nearly a quarter of the community. Drawing from these findings, this paper argues that these vacant lots are due to environmental changes in the community, most likely attributed to climate change. This paper will examine environmental migration in San Mateo as well as availability of resources such as electricity and water. The ultimate goal of this research is to assess how the resettlement of environmental migrants is impacted by the availability of resources. The paper will subsequently illustrate that Central America is especially prone to the effects of climate change, and that this directly impacts the most vulnerable, coastal communities.
Keywords: Belize, case study, climate change.

1 INTRODUCTION

The Disaster Incident Research Team (DIRT), consisting of faculty, staff, and student researchers of the Emergency Management and Homeland Security (EMHS) Program at Florida State University, recently completed field research in San Mateo, Ambergris Caye, Belize, gathering information on demographics and resources within the community. The project was divided into two parts. The first team, comprised of EMHS staff and student researchers, surveyed San Mateo community members to gather data on the community's vulnerabilities to disaster. Community members were asked primarily about infrastructure, utilities, and bills and taxes paid to the local government. The second team of the project involved members of EMHS faculty and staff, making up the small Unmanned Aerial System (sUAS) flight operations team, who flew several missions throughout San Mateo. Employing the use of aerial photography, the team took photos and recorded video of the community so as to construct orthographic photos and gain a better understanding of the physical aspects of San Mateo.

2 SAN MATEO, SAN PEDRO, BELIZE

San Mateo is a small community on the island of Ambergris Caye in Belize. A community composed primarily of immigrants and other marginalized populations, San Mateo has historically been subjected to limited access to clean water, electricity, and economic opportunities for its residents. This is due to a number of factors, though primarily because local government officials claim that San Mateo is 'not [our] priority' [1]. Despite the low

DOI: 10.2495/SAFE-V6-N3-582-588

cost of living for residents and immigrants to the San Mateo neighborhood, there are clear detriments to quality of life and resource security.

Further, and although we cannot provide accurate comparative population numbers, we posit that the high number of vacant lots (about a quarter of the total number of lots in San Mateo), especially those along the coastline, is due to the effects of climate change on this particularly low-lying area. This is an area where climate change is not only visually apparent but experienced firsthand by the community, which is why the Belizean government is taking steps to combat the effects of such detrimental effects to coastal communities. According to the Inter Press Service [2], the recently launched Marine Conservation and Climate Adaptation Project will explore and develop strategies to help coastal communities become more resilient to climate change, and help people learn how they make changes to adapt to environmental factors. Belize, along with many other coastal nations in the Caribbean and the rest of the world, will undoubtedly benefit from such efforts.

2.1 Community profile

As of 2008, the population of the larger city of San Pedro on Ambergris Caye was reported to be 11,600 persons [3]. The community of San Pedro is located on the south side of Ambergris Caye and is the only town on the island [3]. The city is quite small but boasts impressive mercantile abilities for such a small island. The narrow streets necessitate the use of golf carts for most transportation, as average cars and trucks would be cumbersome and bulky in the streets of San Pedro, although a small number are now permitted for use.

San Mateo is an impoverished neighborhood next to the city of San Pedro. San Mateo has roughly 350 private lots, with many lots possessing multiple households, buildings, and various multi-use structures. According to the *Revised Master Development Plan for Ambergris Caye* compiled by Malikah Cardona and Kamilah Cardona in 2000, San Pedro continued to experience problems in overcrowding, an issue still present in San Mateo today [3]. The population of the community is extremely diverse, and individuals and family units in the community come from several different ancestral backgrounds. Some residents of San Mateo keep birds such as chickens and turkeys for subsistence. San Mateo is a relatively new community, with only a few residents having lived there for the past quarter of a century, and even more migrating in the last ten years. The Belizean government allots a plot of land to each Belizean citizen who elects to apply for one, which brought about the arrival of the San Mateo community. Some plots of land are completely underwater, difficult or unable to be reached by road, or are inundated with trash and refuse. Many residents assert that homes or other structures must be built quickly to assure ownership and involvement in the land, or otherwise the Belizean government may reassign the lot to another citizen candidate. Many lots within San Mateo are unable to receive water, electricity, sewer services, or a combination of all three. As a result, both structural and social surveys were completed lot by lot in San Mateo to ascertain which utilities are available at certain locations within the community.

2.2 Access to clean water in San Mateo

Our research in the community shows that many households continue to have inadequate access to necessary resources; most notably, only 40% of residents claimed that they 'always' have access to clean water (Fig. 1), and 81% reported to 'always' treat their water

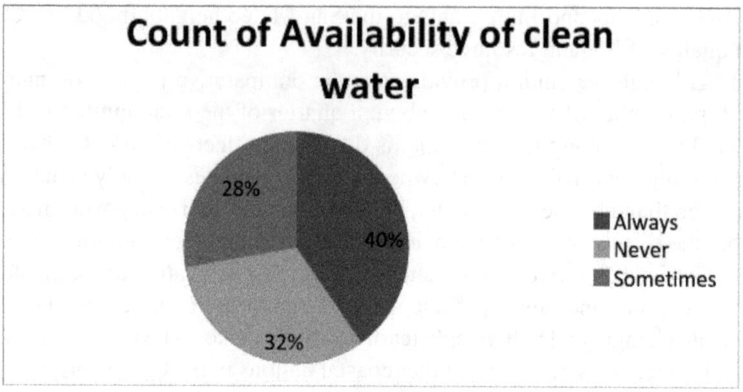

Figure 1: The slight majority of respondents stated that they have clean water available at all times. However, regardless of their response to this question, nearly every person interviewed stated that their drinking water was from bottled water purchased from Crystal Water Co. The concept of the word 'clean' proves difficult to determine.

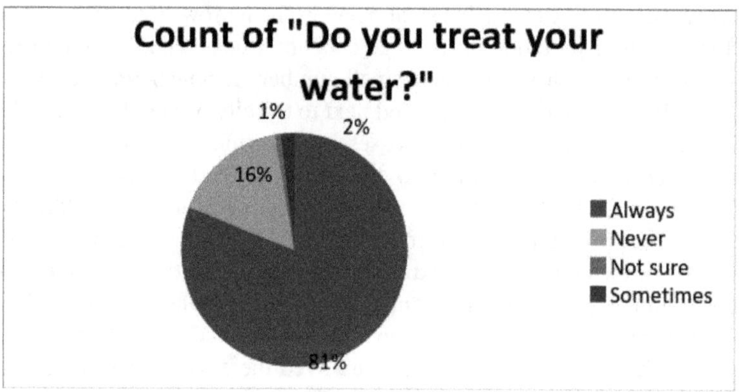

Figure 2: Due to lack of access to clean water from the tap, the residents have found that treating the water is the only alternative to purchasing it by the litre. Many homes have cisterns to collect rainwater but, per recommendation from the Red Cross Belize, even this water should be treated.

with bleach drops or tablets and/or boiling (Fig. 2). Undoubtedly, the issues of San Mateo do not typically concern the local government in San Pedro, but the fact also remains that it is markedly more difficult to route resources like water and electricity to the community. There is just one road in and out of the community, making vehicle access difficult; the dirt roads throughout San Mateo, though they have been improved in the last few years, are damaged and difficult to navigate; and most importantly, not all homes have access to San Pedro's primary water source. These factors, combined with the ever-present risks of living on a coastline in an area prone to various natural hazards, provide for a difficult living situation for San Mateo residents.

Figure 3: Lots colored in red indicate those without access to electricity.

2.3 Electricity

Electricity was present in 236 structures in San Mateo, yet the sources and integrity varied. The bulk of structures with power had lines originating from the main power line. These were installed approximately five to six months prior to our visit, according to the residents. Some lots had improvised power, whether split lines or a simple extension cord running from a neighbor's house. Those without electricity (Fig. 3) did not have power of any kind available in their home. Many of these lots were found on main roads.

Those lots with electricity from the main power lines were provided electrical boxes outside their homes with meters. These were either placed on the structure, on the pole, or, somewhat commonly, on cement slabs that stood along the road just outside the home. However, as stated, the residents were responsible for housing this electrical meter. The majority of the homes with power along the center of San Mateo had electrical boxes, while most of the outer lots with power did not. This indicates that the main line runs through the center of the community.

Although Belize is the recipient of year-round direct sunlight, far fewer than expected structures utilize solar power. A meager 8 structures, located away from those with reliable electricity maintained these panels. Many of those residents stated the solar panels were only used on occasion. Utilization of solar-derived electricity may be a way for the Belize government to provide reliable electricity to those residents who are not on any power line, or to discourage power sharing in the dangerous methods that have already been used. The size and capabilities of the solar panels found in San Mateo varied immensely. The majority of the few systems present had very low capabilities while a few, well-established, homes maintained systems with more extensive capacities.

2.4 San Mateo vacant lots

There were many vacant lots in this area. These were lots that either contained no structure at all or contained partially-built or abandoned structures. We discovered that the majority of lots were completely empty and contained only mangroves, water, or other detritus. Despite

Figure 4: Vacant lots are indicated by grey fill. Some of these lots may have structures, but they are unoccupied. All other lots have residents present.

the absence of a predictable trend of vacant lot orientation, they give insight into the community's origins and future growth trends. The center of San Mateo accommodates the preponderance of the occupied lots, while vacant lots are in greater numbers along the coast or where copious inland water has yet to be displaced. Vacant lots are common along the waterway on the lower constraint (Fig. 4) or along the water in the upper right.

The vacant lots found amongst the central component of the landscape have no discernible pattern to either structure fill or location. There are 74 recorded vacant lots, or about a quarter of the community. A few had dilapidated or abandoned structures, while most were completely empty.

The Belizean government delineates the land and assigns the lots to specific individuals or families, but those who are assigned lots in such undesirable areas are solely responsible for making them livable. As such, in order to live on many of the lots in San Mateo, residents use both organic and inorganic fill by which to displace any standing or, in some cases, moving water. This includes various plastics and metals, along with Sargassum seaweed, an organic, brown macroalgae common to the region. Because of these factors, agriculture and maintaining livestock are much more difficult on San Mateo than on the rest of Ambergris Caye. Some residents of the community were even noted to have been using vacant lots for keeping chickens or goats.

2.5 Migration from or lack of migration to San Mateo as a result of environmental factors

Suitable living conditions and a lack of resources notwithstanding, we argue that the primary reason for the high number of empty and/or vacant lots is due in large part to the receding coastline as a result of the effects of climate change on San Mateo. This argument is strengthened when examining the global warming trends occurring in both Belize and in all of Central America.

An increase in the numbers of hurricanes and floods that have occurred in Belize provide evidence of the growing problem of climate change the country faces [4]. Additionally, rising temperatures are also contributing to coral bleaching. Coral bleaching can occur when reefs begin to die and can sometimes develop because of pollution or storms. However, many scientists agree that the current bleaching that is taking place in Belize is due to rising ocean temperatures [5].

These climatic changes have not only occurred in Belize but throughout Central America. According to research done by the CGIAR Research Program on Climate Change, Agriculture and Food Security, temperatures in Central America have risen since 1901 by around 0.5 [6]. Rainfall has also been more erratic since the 1950s. There have been increased delays in rain and also a greater number of rains that have occurred during the onset season [6].

These factors demonstrate the impact that climate change has had on Central America and Belize. More importantly, they illustrate the real threat to the residents of San Mateo. The numerous vacant lots with structures present point to the fact that many San Mateo residents are most likely leaving their homes due to climate change.

3 VULNERABILITIES AND CLIMATE MIGRATION

A significant geographic area of Belize consists of coastal cities and communities, some of which are the nation's most economically important and tourist-friendly. As such, and as R.B. Richardson notes, climate change will continue to affect these areas through increased incidence of erosion, flooding, inundation, and salinisation [7]. That we have already begun to see the detrimental effects of climate change on many island and coastal nations throughout the world is indicative that we will see such causality in Belize and the Caribbean as well.

It is also important to note that migration due to global warming can leave populations that are already vulnerable even more susceptible to problems such as food insecurity and inaccessible drinking water [8]. According to the United Nations Environment Programme, climate migration occurs when, 'Poverty, failing ecosystems, vulnerability to natural hazards,' lead to the permanent relocation of vulnerable populations [9]. As San Mateo is an impoverished community with a high number of marginalized individuals, some with a severe lack of accessibility to resources, climate migration could prove to be devastating for the community. Climate migration is already occurring in many of the Pacific island nations. It has become such a crucial issue that New Zealand has enacted an initiative known as the Pacific Access Category, in which citizens from Fiji, Kiribati, Tuvalu, and Tonga can apply for residency if their homes are severely affected by the effects of climate change [10]. San Mateo is especially susceptible to environmental factors, and as such, mitigation resources to help the community should be put into place by the Belizean government. If no steps are taken to combat this pressing issue, San Mateo faces a future similar to that of the Pacific island nations, whereby Belizean individuals and families will seek to immigrate to more developed nations.

REFERENCES

[1] Holy Cross Anglican School, available at: http://holycrossbelize.org/about-us/san-mateo/

[2] Humes, A., "In Belize, Climate Change Drives Coastal Management." Inter Press Service, available at: http://www.ipsnews.net/2015/04/in-belize-climate-change-drives-coastal-management/, 2015.

[3] Cardona, K. & Cardona, M., *Revised Master Development Plan for Ambergris Caye*. Produced for the Coastal Zone Management Authority & Institute, 2009.

[4] UNICEF Belize-Climate Change Prevention and Emergency Response, UNICEF, available at: http://www.unicef.org/belize/climate_change_22129.htm

[5] Handwerk, Brian, "Belize Reef Die off due to Climate Change?" available at: http://news.nationalgeographic.com/news/2003/03/0325_030325_belizereefs.html
[6] Marengo, Jose, etc. all. "Climate Change in Central and South America: Recent Trends, Future Projections, and Impacts on Regional Agriculture," CGIAR Research Program on Climate Change, Agriculture and Food Security, available at: https://cgspace.cgiar.org/bitstream/handle/10568/41912/Climate%20change%20in%20CA%20and%20SA%20final_.pdf?sequence=2
[7] Richardson, R.B., Belize and climate change: the costs of inaction, available at: http://www.bz.undp.org/content/dam/belize/docs/UNDP%20BZ%20Publications/Belize-and-Climate-Change-The-Costs-of-Inaction.pdf
[8] "UNHCR - Climate Change, The Storm Ahead," The UN Refugee Agency, available at: http://www.unhcr.org/pages/49e4a5096.html
[9] "Climate Change and Environmentally Induced Migration," United Nations Environment Programme, available at: http://www.unep.org/conflictsanddisasters/Policy/DisasterRiskReduction/ClimateChangeAndMigration/tabid/282/language/en-US/Default.aspx
[10] "Pacific Access Category," Immigration New Zealand, available at: http://www.immigration.govt.nz/migrant/stream/live/pacificaccess/

TWO RECENT STORM-SURGE EPISODES IN THE ADRIATIC

I. MEĐUGORAC, M. PASARIĆ, Z. PASARIĆ & M. ORLIĆ
Department of Geophysics, Faculty of Science, University of Zagreb, Croatia.

ABSTRACT

In the recent past, two intensive episodes of high sea level occurred in the Adriatic, which affected particularly the eastern coast of the basin. The events happened on 1 December 2008 and on 1 November 2012 when the eastern-coast station Bakar recorded the two highest water levels in its operational history. Interestingly, the episodes were of milder intensity in Venice, situated at the opposite coast. The question is: why these events were so pronounced along the eastern shore and was there specific meteorological background favouring stronger flooding of the eastern Adriatic coast? The study was carried out using hourly sea-level series along both sides of the basin and ECMWF reanalysis of air pressure and wind fields. Sea level was attributed to four processes: low-frequency variability, tide, pre-existing basin-wide seiche and storm-surge activity and respective contributions to high sea-level formations were evaluated. The effects of meteorological forcings over the basin were analysed by calculating along- and across-basin wind stress with respect to local bathymetry. The two marine storms were outcome of fine interplay between the four mentioned processes. During the 2008 flood, storm-surge activity was dominant while the other components reinforced the maximum. In the 2012 episode, storm surge was intensive at most stations, but relative timing of all the components was more complex. Results also revealed one common feature that was crucial for both floods being stronger along the eastern Adriatic coast. Specifically, presence of Sirocco-wind shear induced higher-than-usual water level rise along the eastern coast while transverse component of wind had a relatively small effect. This means that, at least in these two extreme episodes, transverse sea-level slope in the Adriatic was more sensitive to shear of longitudinal component of wind than to wind's direction.
Keywords: acqua alta, Adriatic Sea, Sirocco-wind shear, storm surge.

1 INTRODUCTION

High-sea-level events in the Adriatic, known as the acqua alta phenomena, occur several times a year, usually during late autumn and winter [1]. They cause flooding of the northern Adriatic cities, many of which are of historical significance. The city of Venice and its surrounding low-lying area, situated at the very north of the basin, are particularly vulnerable to these flooding episodes. Some events in the past, however, were more pronounced along the eastern Adriatic coast than in Venice, where their effect was of milder or even weak intensity. This was observed in two recent exceptional events.

In the early morning of 1 November 2012, the oldest Croatian tide-gauge station Bakar has recorded the highest water level in its operational history, when the sea level rose to 99.3 cm above the mean. Absolute maxima were recorded in the middle Adriatic as well, by tide-gauge stations Zadar and Split. During the same event, however, Venice recorded a weaker sea-level maximum that was ranked as eleventh in recorded history [2, 3]. Mediterranean cyclone active at the time brought heavy rain and strong winds over Istria in the north and Dalmatia in the middle Adriatic. Istria suffered the heaviest rainfall on the occasion.

DOI: 10.2495/SAFE-V6-N3-589-596

According to data [4], Plomin meteorological station recorded 154 mm of rain accumulated from midnight to the morning hours of 1 November. In the area of the Raša river, a state of emergency was declared because of an alarming increase of its upper part water levels. A tornado occurred as well, devastating villages of Šugari and Filini in eastern Istria. During the night of 1 November, the strongest wind velocities were observed in the middle Adriatic – in coastal Split and at islands of Lastovo, Palagruža, Mljet and Vis. Wind gusts in Split reached a speed of 112 km/h while at Vis island the observed velocity was even higher – it reached 133 km/h [4]. Normally, strong wind and low air pressure raise sea level only in the northern Adriatic, but on this occasion they also caused exceptionally high sea levels along the Dalmatian coast. The sea water was reported to enter houses, cellars, restaurants, shops and cars in several Croatian coastal cities (Rijeka, Crikvenica, Bakar, Cres, Rab, Zadar, Split...). It partially flooded the Adriatic highway in Sukošan slowing the traffic through the city [5]. Altogether, the flood of 1 November 2012 left Croatia with considerable material damage (Fig. 1, left).

Four years earlier, on 1 December 2008, another severe storm hit the Adriatic. This flood was already analysed in detail [6], and therefore only a short summary will be given in this paper. During this episode, the oldest Adriatic tide-gauge station Trieste recorded the second highest sea level in its operating history. The oldest Croatian tide-gauge station Bakar has recorded the second highest level as well. Venice experienced a less dramatic event, with sea level ranking the fourth highest in history. Similarly as in 2012, Dalmatia was flooded together with Istria and Kvarner area. Croatian contemporary newspapers reported on troubles brought by severe weather to residents of the coastal area. The old town of Trogir was flooded together with the centre of Rijeka (Fig. 1, left), as were parts of Pula, Vodice, Šibenik and Zadar. As examples of intensity of the storm, some of the ferry routes from Zadar were temporarily cancelled, and a fishing boat was thrown ashore near Kraljevica [7].

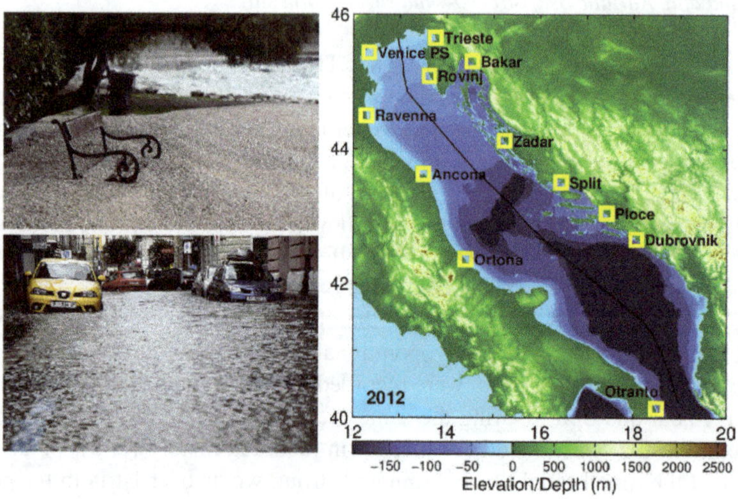

Figure 1: A scene from Zadar after the flood of 1 November 2012 (left, top). City of Rijeka during the flood of 1 December 2008 (left, bottom). Right: location of tide-gauge stations from which the sea-level data were collected for analysis of the 2012 flood. The along-basin line is needed for the purposes of Fig. 4. Same colour is used for all depths greater than 250 m.

In this paper, we closely analyse the flood of 1 November 2012, using hourly sea-level series, recorded at the eastern and the western coast, and complemented by ERA-Interim meteorological dataset. The results are then compared to those obtained from the study of the 1 December 2008 flood [6]. This comparison should indicate common features of the two meteorological backgrounds that led to stronger flooding of the eastern as opposed to the western Adriatic coast.

High sea levels in the northern Adriatic (acqua alta) are result of superposition of several processes: storm surge, tide, pre-existing Adriatic-wide seiche and low-frequency variability.

Storm surge has a dominant role in formation of high sea level, whereas the other processes can either amplify or mitigate the effect. The meteorological background favouring storm surges in the northern Adriatic is well established [1, 8, 9]. It develops when cyclone approaches the basin from the west, resulting in air-pressure gradient along the Adriatic which causes south-easterly wind known as Sirocco. Low air pressure in the north together with Sirocco wind accumulates water in the closed and relatively shallow north end of the basin (Fig. 1, right).

Adriatic tide reinforces flooding event if its culminating phase is superimposed on the induced storm surge [10]. The tide in the Adriatic is of mixed type and has traditionally been modelled as a superposition of seven constituents (M2, S2, N2, K2, K1, P1, O1). Its amplitude reaches the maximum at the closed north end of the basin so that tidal range in the Gulf of Trieste is larger than 1 m.

To a much lesser degree, pre-existing Adriatic-wide seiche, which is usually triggered by a sudden weakening of Sirocco or its change to Bora wind, can take a part in an acqua alta formation as well [8]. It has period of 21–22 hours and a decay time of 3.2 days [11, 12]. Since it is a slow decaying oscillation, it could last for several days and could influence a next upcoming storm surge.

Low-frequency sea-level oscillations, which are induced by slow atmospheric planetary waves travelling to the east, are known to contribute to all coastal floods [10]. Whereas atmospheric planetary waves are prominent in the middle troposphere, they can also be observed in surface air pressure and wind velocities at temporal scales of weeks to several months.

2 DATASET AND METHODOLOGY

The flood of 1 November 2012 was analysed using 1-hourly sea-level series at the stations along both sides of the Adriatic (Fig. 1, right). The time interval used is four months long: from 2 September to 30 December 2012. The verification of the time series was performed with Tukey 53H filter [13] and no outliers were found. Occasional single missing values were linearly interpolated.

First, the data were reduced to anomalies by subtracting the respective four-month means. After that, the four contributing processes have been determined as previously described [6]: first the tide was synthesised using seven constituents, then the low-frequency variability was extracted by applying low-pass filter, upon which a pre-existing Adriatic seiche was estimated. Storm-surge activity was defined as the residual sea level after the three processes were removed from the series.

Meteorological dataset consists of ECMWF reanalysis [14] – the ERA-Interim surface wind and mean sea-level pressure fields given with 6 hour time step and 0.25 × 025 deg spatial resolution. The wind field was enhanced by a factor of 1.47 as suggested in the literature [6]. In order to examine forcing on longer spatial and temporal scales, we inspected geopotential hight of 500-hPa surface in addition to surface fields. All times in this paper are given in UTC.

Similar dataset and methods were used in analysis of the 1 December 2008 flood [6].

3 RESULTS

In the morning hours of 1 November 2012, tide-gauge stations in the whole Adriatic recorded raised sea level (Fig. 2). The maximum in Bakar was observed at 6:00 am and it was 99.3 cm above respective mean. At northern stations, Trieste and Rovinj, maxima were reached later, at 7:00 and 8:00 am respectively, and were lower than in Bakar – sea level was 93.6 cm high in Trieste and 88.6 cm in Rovinj. In Venice, the maximum was recorded earlier, at 1:00 am.

According to ECMWF charts, before the decisive cyclone crossed the basin, a low-pressure system born in the western Mediterranean, propagated towards the Gulf of Genoa where it remained from 27 to 29 October. It caused a rise of the water in the Adriatic and probably triggered the Adriatic seiche. The system diminished on 29 October while moving towards the south. After that, on 31 October, a cyclone developed in the western Mediterranean as a part of a larger low-pressure system centred above Scotland. From there, it moved towards the Adriatic, crossing its northern border at 6:00 am on 1 November.

The cyclone induced an air-pressure gradient over the basin, oriented in the west-to-east direction, and giving rise to Sirocco wind of non-uniform speed (Fig. 3, left). In the velocity field, south-easterly direction dominated over the larger portion of the basin, whereas north-eastern direction occurred over the northern Adriatic. The velocity field reached maximum speed, about 20 m/s, around the latitude of Zadar, with maximum slightly closer to the eastern coast. In the northernmost part, in front of Venice and in the Gulf of Trieste, the speed was relatively low, reaching 12 m/s or less.

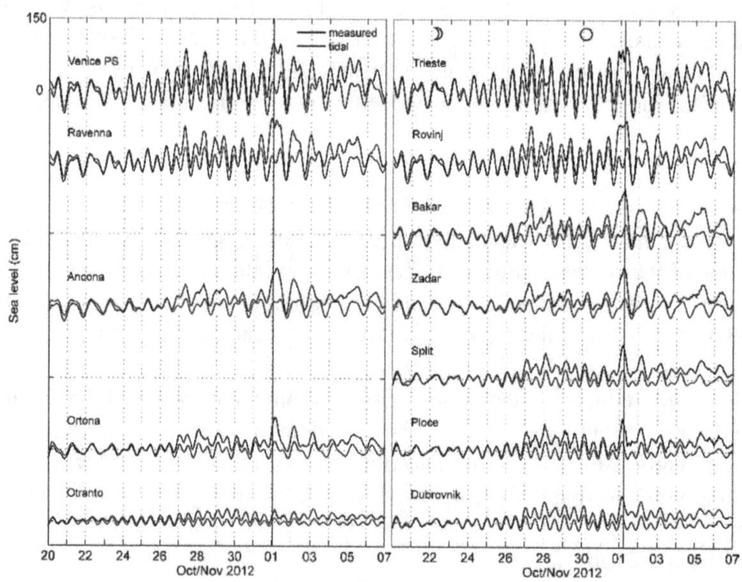

Figure 2: Measured sea-level anomalies, referred to the respective 4-month means, and tidal predictions in the Adriatic for the 2012 flood. Vertical lines indicate onset of maxima at Venice Punta Salute (left) and Bakar (right). The moon phases are also indicated in the figure.

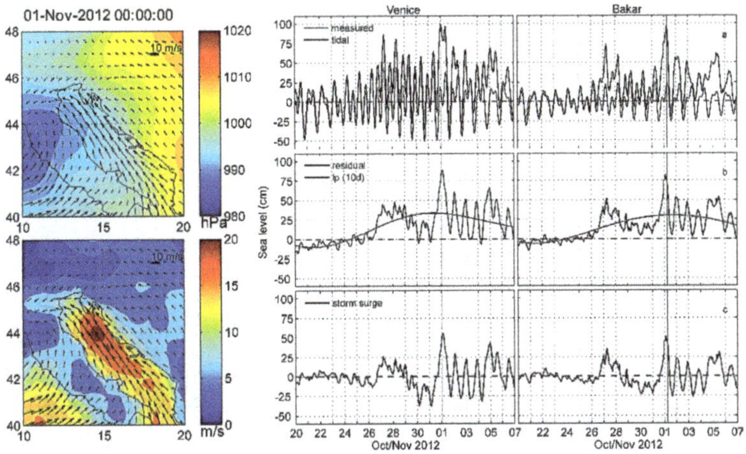

Figure 3: Left: ERA-Interim field of air pressure and enhanced wind for 1 November 2012 at 00:00 UTC. Right: sea-level evolution in Venice and Bakar - measured sea level and tidal prediction (a), sea-level residual and low-frequency sea-level change (b), storm-surge activity (c).

Preconditioning of the flooding event started a week before when the sea level began to slowly rise (Fig. 3, right) as a result of atmospheric large-scale activity. In fact, this disturbance was related to the passage of atmospheric planetary waves which can clearly be seen from the geopotential height of 500-hPa surface. Its contribution to high sea level was almost uniform along the two Adriatic coasts (Fig. 4), as expected considering that this is a large-scale process. At the northern stations it contributed about 30 cm to the local maxima.

Residual sea-level series (Fig. 3, right) indicate a large energy input on 26 and 27 October, which coincided with the Genoa cyclone. This abrupt sea-level change generated the Adriatic-wide seiche of significant amplitude. However, this seiche was generated five days before the main event and therefore was mostly damped at the time of flooding and had almost no effect on the event.

The Moon phase was between full moon and the last quarter (Fig. 2), therefore the semi-diurnal tide prevailed. Onset of maxima at all stations coincided with the positive tidal contribution, but the eastern stations were more affected by it (Fig. 4). Sea-level maxima at Trieste and Rovinj have coincided with the tidal maxima that were reached when storm surge was subsiding. On the other hand, the maximum at Venice was formed earlier in the night of 1 November, during the storm-surge peak and small tide (Figs 3 and 4). The situation in Bakar was something in between – the maximum emerged while the storm surge was subsiding and the tide was heading towards its maximum.

It is interesting to note (Fig. 3, right) that the storm surge was of similar intensity in Bakar and in Venice. One would expect a much higher storm surge in Venice because the Sirocco-related fetch is longer and the depths are smaller off it. Comparison of the storm-surge series at the eastern stations with corresponding series at the western stations indicates that the along-basin water-surface slope was not very steep while in an across-basin direction water surface was horizontal. Obviously, the wind forcing (Fig. 3, left), characterised by low speeds over the northernmost part of the Adriatic and by a maximum in the vicinity of Zadar, considerably raised water in the middle Adriatic producing the storm surge of a moderate along-basin

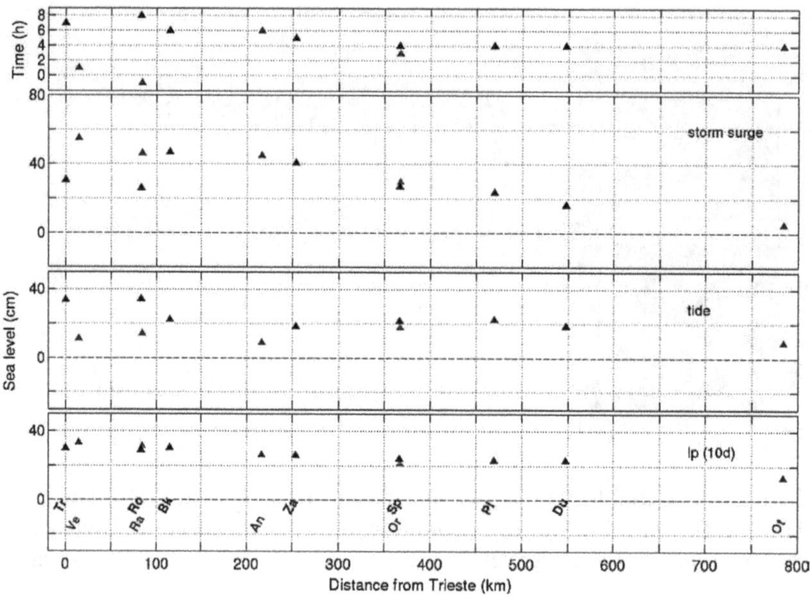

Figure 4: Contribution of each component at the onset of local sea-level maxima to the 2012 flood. Eastern Adriatic stations are denoted by black and western by grey triangles: Trieste (Tr), Venice (Ve), Rovinj (Ro), Ravenna (Ra), Bakar (Bk), Ancona (An), Zadar (Zd), Split (Sp), Ortona (Or), Ploče (Pl), Dubrovnik (Du), Otranto (Ot). Abscissa denotes a distance of a station from Trieste when a station is projected onto the along-basin line indicated in Fig. 1. Time of the occurrence of sea-level maxima is indicated at the top of the figure.

slope. Interestingly, the wind managed to accumulate water equally along the two coasts even though the eastern Adriatic is much deeper. Hence, although the eastern Adriatic maxima were influenced by the tide, they would not have been so pronounced if the storm-surge level had not been higher along the eastern coast than it usually is during Sirocco-wind episodes.

To summarise, the flooding of the north-western coast during the Adriatic marine storm of 1 November 2012 was a result of high storm surge and moderate tide, all superimposed on raised sea level due to low-frequency variability. Flooding of the eastern Adriatic coast was a result of an interplay between storm surge and tide, superimposed on prolonged interval of raised water level.

The flooding episode of 1 December 2008 was induced by Genoa cyclone that was not particularly pronounced [6]. The cyclone supported air-pressure gradient along the basin and south-easterly wind of non-uniform speed. Intensity of the wind field was generally higher along the eastern side and in the south where southerly direction was prevailing. Sea-level maxima emerged as a result of a high storm surge accompanied by a large along-basin slope, as well as of a high tide and a low-frequency sea-level rise. Pre-existing seiche was not especially strong and its contribution was rather small. An interesting fact about this event is that all components, except the storm surge, had a similar effect at both sides of the basin. This means that severe flooding of the eastern shore in 2008 resulted from rarely experienced eastward storm-surge surface slope.

4 DISCUSSION AND CONCLUSION

In recent years, on two occasions (1 December 2008 and 1 November 2012) the eastern coast of northern Adriatic experienced extremely high sea level – in Bakar on both occasions it surpassed its ever recorded value – while the events were of a lower intensity at the opposite coast. The aim of our analysis was to determine specific conditions under which the floods were more pronounced along the eastern coast than at the western coast. The results showed that in both episodes the across-basin storm-surge slope was decisive for flooding of the eastern coast. The two episodes were governed by similar meteorological backgrounds. Air-pressure gradient was of along-basin (2008) or west-to-east (2012) direction, while consequent Sirocco-wind speed was spatially non-uniform with higher values near the east coast. Air-pressure forcing was in favour of higher sea level at the western coast and in both cases its contribution was estimated to be less than 10 cm. In order to evaluate the wind-stress effect on sea level over the variable bathymetry, we mapped wind-stress-to-depth ratio shown in Fig. 5. The wind stress, calculated as described before [6], was decomposed into along- and across-basin component. Evidently, the transverse water-level slope emerged as a result of dynamics related to longitudinal wind. The Sirocco-wind shear related to stronger winds along the eastern coast surpassed the bottom-slope effect and accumulated water along the deeper eastern coast. On the other hand, the impact of transverse wind on across-basin water slope was of secondary importance since its contribution was rather small, probably due to the wind component being small and the related fetch being short.

Therefore, we can conclude that two recent episodes of severe flooding along the eastern Adriatic coast, the events of 2008 and 2012, were the result of storm surge controlled by Sirocco-wind shear superimposed on the tide and the low-frequency sea-level rise.

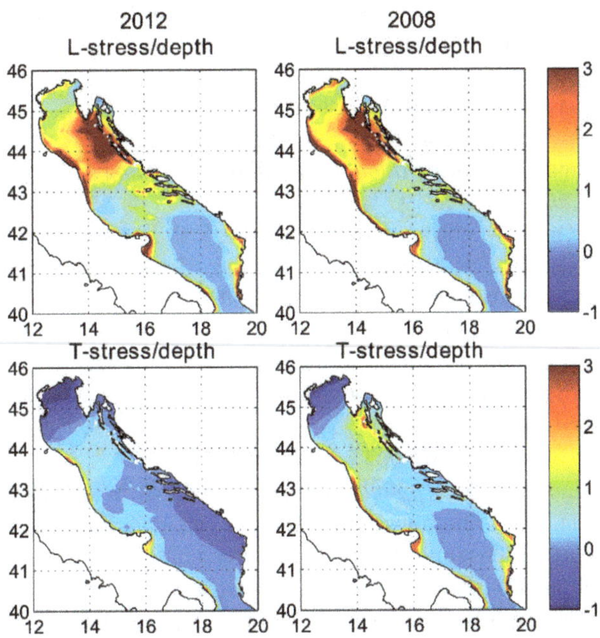

Figure 5: Along-basin (top) and across-basin (bottom) wind stress with respect to local bathymetry and normalized by average wind stress over the basin (in units 10^{-2} m^{-1}) for the 2012 (left) and 2008 (right) floods. Wind stresses are related to the calibrated ERA-Interim wind fields of 1 November 2012 at 00:00 h and 1 December 2008 at 00:00 h.

ACKNOWLEDGMENTS

The authors are thankful to the following institutions for providing the data: Consiglio Nazionale Delle Ricerche, Hydrographic Institute of the Republic of Croatia, Istituto Superiore per la Protezione e la Ricerca Ambientale and Istituzione Centro Previsioni e Segnalazioni Maree. This work has been fully supported by Croatian Science Foundation under the project 2831 (CARE).

REFERENCES

[1] Lionello, P., Cavaleri, L., Nissen, K.M., Pino, C., Raicich, F. & Ulbrich, U., Severe marine storms in the Northern Adriatic: characteristics and trends. *Physics and Chemistry of the Earth*, **40–41**, pp. 93–105, 2012. http://dx.doi.org/10.1016/j.pce.2010.10.002

[2] Bonato, N., Egiatti, G., Ferla, M. & Filippi, M., Tidal observations in the Venetian Lagoon. Update on sea level change from 1872 to 2000. *Proc. of the Final Workshop of COST Action 40*, eds. I. Vilibić, H.P. Plag & M. Orlić, Hydrographic Institute of the Republic of Croatia, Split, pp. 97–105, 2001.

[3] Venice Municipality – Tide Monitoring and Forecast Center, available at http://www.comune.venezia.it/flex/cm/pages/ServeBLOB.php/L/IT/IDPagina/25419

[4] Crometeo – monitoring and forecasting the weather, available at www.crometeo.hr

[5] www.dnevnik.hr

[6] Međugorac, I., Pasarić, M. & Orlić, M., Severe flooding along the eastern Adriatic coast: the case of 1 December 2008. *Ocean Dynamics*, **65**, pp. 817–830, 2015. http://dx.doi.org/10.1007/s10236-015-0835-9

[7] www.jutarnji.hr

[8] Robinson, A.R., Tomasin, A. & Artegiani, A., Flooding of Venice, phenomenology and prediction of the Adriatic storm surge. *Quarterly Journal of the Royal Meteorological Society*, **99**(422), pp. 688–692, 1973. http://dx.doi.org/10.1002/qj.49709942210

[9] Trigo, I.F. & Davies, T.D., Meteorological conditions associated with sea surges in Venice: a 40 year climatology. *International Journal of Climatology*, **22**, pp. 787–803, 2002. http://dx.doi.org/10.1002/joc.719

[10] Pasarić, M. & Orlić, M., Long-term meteorological preconditioning of the North Adriatic coastal floods. *Continental Shelf Research*, **21**, pp. 263–278, 2001. http://dx.doi.org/10.1016/S0278-4343(00)00078-9

[11] Raicich, F., Orlić, M., Vilibić, I. & Malačič, V., A case study of the Adriatic seiches (December 1997). *Il Nuovo Cimento C*, **22**, pp. 715–726, 1999.

[12] Cerovečki, I., Orlić, M. & Hendershott, M.C., Adriatic seiche decay and energy loss to the Mediterranean. *Deep-Sea Research*, **44**(12), pp. 2007–2029, 1997. http://dx.doi.org/10.1016/S0967-0637(97)00056-3

[13] Otnes, R.K. & Enochson, L., *Applied Time Series Analysis*, Wiley: New York, 1978.

[14] Dee, D.P., Uppala, S.M., Simmons, A.J., Berrisford, P., Poli, P., Kobayashi, S., Andrae, U., Balmaseda, M.A., Balsamo, G., Bauer, P., Bechtold, P., Beljaars, A.C.M., Van de berg, L., Bidlot, J., Bormann, N., Delsol, C., Dragani, R., Fuentes, M., Geer, A.J., Haimberger, L., Healy, S.B., Hersbach, H., Holm, E.V., Isaken, L., Kallberg, P., Kohler, M., Matricardi, M., Mcnally, A.P., Monge-Sanz, B.M., Morcrette, J.J., Park, B.K., Peubey, C., De Rosnay, P., Tavolato, C., Thepaut, J.N. & Vitart, F., The ERA-Interim reanalysis: configuration and performance of the data assimilation system. *Quarterly Journal of the Royal Meteorological Society*, **137**, pp. 553–597, 2011. http://dx.doi.org/10.1002/qj.828

COPING MECHANISMS OF SME IN RESPONSE TO 2011 FLOODS IN PATHUMTHANI, THAILAND

S. PATHAK[1] & M.M. AHMAD[2]
[1]Disaster Preparedness, Mitigation and Management, Asian Institute of Technology, Thailand.
[2]Associate Professor, Asian Institute of Technology, Thailand.

ABSTRACT
Disaster recovery for SMEs are related to bouncing back after the disaster event and attaining resilience and self-dependency. This study focuses on the flood recovery coping mechanism adopted by the small and medium enterprises (SMEs) in Pathumthani province of Thailand post 2011 floods. Thailand is a middle income country prone to natural hazards with a major risk from the floods. The affected population includes all sectors of the society such as business. The SMEs contribute a major share in the GDP of Thailand and are have to be protected from the disastrous impacts of the re-occurring floods. This paper analyses the capacities adopted by the manufacturing SMEs to recover from the floods in the Pathumthani province in Thailand. It is based on extensive literature review synthesis and a methodological approach to ascertain the impacts and coping mechanisms of flood disaster to vulnerable SMEs in the Pathumthani province, Thailand.
Keywords: coping mechanisms, disaster, disaster recovery, disaster risk reduction, floods impact, small and medium enterprises (SMEs).

1 INTRODUCTION

The 2011 floods event, which is also termed as 'Great Floods', is chosen to ascertain the maximum impacts that were faced by the small and medium enterprises (SMEs) in the recent times. Thailand is a developing economy and SMEs contribute significantly to the Thai growing economy by generating employment and adding to the GDP and alleviate poverty [1, 2]. There was a tremendous growth in the SMEs during the years 1985 to 1995 in Thailand [3]; however, a declining trend is observed after this period [4]. Still almost 70% of the total employment in Thailand is generated through SMEs [1].

The study focuses on the recovery strategies adopted by the SMEs in the Pathumthani province of Thailand to cope up with the reoccurring floods. The flood event of 2011 is taken as the sample year of impact, which affected majority of the SMEs in Thailand. The paper analyses the various factors that contribute to recovery of these SMEs from the floods and how could they be managed in a way to achieve faster recovery from future flood events.

2 FLOODS IN THAILAND

The Floods affect almost all walks of life as a slow receding disaster. Flood impacts directly on the human lives, infrastructure, plant and machinery, transportation and warehouse storage capacities [5]. The floods impact indirectly on the social and psychological set up of any

DOI: 10.2495/SAFE-V6-N3 -597-606

community, as production is hurdled, negative impacts on market values of property and stock market [6].

The coping mechanism of the SMEs from the floods involves various factors that work together to attain resilience and sustainability from catastrophic flood events. These factors include SME's size, type, location [7], disaster risk perception and management techniques adopted [4, 8], financial resilience [4], flood risk mapping, community support, research and development on SMEs [1], governmental aid on both local as well as national level and aid from various non-governmental organisations. SMEs are severely affected in the absence of disaster risk reduction, business risk assessment and business continuity planning [9, 10]. The limitation of the SMEs in terms of location, small workforce and reach to limited market enhances the disaster risk [11, 12]. The study found out that apart from these factors absence of insurance is one of the major factors affected the recovery time period for the SMEs [9, 13, 14].

3 DISASTER RECOVERY IN THAI SMES

Disaster recovery studies in the past have overlooked the SMEs affected by disasters [15, 16] as most of them are focused on the household or community level, government bodies and departments dealing with emergency services [17, 18]. Few studies have been found on disasters and their impact on SMEs [19–21]. There is a dire need to study the SMEs as large firms have adequate resources and access to finances that lead to faster recovery than the SMEs [19].

In Thailand during the floods of 2011, the SMEs were heavily impacted by floods and recovery period extended from few hours to over a year depending upon the size, financial resources and managerial decision making. The overall sample size of 200 SMEs revealed (refer Table 1) that flood preventive measures adopted by the SMEs prior to the floods resulted in faster recovery (2%) in comparison to the others especially two SMEs which have been unable to commence the business till date (1%). Such huge time difference in the recovery period influence the overall industrial output due to linkages of supply chain from the SMEs, which provide the raw materials.

4 THAI SMES

SMEs in Thailand have been defined on the basis of capital invested or number of employees. SMEs in Thailand are defined as those enterprises, which employ less than 200 employees,

Table 1: Recovery time taken by sample SMEs in Pathumthani district.

Recovery Time period (in month/s)	Frequency (Number of SMEs)	Percent
0	4	2.0
0.1–1	41	20.5
1.1–3	93	46.5
3.1–6	52	26.0
6.1–12	8	4.0
Never	2	1.0
Total	200	100.0

Source: Field Survey.

having investment capital of less than 100 million baht, and fixed assets of less than 100 million baht [22, 23]. SMEs are termed as '*boresat*' in the native Thai language and contribute to the second largest employment opportunity after agriculture [24].

The SMEs in Thailand are categorised on the basis of three primary sectors including production/manufacturing, trading and services. The major portion of the SMEs are in the manufacturing sector and contributed to the selection of these enterprises to be studied as the flood impact on these firms affects further supply chain and export/import of the Thai economy as a whole. These small business enterprises are more vulnerable towards the adverse impacts of floods [5]. The study is focused to prevent the SMEs damage from future flood events.

5 METHODOLOGY

The study was carried out in seven districts of the Pathumthani province of Thailand. The province has a total of 401 SMEs [23] out of which a sample size of 200 companies [25] was selected. The study focused on the coping mechanisms adopted by the affected SMEs in the study area in order to recover from the devastating floods of 2011.

The methodology involved qualitative and quantitative methods by the usage of survey questionnaires, key informant interviews and focused group discussions along with secondary data gathered from various concerning institutions, bodies and previous literatures on flood recovery for SMEs in Thailand. The data was collected in three distinct phases. The first visit provided a detailed information of the study area which assisted the second phase in collection of survey data according to the 2011 flood affected SMEs. The survey questionnaire involved multiple choice and open-ended questions to gather information related to SME's profile, management, and impacts incurred from the 2011 floods, disaster risk perception and adopted coping mechanisms. The survey questionnaire respondents included the hired managers and owners/managers who are managing a company in the Pathumthani district.

The respondents consisted of 70.5% males and 29.5% females. The respondents were among the age group of 21–30 years (20%), 31–40 years (35.5%) and beyond 40 (44.5%) and the education level comprised of graduates (66%), post graduates (11%) and below graduation level (23%).

6 FINDINGS AND DISCUSSIONS

The study finds out the severe impacts from floods had been borne by the SMEs during the floods of 2011.The damages included financial, social as well as psychological losses. The government aid and policies lack the force for faster recovery of these enterprises. The non-governmental sector is almost negligible in terms of assistance during the floods for the SMEs. The major recovery was brought through the financial assistance from the owners own equity or through the profits of future business, which pushed the recovery process for over a period of months. The growth is hampered as recovery time period is elongated due to lack of supporting mechanism from various governmental or non-governmental bodies. The recent political turmoil and droughts (2015) have added to the already affected recovery, created hurdles in the growth and development of the SMEs.

6.1 Socio-economic impact

The production was on standstill for a period of 2 months (mid-October to mid-December), however, after the floods receded, it was hard for the SMEs to jump start the production due to non-existence of adequate finances. It took around five months (March 2012) on an average for the SMEs to restart the production of the enterprise. However, two SMEs were unable to recover from the 2011 floods in the study area and finally winded up their business activity leading to loss of employment for 67 employees.

The financial losses were enormous as almost 40% of the total sample size had to bear an economic loss of more than 1 million baht. These economic losses included production days, sales, stock and other miscellaneous expenses to claw back to the normalcy of the business activities. However, the costs of repair and purchasing of new machineries and equipment is not included. The total financial impact is almost double when the total expenses incurred are calculated.

Thailand being an upper middle income country, largely depends upon the SMEs for future development and sustainable growth in the economy. The SMEs in the developing countries have general tendency to operate in hazard prone and vulnerable locations, lack of human and financial resources, limited institutional capacity and inadequate futuristic planning in terms of disaster risk [13]. The study found the limitations of financial resources for SMEs in the study area along with misconceptions regarding flood risk and importance of mitigation tools such as flood insurance and medical insurance policies for the workforce.

Almost 85% of the sample size in the study depicted lack of financial resources and sustainability through preparedness and mitigation strategies to cope from a natural hazard. SMEs lack planning for business continuity and risk transfer mechanism during an unforeseen natural calamity [5, 11, 12, 13]. The flood disaster recovery has a positive association with the business risk perception and the financial resources at the management's disposal during a disaster event. Therefore, the SMEs should try to identify adequate financial security and sustainable mechanisms to recover faster from the recurring Thai floods.

6.2 Flood risk

Flood risk is a recurring risk in the Pathumthani province as it lies in the catchment area of the Chao Phraya River, the largest river running through the northern and central provinces. The respondents were vulnerable to the flood risk and had adopted various mitigation measures using their indigenous knowledge. However, they lack the motivation and interest for adopting preparedness and mitigation tools such as structural and non-structural measures such as insurance and flood recovery planning.

Risk perception is a predominant factor in the formulation of flood disaster risk reduction policies and planning [26]. The risk perception would formulate the managerial decision making in SMEs during disaster scenarios. This study finds out that the SMEs have been reluctant in acknowledging the flood risks. We found that they have a laid back and carefree attitude towards future flood risk. A severely affected respondent from a leading noodle company Mr. Nattaporn Noree adds

> 'There is no flood risk in our area. We did get flooded in 2011 but I think the government will take care of it in case it happens in future. Similarly, we don't need insurance also, we will cross the bridge when we reach there!'

Such attitudes toward the flood risk need to be revisited both at governmental and individual level. The flood disaster risk is prevalent in the area, therefore, the risk prevails but in terms of perception, the SMEs are safe from flood risk. The negligence of the risk further reduces the adoption of coping mechanisms while enhancing vulnerability.

6.3 Government aid and assistance

The Thai government has been taking various steps at the national as well as local levels to assist the population from the impacts of the floods, however in 2011, the management of floods by the government was heavily criticized. Around 88% of the respondents were not satisfied with the government policies regarding the flood and water management during 2011 floods. SMEs stood alone during the floods and efforts were focused on relief rather than commencement of the recovery process. It was found that there was an acute lack of SME database, which in turn affected the recovery process and policy formulation [27]. With the recent political instability in the country, disaster recovery initiatives have taken a back seat both at SME and government level.

The government was also heavily criticized for untimely and inaccurate early flood warning (refer Table 2. In spite of various local level bodies working on 24*7 basis, the SMEs were unable to receive flood early warning from the government. The major source was through social media and word of mouth information dissemination during the floods. We found that only 72% of SMEs received the early warning, out of which only 38% had more than a day notice. This was a crucial factor as the SMEs were unable to cope and prepare for the coming flood waters.

The government has a crucial role in disaster recovery. The governing of several factors such as disaster management, aid and assistance programs forms the foundation of faster disaster recovery and enables the economy to bounce back to normalcy. We found out that there is lack of emphasis from the government on Pre-disaster mitigation, inclusion of NGOs, disaster management and recovery planning and adopting insurance as a necessary requirement to curtail the flood losses and provide sustainability to the business unit in itself for speeding up their recovery process. A key informant from an NGO working for SMEs in the province added

> *"The government should include us in the disaster management infrastructure. We have the human resources and volunteers who are willing to assist in flood times. A proper pre-disaster management planning could enable the SMEs to reduce losses and cope up faster from floods as was the case in 2011."*

Table 2: Flood early-warning time taken to disseminate warning to SMEs.

Warning time	Frequency	Percent
No warning	56	28
Less than 1 h	21	10.5
1–6 h	26	13
7–12 h	9	4.5
13–24 h	12	6
+24 h	76	38
Total	200	100

Source: Field Survey.

6.4 Coping mechanisms

The coping mechanisms adopted by the SMEs during the 2011 floods depicts a lack of adequate risk assessment, planning and managerial approach. The subsequent results were huge economic damages and long recovery time periods (refer Table 1). The following is the SWOT analysis on the basis of data collected from SMEs and other stakeholders including government and non-government organizations.

The SWOT analysis of the existing coping mechanisms involves:

Strengths: The current coping mechanisms are related to various aids and assistance provided by the Thai government. The political structure provides for basic disaster response and relief to the affected communities. The local level government provides immediate assistance in terms of food and other basic amenities. However, there is a lack of focused assistance towards SMEs and business community.

An example from a SME manager Mr. Nattapol who quoted

'The government was very prompt in providing food, water and basic amenities. The staff also helped in recovering from losses and assisted even during flood time. The main problem arose once the company was to restart its daily business.'

Weaknesses: The major setback was in the managerial decision making both at SME and governmental level. The tremendous impact of 2011 floods was largely due to the misperception of the flood risk. The SMEs were not prepared and government lacked in providing early warning. The financial resources of SMEs were limited and the government focused on disaster relief and response as the scale of 2011 floods was enormous and beyond measures for the Thai government.

Opportunities: There were ample opportunities for the SMEs if they would have prepared for the uncertainties and perceive the flood risk prior to the 2011 floods. The existing opportunities involved tax rebates and other incentives given to the SMEs during the next three years subsequent to the 2011 floods. The business continuity planning and disaster risk management planning proved to be effective in the disaster scenario for some SME's that had prepared adequately.

In the words of SME owner and manager Mr. Tantripol Pornsiri

*"We were not affected by the 2011 floods at all. The planning was in place and we carried out business activities from our Bangkok branch. The factory was secured with insurance, disaster management plans and measures such as boats, flood dykes around the premises and 24*7 security guards on duty during the flood months."*

Threats: The absence of participatory approach from all the stakeholders involved resulted in enhanced vulnerabilities and threats for the SMEs and eventually to the economy as a whole [28]. The perception of risk was a major threat towards sustaining the floods [29]. Pathumthani province is vulnerable to reoccurring floods. There is a dire need to reduce the vulnerabilities through effective flood early warning system and policy level implications and implementations. The effective use of information and available resources would assist in curtailing enormous impacts of floods among the SMEs.

The overall approach has been towards achieving faster flood disaster recovery for SMEs. The future success depends on the improvisations and additions of new and better coping mechanisms on national, local as well as SME level.

6.5 Disaster risk management & disaster recovery

Disaster risk management deals with the mechanisms involving reduction of vulnerabilities of the unforeseen disasters and provide sustainable and resilient development. In terms of the business sector, it is essential to perceive the risk in its actual state and provide for its adequate and effective disaster management planning and mitigation measures. The overall process of flood disaster management and rapid recovery involves four distinct stages as described below.

Stage 1: Flood Risk Assessment – The first stage involves the able and effective managerial decision making, accurate risk perception to develop and frame the business continuity planning in accordance with the disaster risk reduction planning for future hazards. Through proper governmental policies and utilizing new technologies such as remotely sensed data [30], the flood risk assessment process would provide for a robust and holistic policy framework for coping from disasters such as floods.

Stage 2: Flood Risk Planning – Planning ahead of a hazard is a key to better managing the disaster event. The flood risk planning process involves undertaking every aspect of disaster, risk and ensuring faster recovery after the disaster event. It is essential to plan ahead to be resilient towards the future uncertainty [31]. The climate change impacts are to be included in the planning, as the effects are not uniform throughout an economy [32]. The planning must ensure reduced risk through tools including financial resources, risk transfer tools and sustainable progress of the enterprise.

Stage 3: Flood Disaster Management – The success of flood disaster risk management is dependent upon the accuracy of the flood risk assessment and planning [33]. The external factors such as political and government policies and assistance programs must be fully utilized by the flood risk management. Disasters might open new horizons for the enterprise to grow; however, the chances are reduced if proper risk management is absent.

Stage 4: Flood Recovery – The post flood recovery is of utmost importance due to the fact that SMEs provide for social and economic development. The faster the SMEs recover from the disaster event, the faster would be economic growth of the economy. The vulnerabilities are reduced with adequate implementation of the disaster recovery mechanisms. It is essential to provide for faster recovery to the SMEs as they act as a lifeline to the economy and contribute towards sustainable development.

7 CONCLUSION

The study has identified the loopholes in disaster management and recovery mechanism adopted among the SMEs during the 2011 flood in the affected areas of Pathumthani Province. The lack of government planning and policy implementation, non-utilization of adequate financial and managerial disaster recovery tools, such as insurance and business continuity planning are the major factors responsible for the longer recovery of SMEs after the natural disaster.

The existing coping mechanisms are unable to reduce the long recovery time. The disaster risk reduction and disaster recovery had to be involved at all levels of policy formulations. The government should focus on recovery as well as disaster preparedness among the SMEs that are crucial for the provincial growth and development, which in turn increases the national productivity and strengthen the economy.

We have found that with adequate flood preparedness and business continuity strategies adopted at SME and government levels the total recovery period after a flood event would be optimally reduced. A comprehensive disaster risk reduction and business continuity planning would enhance the sustainability of the SMEs in Pathumthani province from future flood hazards. The government could assist before the disaster strikes in terms of risk awareness campaigns, training exercises and propagating self-sustainability through insurance. The post disaster relief and recovery strategies involve restoration of critical facilities, tax subsidies and risk reduction workshops and follow up.

SME forms the major contributor to middle income countries economy. Thai SMEs need to be flood resilient in order to maintain their continuous contributions towards the Thai economy. There are several loopholes in the current system of SME in the Pathumthani and other provinces in Thailand. There is a dire need to further investigate the factors contributing towards SME vulnerability towards disasters. The joint efforts of all stakeholders involved would result in sustainable and resilient SMEs from any future disasters.

REFERENCES

[1] OSMEP, "Situation and Structural Indicators of SMEs in 2008 and 5-Year Changes," White Paper on SMEs 2008 and Trends 2009, 2008.

[2] Tannock, J., Krasachol, J. & Ruangpermpool, S., The development of total quality management in Thai manufacturing SMEs: A case study approach. *International Journal of Quality & Reliability Management,* **19**(4), pp. 380–395, 2002. http://dx.doi.org/10.1108/02656710210421562

[3] Régnier, P., *Small and Medium Enterprises in Distress: Thailand, the East Asian Crisis and Beyond*, Hampshire: Gower, p. 181, 2000.

[4] Sleuwaegen, L. & Goedhuys, M., Growth of firms in developing countries, evidence from Co^te d'Ivoire. *Journal of Development Economics,* **68**, pp. 117–135, 2002. http://dx.doi.org/10.1016/S0304-3878(02)00008-1

[5] Alesch, D.J., Holly, J., Mittler, E. & Nagy, R., *Organizations at Risk: What Happens when Small Businesses and Not-for-Profits Encounter Natural Disasters*. Public Entity Risk Institute, Fairfax, VA, 2001.

[6] Rose, A., A framework for analyzing the total economic impacts of terrorist attacks and natural disasters. *Journal of Homeland Security and Emergency Management,* **6**(1), pp. 2009.

[7] Brush, C.G., Ceru, D.J. & Blackburn, R., Pathways to entrepreneurial growth: The influence of management, marketing, and money. *Business Horizons,* **52**, pp. 481–491, 2009.
http://dx.doi.org/10.1016/j.bushor.2009.05.003

[8] Lind, P., Artist, *The SME in a Global Context – Cultural Impact.* [Art]. Gotland University, 2009.

[9] Zhang, Y., Lindell, M.K. & Prater, C.S., Vulnerability of community businesses to environmental disasters. *Disasters,* **33**, (1), pp. 38–57, 2009.
http://dx.doi.org/10.1111/j.1467-7717.2008.01061.x

[10] Asgary, A. & Sadeghi Naini, A., Modelling the adaptation of business continuity planning by businesses using neural networks. *Intelligent Systems in Accounting, Finance and Management,* **18**, pp. 89–104, 2011.
http://dx.doi.org/10.1002/isaf.326

[11] McClure, D.L., Disaster recovery for small businesses. Central U.S. *Earthquake Consortium Journal,* **7**(1), p. 10, 2000.

[12] Webb, G., Tierney, K. & Dahlhamer, J., Business and disasters: empirical patterns and unanswered questions. *Natural Hazards Review,* **1**(2), pp. 83–90, 2000.

http://dx.doi.org/10.1061/(ASCE)1527-6988(2000)1:2(83)

[13] Green, R., Miles, S., Gulacsik, G. & Levy, J., Business recovery related to high-frequency natural hazard events. Quick Response Report, Natural Hazards Centre, 2008.

[14] Yoshida, K. & Deyle, R., Determinants of small business hazard mitigation. *Natural Hazards Review,* **6**(1), pp. 1–12, 2005.
http://dx.doi.org/10.1061/(ASCE)1527-6988(2005)6:1(1)

[15] Quarantelli, E.L., Disaster crisis management: a summary of research findings. *Journal of Management Studies,* **25**(4), pp. 373–385, 1988.
http://dx.doi.org/10.1111/j.1467-6486.1988.tb00043.x

[16] Hale, J.E., Dulek, R.E. & Hale, D.P., Crisis response communication challenges: building theory from qualitative data. *Journal of Business Communications,* **42**(2), pp. 112–134, 2005.
http://dx.doi.org/10.1177/0021943605274751

[17] Jones, B.G. & Chang, S.E., Economic aspects of urban vulnerability and disaster mitigation. In *The Role of Engineering and Technology*, Oxford: Elsevier Science Ltd, 1995.
http://dx.doi.org/10.1016/b978-008041920-6/50030-0

[18] Dahlhamer, J.M., Artist, *Small Business and the Whittier Narrows Earthquake: Loan Request Outcomes in the US Small Business Administration Disaster Loan Program for Businesses.* [Art]. Department of Sociology, University of Delaware, 1992.

[19] Runyan, R.C., Small business in the face of crisis: Identifying barriers to recovery from a natural disaster. *Journal of Contingencies and Crisis Management,* **14**, pp. 12–26, 2006.

[20] Corey,C.M. & Deitch, E.A., Factors affecting business recovery immediately after Hurricane Katrina. *Journal of Contingencies and Crisis Management,* **19**(3), pp. 169–181, 2011.
http://dx.doi.org/10.1111/j.1468-5973.2011.00642.x

[21] Asgary, A., Anjum, M. & Azimi, N., Disaster recovery and business continuity after the 2010 flood in Pakistan: Case of small businesses. *International Journal of Disaster Risk Reduction,* **2**, pp. 46–56, 2012.
http://dx.doi.org/10.1016/j.ijdrr.2012.08.001

[22] OSMEP, "SMEs Trend," 2005. [Online]. available at: http://www.sme.go.th.

[23] OSMEP, Situation and structural indicators of SMEs. In *White Paper on Small and Medium Entreprises in Thailand in 2014*, pp. 27–42, 2014.

[24] NSO, Statistical yearbook Thailand 2012. National Statistics Office. Ministry of Information and Communication Technology, Bangkok, 2012.

[25] Yamane, T., *Statistics, An Introductory Analysis*, 2nd ed., Harper and Row: New York 1967.

[26] Tran, P., Marincioni, F., Shaw, R., Sarti, M. & Van An, L., Flood risk management in Central Viet Nam: challenges and potentials. *Natural Hazards,* **46**, pp. 119–138, 2008.
http://dx.doi.org/10.1007/s11069-007-9186-2

[27] Waiyahong, N., Information provision for thai small and medium sized enterprises (SMEs). *Asia Pacific Journal of Library and Information Science,* **2**(2), pp. 117–124, 2012.

[28] Plate, E., Flood risk and flood management. *Journal of Hydrology,* **267**, pp. 2–11, 2002.
http://dx.doi.org/10.1016/S0022-1694(02)00135-X

[29] Renn, O., Perception of risks. *Toxicology Letters,* **149**, pp. 405–413, 2004.
http://dx.doi.org/10.1016/j.toxlet.2003.12.051

[30] Hall, J.W., Dawson, R.J., Sayers, P.B., Rosu, C., Chatterton, J.B. & Deakin, R., A methodology for national-scale flood risk assessment. *Water Maritime Engineering,* **156**, pp. 235–247, 2003.
http://dx.doi.org/10.1680/wame.2003.156.3.235

[31] Wilby, R.L., Beven, K.J. & Reynard, N.S., Climate change and fluvial risk in the UK: more of the same? *Hydrological Processes,* **22**, pp. 2511–2523, 2008.
http://dx.doi.org/10.1002/hyp.6847

[32] Hattermann, F.F., Kundzewicz, Z.W., Huang, S., Vetter, T., Kron, W., Burghoff, O., et al., Flood risk in holistic perspective – observed changes in Germany. In: *Changes in Flood Risk in Europe*, Special Publication No. 10, ed. Z.W. Kundzewicz, IAHS Press, pp. 212–237, 2012.
http://dx.doi.org/10.1201/b12348-14

[33] Kundzewicz, Z., Flood risk assessment–how certain are we? *Climate-Impacts-2013. Org,* pp. 1–10, 2013.

IMPROVING THE UPTAKE OF FLOOD RESILIENCE AT THE INDIVIDUAL PROPERTY LEVEL

CARLY ROSE[1], JESSICA LAMOND[1], MARY DHONAU[2], ROTIMI JOSEPH[3] & DAVID PROVERBS[4]
[1]University of the West of England, UK.
[2]MDA Community Flood Consultants, UK.
[3]Cunningham Lindsey, UK.
[4]Birmingham City University, UK.

ABSTRACT

One of the drivers for improving resilience to flooding at property level (also known as adoption of a 'water entry strategy') was demonstrated by the overtopping of hard engineered flood defences across Cumbria, UK during 'Storm Desmond' in December 2015. Although the uptake of water exclusion strategies (also termed 'flood resistance) is gradually improving in the UK, the longer term resilience options that permit water entry are less popular. Findings from an evidence assessment of the barriers and drivers to uptake of low-cost water entry strategy options are presented. Evidence was collected using a Rapid Evidence Assessment of published materials, semi-structured interviews and workshops with professionals, and a series of case studies of properties that had been adapted to flood risk. Factors leading to successful implementation are identified and illustrated by case study material from homes and small businesses. Water entry and water exclusion approaches were found to be seen as two methods used as part of an integrated approach that can bring about important co-benefits such as improved thermal efficiency and lower energy bills.
Keywords: flooding, flood-proof, flood-repairable, resilience, water entry strategy.

1 INTRODUCTION

One of the drivers for improving resilience to flooding at property level (also known as adoption of a 'water entry strategy') was demonstrated by the overtopping of hard engineered flood defences across Cumbria, UK during 'Storm Desmond' in December 2015 [1]. The uptake of water exclusion strategies (also termed 'flood resistance) is gradually improving in the UK, in large part due to the demonstration projects funded by Defra and the Environment Agency between 2009 and 2012; the longer term resilience options that permit water entry, however, are less well known [2]. Water entry strategy has been defined [3] as:

> *Allow water through property to avoid risk of structural damage* but noting that, for low depths, this strategy can also include *Attempt to keep water out for low depths of flooding*

Flood resilience within a building can be achieved in different ways: for example, avoidance of inundation by raising items such as electrical sockets above expected flood levels; or using resilient materials that do not deform/disintegrate on contact with floodwater, such as cementitious materials. In all cases, it is important to ensure that water can be expelled from the building quickly, and an adequate circulation of air around the exposed elements for reasonably rapid drying must be maintained. For any given building, therefore, there may be multiple possible water entry strategic solutions.

DOI: 10.2495/SAFE-V6-N3-607-615

Despite efforts by multiple agencies in the UK, the tendency of communities at risk to adopt measures to protect their property from flooding is generally low. Various aspects of flood adaptation behaviours have been studied [4–6]; while a study specific to 'resilient' or 'flood repairable' adaptation was conducted [7]. In each case, a complex set of constraints was found to exist that inhibited the necessary behavioural changes. In a review of the international literature on resilience, the findings are grouped in terms of four barrier types: informational, financial, emotional and timing issues [8]; these, in turn, impacted variously on components within the decision making process such as the necessary awareness and perception of risk; ownership of the risk; knowledge of solutions; resources to implement solutions (including financial issues); and belief that the measures would work. It has been found that, although resilient reinstatement did cost more than traditional repair, this approach did significantly reduce future repair costs (assuming a subsequent flood were to take place) [9].

The Department for the Environment, Food and Rural Affairs (Defra) in the UK therefore sought to more fully explore the research gap around the cost and performance of resilient measures [10]. The primary question for this research was formulated as:

> 'How can low-cost adaptation approaches be used in existing residential and small business properties to limit the damage from flood water'.

2 METHOD

First, a scoping exercise was undertaken, followed by consultation and interviews with key stakeholders. Informed by the findings from these stages, a Rapid Evidence Assessment (hereafter REA) of published material on resilience approaches was then conducted using a structured protocol, and encompassing both academic and 'grey' literature sources (such as industry/government publications and guidance). This initial work identified 89 resilience measures in all; costing information was, however, found to be scanty and contradictory in some cases, making it difficult to identify low-cost approaches definitively. Robust scientific evidence on actual performance of measures in flood events was also found to be lacking in many areas, although industry experts have consistently recommended a large number of the approaches concerned, based on their experiential evidence.

In the second phase, input from the professional and practitioner communities was gained by means of semi-structured interviews with 18 individuals. This method captured additional evidence about emerging approaches not represented in published sources, as well as examples of leading practice and reflection on the effectiveness of approaches. Potential literature sources identified by the respondents were located, where possible, and considered for inclusion in the final literature review. A second series of semi-structured interviews was conducted with 13 householders/small business owners, all of whom had already made adaptations to their properties. These interviews explored the experience of individuals at different stages in their adaptation journeys: the process of deciding to adapt; the adaptation activities themselves; and the learning points arising from living in their adapted property, particular the performance of different measures in subsequent flood events.

Third, an interim report summarising the outcomes of the above processes was submitted for critical appraisal by the industry experts who comprised the Project Board overseeing the research. This took the form of a workshop structured around key questions which would be used to inform the subsequent stages of the project. For example, an important issue was how 'low-cost' should be defined: for many of the measures, the financial outlay required could be very different if the work was undertaken as a part of the reinstatement process following flood damage, as opposed to the same work undertaken as a retrofitting option. This, in turn, generated many additional publications which were added to the REA database. The evidence collected from each of the householder/small business interviews were combined with photographic evidence (where available) to form summaries of the key learning points in a case study format.

3 RESULTS

The final list of measures identified numbered 139 in total and covered a wide range of water compatible building elements including: internal walls; flooring/floor coverings; kitchen and bathroom fixtures and fittings; various timber fixtures and fittings; approaches relating to building services and contents protection; and a miscellany of tried and tested hints and tips, largely provided by the householder and small business community. For example, less than £40.00 (48 Euros) is needed to purchase a 'sack truck' that allows items of furniture to be moved more easily when a flood warning has been issued. The review also highlighted the critical importance of understanding the specific detail of proposed measures: materials such as wallboards may bear superficially similar descriptions such as 'moisture resilient' or 'water resilient' but neither of these types is actually designed to be water-proof (boards designed to be fully water-proof exist, but can be considerably more expensive).

Where published studies included performance data, these largely reported the results of experiments on building assemblies, or building components subject to submersion or simulated flooding, on a limited number of materials. However, the properties of any material considered in isolation do not adequately predict resilience within a building setting: the nature and quality of construction, the interfaces between materials, and the drying spaces around them are equally relevant. Whilst there was anecdotal and testimonial evidence suggesting that a number of resilient interventions would be successful in preventing damage, the possibility of bias (towards the reporting of positive results) was likely to be present in sources such as this, as failures are unlikely to be reported.

Evidence of positive performance of some materials was provided by many of the professional interviewees, which largely concurred with current recommendations and is therefore unlikely to be subject to bias. The resilience of some materials traditionally subject to 'stripping out' was highlighted by some of these respondents: for example, several questioned the necessity of replacing (hardwood) timber floors with concrete for most properties. They also reported instances where normally resilient materials had deteriorated, usually after prolonged inundation. This group also emphasised the importance of considering building assemblies, joints and interfaces, rather than individual materials in isolation.

From the households and small business interviews, kitchens units made from both marine-grade plywood and plastic had been found to be resilient. Lime plaster, and sand and cement render, had been successfully used by some respondents. Here again, hardwood elements, together with (heavily) varnished and painted wood were reported to have survived repeated flood events, echoing the professionals' comments regarding the resilience of much timber, which calls into question the commonly accepted practice of 'stripping out' such elements. Households also reported on the success of some of the less frequently recommended measures, such as isolating ground floor circuits from those in the rest of the building (thereby allowing residents to live upstairs during and immediately after a flood); and, as well as moving contents, protecting items by, for example, wrapping in plastic, or using proprietary waterproof bags were recommended. Some issues were also reported, however, including concerns over sanitising contaminated kitchen surfaces; higher than expected flood depths; and failure under hydrostatic pressure. Other issues, such as minor salt accretion, compromised breathability and rusting issues with one steel kitchen appear to be related to existing building conditions rather than flood damage.

The most commonly reported factor that contributes to the desire to adapt property to flooding is flood experience, usually direct experience of flood damage to the property [11]. Householder interviews confirmed that their motivation for adaptation was typically triggered by the desire to avoid further flood damage, with its attendant distress and often lengthy displacement from their homes. It has been found this behavioural driver is strongest in the period immediately following a flood [12]; some professional interviewees, however, had noted a reluctance to slow down the reinstatement/reoccupation process by incorporating adaptations. This means that, even though adaptation has been found to be most cost effective at reinstatement [13] there may be a need to consider additional intervention points within the property life-cycle, such as pre-planned maintenance, or renewal of fixtures and fittings [14]. It has also been suggested that businesses, in particular, may be driven to install measures during insurance renewal [15]. General awareness campaigns, however, were not mentioned as a factor by any of the interviewees: this underscores the importance of utilising innovative awareness raising activities, targeted at the appropriate windows of opportunity.

There are a number of other issues directly impacting upon the most appropriate resilience choices for a specific property which were highlighted by some of the interviewees. These included the likely speed of onset, as surface water or 'flash' floods typically occur very rapidly and often lack any prior warning system; in such circumstances, passive measures represent the most appropriate solutions, as the building's occupants may be unable to put more planned measures into effect. This was reported by one householder, whose flood plan involved actions such as inserting proprietary toilet bungs, and removing internal doors; when the local flood warning system failed to operate late one night, the property was inundated far too rapidly for these measures to be deployed. The type and age of the property is also of importance: for example, historic buildings in the UK are typically subject to legal restrictions regarding alterations to the fabric of the structure, as well as its external appearance, which may mean making resilient alterations poses a greater challenge. Finally, the capacities and preferences of the building occupiers themselves must be taken into account: those with existing mobility issues may have difficulty deploying some measures, while aesthetic considerations, such as the desire to maintain the building's 'normal' appearance, cannot be disregarded.

4 CASE STUDIES

Most of the householders had been flooded more than once, with the worst case being 14 times in 12 years. For some, however, a single flood, or knowledge of the flood risk, had prompted them to install measures; these individuals appear to be atypical and more proactive than the average population at risk, as 6% is the typical proportion found to act in these circumstances [16]. The commonest adaptation among the householders interviewed was the raising of electrical sockets above the anticipated flood levels, with the supply cables running down from above. Tiled floors were frequently adopted (including ceramic, porcelain, reconstituted marble and encaustic); one had chosen solid wood flooring that was 'heavily varnished' (the latter having already survived three inundations unscathed), while another had selected very cheap (sacrificial) carpet tiles to cover a concrete floor. In other respects, however, both the variety and ingenuity of adaptation measures was remarkably wide ranging.

4.1 Case Study 1 – Worcestershire, UK

The first case concerns a property with extensive and comprehensive adaptations, being an 18th century half-timbered cottage which had suffered fluvial flooding twice during the owner's period of occupancy. On the first occasion around 20 cm of water entered the building causing some minor damage and an insurance claim was made. Seven years later, a major flood occurred, with over 1 m of water inside the house: some structural damage resulted, and the owner was advised that extensive rebuilding work would be required. This prompted the owner to investigate what resilience measures could be incorporated during this process, not only to avoid having to go through the 'upheaval and trauma' again, but also due to well-founded fears that insurance cover would become prohibitively expensive in the future.

In addition to many of the common solutions shown in Fig. 1, some innovative measures have been adopted: for example, in the kitchen the dishwasher is not permanently plumbed in, so that it can easily be removed to a safer location when flooding is forecast. An extension to the oiled oak work surface forms a breakfast bar supported by stainless steel legs (as opposed to the more usual chipboard or MDF end panel); the accompanying bar stools are an elegant Italian design, but are made of fully washable plastic material. In the lounge, the owner's extensive book collection is accommodated on shelving supported by wall brackets (rather than traditional free standing bookcases); furthermore, the lowest shelves are easily removable, once the books themselves have been packed away and moved to safety. Many items of furniture on the ground floor are of a lightweight design so they can easily be moved upstairs upon receipt of a flood warning. Long-term sustainability was also an important factor for the owner, with the central heating radiators being not only removable, but also the most energy-efficient design available, to minimise future heating bills. In order to reoccupy the property as soon as possible following a future flood, the upstairs rooms are now on

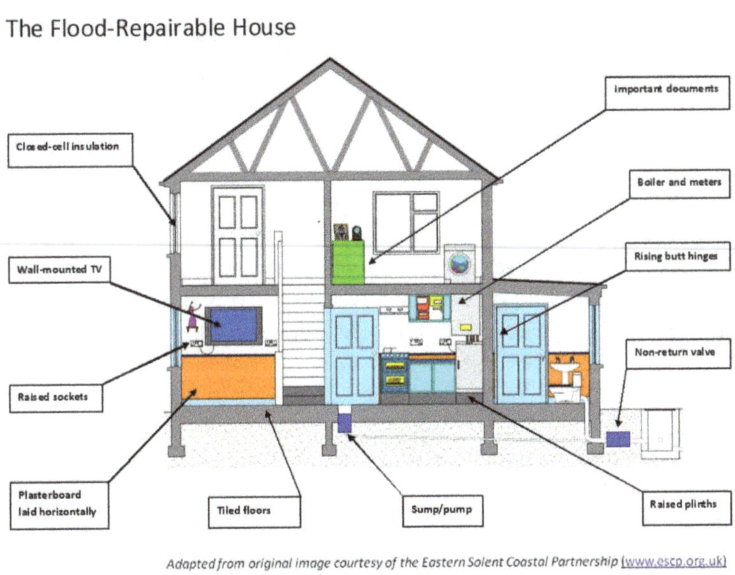

Figure 1: The Flood Repairable House (as adapted from original image, courtesy of the Eastern Solent Coastal Partnership - www.escp.org.uk).

separate circuits from the ground floor: this allows the owner to live in relative comfort upstairs, with heating and lighting unaffected, while the ground floor rooms are being dried out and cleaned.

The owner in this instance had undertaken a considerable amount of research via the Internet in order to identify the most appropriate solutions in costing terms: for example, an open tread style of solid wood staircase was selected, as this cost the same as the standard design made of inferior quality materials, a change that was, therefore, cost neutral to the insurance company. The replacement kitchen units chosen were of powder-coated steel, as this was not only water resilient, but also had an advertised life-span of around 30 years, which the owner felt justified the higher initial outlay. This property, therefore, includes several measures that have co-benefits, as well as meeting the needs of the owner in terms of aesthetic acceptability.

4.2 Case Study 2 – Buckinghamshire, UK

A four-floor 19th century semi-detached property situated in a town with a long history of flooding. The owner was not aware of the flood risk at the point of purchase, however, and as part of the initial renovation work converted the basement level to create a living room and a hi-tech kitchen. The property subsequently suffered two floods within 5 years, and the owner (belatedly) realised the appropriateness of the original Victorian design, in which the basement had housed the servants' quarters, with stone-flagged floors laid on sand. This had, in effect, been flood resilient as water was allowed in, but it also had an escape route following a flood. After the second flood event the owner, who had some practical building experience, then decided to make the house resilient, using what they described as a 'common-sense approach'.

The kitchen floor was re-covered using encaustic tiles affixed with waterproof cement, which can be easily washed down. The kitchen units were constructed from marine ply, which can also be washed out and disinfected; removable kickboards permit the voids below to dry out post flooding. A sump and pump installation is designed to keep the water level within the property as low as possible, and the electricity supply for this comes from the separately zoned upper floor. The free-standing items of furniture in the affected rooms have been selected to be easily raised on stacks of bricks; as a variation on this common method; however, in this case, the bricks are wrapped in plastic bags, to prevent water permeating through bricks and thus affecting the wooden legs of the furniture.

The owner of this property had successfully applied for a government grant to cover the costs of some of the changes made following the second flood, but, as with Study 1, they too had carefully considered the long-term resilient options they wished to adopt. For example, using marine ply to construct kitchen units involves a relatively high initial layout, but was expected to repay the cost in the course of subsequent flood events.

4.3 Case Study 3 – Yorkshire

A small Victorian property which the owner has rented to the current tenant for four years, during which period four 'flash 'floods have occurred (typically with very little warning time, owing to the steepness of the local catchment). One of the reasons for opting for a resilient approach is that the occupant has several pets, making it difficult to obtain alternative accommodation following flooding (as many landlords will not allow animals in their properties). Both parties agreed upon a programme of alterations that was acceptable both in cost terms (to the owner) and on aesthetic and practical terms (to the tenant). As with Study 1, the house has been rewired with a separate circuit serving the upper floor, enabling the occupant to continue living in the property while the lower floors are being dried and cleaned. A modestly priced (and therefore sacrificial) carpet has been used to cover the concrete floor of the living room, but the tenant uses a good quality, easily removable rug on top of this. As the original front door had been coming towards the end of its useful life, it was replaced by a low maintenance and draught-free UPVC version, benefitting both the tenant and property owner alike. Finally, the tenant chose to change the way in which some of the rooms were used, including ceasing to store items in the cellar.

5 DISCUSSION

There seemed to be a consensus, from both professionals and householders, that the water entry approach is about damage limitation, with water *exclusion* strategies remaining the preferred option. Respondents from both groups favoured a combination of measures to limit water entry, as well as incorporating repairable internal features. This suggests that repairable approaches might usefully be reappraised: instead of being seen as the alternative of last resort, adoption could be promoted as being useful within any property level scheme. Stressing co-benefits could be helpful when recommending adaptations: for example, the householder who had purchased stylish Italian furniture, enhanced her interior décor, as well as ensuring the items were lightweight and resilient. Similarly, raised sockets can be convenient in the longer term, as they are more easily accessed by older people, or those with mobility issues; both waterproof insulation materials and UPVC door and window frames offer improved thermal efficiency and, thus, lower energy bills.

6 CONCLUSIONS

The weight of evidence suggests that low-cost approaches can be used to prevent some of the damage from floodwaters entering the home and increasing their uptake could save money for households, small businesses and their insurers. Of the 138 measures identified, over 70 were deemed to be ‘low-cost’ or ‘affordable’, depending upon the point in the property life-cycle when the adaptations are made (for example, during post-flood repair work, or as part of other property renovations or repairs). Although scientific evidence of performance is lacking in some cases, industry experts have consistently recommended many of the approaches examined; more research is, therefore, needed to gather improved evidence of effectiveness. This will need to combine laboratory testing, testimonial and documentary evidence and post flood damage assessments, not only of individual materials, but with multiple components as used in building assemblies. The majority of the householder interviewees, and many professionals, emphasised that more information and guidance was needed in this area, supporting the view that overcoming informational barriers will be critical in improving uptake. Although the extant literature and guidance often make a binary distinction between water entry and water exclusion, interviewees saw the two methods as forming important elements of an integrated approach.

ACKNOWLEDGEMENT

This research is funded by Defra under project reference FD2682.

REFERENCES

[1] Balmforth, D., Why did flood defences fail against Storm Desmond? 2015 16 December, 2015 [cited 2015; available at: http://www.ice.org.uk/media-and-policy/the-infrastructure-blog/december-2015/why-did-flood-defences-fail-against-storm-desmond.

[2] JBA Consulting, *Establishing the Cost Effectiveness of Property Flood Protection: FD2657 - Final Report*, Defra: London, 2013.

[3] Tagg, A., Escarameia, M. & Ortiz, J.M., Improving the flood resilience of buildings through improved materials, methods and details - Report WP6 - collation and analysis of post-flood observational data, in DCLG Building Regulations (Sanitation) Framework. 2007.

[4] Thurston, N., Finlinson, B., Breakspear, R., Williams, N., Shaw, J. & Chatterton, J., *Developing the Evidence Base for Flood Resistance and Resilience - R&D Technical Report FD2607/TR*, Defra: London. p. 24, 2008.

[5] Spence, A., Poortinga, W., Butler, C. & Pidgeon, N.F., Perceptions of climate change and willingness to save energy related to flood experience. *Nature Climate Change*, **1**, pp. 46–49, 2011.
http://dx.doi.org/10.1038/nclimate1059

[6] Bubeck, P., Botzen, W.J.W. & Aerts, J.C.J.H., A review of risk perceptions and other factors that influence flood mitigation behavior. *Risk Analysis*, **32**(9), pp. 1481–1495, 2012.
http://dx.doi.org/10.1111/j.1539-6924.2011.01783.x

[7] Wassell, P., Ayton-Robinson, R., Robinson, D., Joseph, R., Hack, K., Butler, D., et al., *Resilient Reinstatement – the Cost of Flood Resilient Reinstatement of Domestic Properties (ABI Research Paper no 14, 2009)*. In *Research Paper*, Association of British Insurers: London, 2009.

[8] Lamond, J.E. & Proverbs, D.G., Resilience to flooding: lessons learned from an international comparison. *Proceedings of the Institution of Civil Engineers – Urban Design and Planning*, **162**(2), pp. 63–70, 2009.
http://dx.doi.org/10.1680/udap.2009.162.2.63

[9] Joseph, R., Proverbs, D., Lamond, J. & Wassell, P., An analysis of the costs of resilient reinstatement of flood affected properties. *Structural Survey*, **29**(4), pp. 279–293, 2011.
http://dx.doi.org/10.1108/02630801111162350

[10] Defra. *Low Cost Resilience – FD2682*. 2015, available at: http://sciencesearch.defra.gov.uk/Default.aspx?Menu=Menu&Module=More&Location=None&Completed=0&ProjectID=19221.

[11] Rose, C.B., Proverbs, D.G., Booth, C.A. & Manktelow, K.I., Three times is enemy action – flood experiences and flood perceptions. In: *FRIAR III (3rd International Conference on Flood Recovery, Innovation and Response)*, eds. D. Proverbs, S. Mambretti, C.A. Brebbia & D. de Wrachien, WIT Press: Dubrovnik, Croatia. pp. 233–242, 2012.

[12] Steinführer, A., Local communities at risk from flooding: social vulnerability, resilience and recommendations for flood risk management in Europe, Helmholtz centre for environmental research – UFZ Leipzig, 2009.

[13] Joseph, R., Proverbs, D. & Lamond, J., Resilient reinstatement: what can we learn from the 2007 flooding in England? *Flood Recovery, Innovation and Response IV*, **184**, p. 175, 2014.

[14] Soetanto, R., Proverbs, D.G., Lamond, J. & Samwinga, V., Residential properties in England and Wales: an evaluation of repair strategies towards attaining flood resilience. In *Hazards and the Built Environment: Attaining Built-in Resilience*, ed. L. Boscher, Taylor and Francis, pp. 124–149, 2008.

[15] Lamond, J., Dhonau, M., Rose, C. & Proverbs, D., Overcoming the barriers to installing property level flood protection – an overview of successful case studies. In *Road Map*

Towards a Flood Resilient Urban Environment, Urban Flood Management Cost Action Network C22: Paris, 2009.

[16] Harries, T., Public perceptions and adaptations to flood risk. In *Flood Repair Network Workshop no 4, 08/05/08*, ed. D.G. Proverbs, Embassy House: Birmingham UK, 2008.

DESIGN *WITH* FLOODS: FROM DEFENCE AGAINST A 'NATURAL' THREAT TO ADAPTATION TO A HUMAN-NATURAL PROCESS

L. HOBEICA[1] & P. SANTOS[2]
[1]Institute for Interdisciplinary Research, University of Coimbra, Portugal.
[2]Centre for Social Studies, University of Coimbra, Portugal.

ABSTRACT

Recognizing that traditional flood management interventions focus on defence, attempting to eliminate contingencies in the urban relationship with rivers, an emergent perspective, spearheaded by spatial design, seeks to deal with floods through a more holistic framework. In contrast to the prevalent 'design *against* floods' approach that targets either the hazard or the exposure components of flood risk, 'design *with* floods' focuses as well on the assets at stake (including the built envelopes of exposed people and activities, usually covered under the term vulnerability), duly acknowledging the intertwining of natural and human processes. Using a multiple case study comprising three European flood-prone urban projects, we explore potentials of spatial design as an adaptation tool that goes beyond flood protection to foster wider societal gains. Our analyses have so far suggested that 'design *with* floods' requires a positive stance through which problem-solving and sense-making approaches are merged to provide both safety and urbanity (enriched urban realm and experience), without eliminating floods per se, accepted as a complex hybrid process.

Keywords: flood adaptation, riverine urban projects, spatial design, urban regeneration.

1 INTRODUCTION

Giving space for river waters as a flood mitigation strategy is a noticeable growing trend [1], which partially recognizes the limits of traditional defensive approaches that restrain either the river (as the hazard source) or the urban occupancy of floodplains (exposure), in the context of greater uncertainties heralded by climate change. At the same time, other environmental concerns (in part linked to the mitigation of climate change itself), such as preventing urban sprawl, have been pressing the redevelopment of floodable urban areas that have become underused in the post-industrial era [2]. In this sense, the adaptation of the new exposed structures, as a more flexible, long-term and no-regret strategy, has been advocated to reconcile the demands of flood risk management and urban development, which have clear spatial expressions [2, 3]. With its general aim of enhancing existing conditions through the manipulation of form and substance of space at different scales, spatial design, meant to be an interdisciplinary professional field, can thus be a valuable tool to reach that end. But what can make spatial design successful in pursuing such a task? In this paper, we explore the relationships between spatial design and flood risk adaptation in the redevelopment of riverine urban areas, based on the analysis of three European flood-prone urban regeneration projects.

DOI: 10.2495/SAFE-V6-N3-616-626

After briefly introducing the challenges of regenerating flood-prone urban areas (Section 2), the paper presents adaptation as a flood conciliation strategy (Section 3) and spatial design as a means to foster a broader flood-adaptive approach (Section 4). The three studied cases are then briefly introduced (Section 5), followed by some general lessons that can be learned from them (Section 6). The raised insights lead us to propose a scale of flood adaptation by design (Section 7), before ending the paper with some general implications of and prospects for 'design *with* floods' (Section 8).

2 URBAN REGENERATION AND THE NEED TO DEAL WITH INTRA-URBAN FLOODABLE SPACES

As a contemporary feature of the post-industrial world, urban regeneration refers to intentionally organised initiatives to boost new life into city segments when their intrinsic vitality has given signs of decline. Despite their great inertia, cities are indeed dynamic systems that are always evolving and changing their scope and atmosphere, while some basic hard structures remain as their most long-lasting elements (even though the use and meaning of these also shift across time). Yet, when the inherent urban vigour is hampered by adverse socioeconomic trends, the regeneration of urban areas requires ad hoc strategic interventions that often take the form of comprehensive urban projects.

Being closely linked to the identity of riverine cities, riverfronts play a special role in urban regeneration initiatives, since the open horizons and the sense of naturalness attached to them currently represent valuable urban assets. At the same time, many riverine urban spaces were once a privileged location for industries and transport infrastructures now in decline, configuring actual 'bad places' in terms of remaining risks (notably floods and pollution) [4], which nevertheless need to be brought to urban life again. In fact, such brownfields are probably instead *excellent* places if other economic, landscape and social potentials are also considered; and this is perhaps the main reason why their redevelopment, even when they are prone to floods, is often considered as necessary [2, 3, 5]. Revamping such areas is definitely in line with the human historic settlement pattern of living by water bodies, which constitutes an important cultural heritage in itself. The regeneration of riverine urban spaces is thus a complex issue; yet it should be noted that flood proneness represents only one feature of these ever changing spaces, in which multiple and conflicting demands frequently overlay. Floods indeed represent one-off contingencies (although with probable serious repercussions) when reasonably compared with other more immediate and regular urban needs [3, 6]. This does not mean that urban redevelopment should be carried out ignoring floods as a real threat, but that it should be pursued in an integrated fashion, in which overall urban life enhancement and flood protection merge into a single goal.

Standing 'at the point where nature and artifice meet' [7], cities are much more than mere concentrations of people and assets, as traditionally seen through strict flood management lenses. Presently home to half of humanity, cities – '*the* human invention *par excellence*' [7] – should provide inspiring environments in which human life can positively unfold. Therefore, at stake in cities are not only the technical issues related to life protection in a narrow sense, but also wider necessities of the being that include experiencing life in its wholeness. In this connection, the greatest challenge in redeveloping flood-prone riverine spaces is to balance existing and envisaged short- and long-term constraints and opportunities, while keeping cities' identity and sense of place (their human face), without compromising human life in its broadest sense. The urban regeneration of such spaces thus conforms a timely opportunity to review current concepts and practices regarding dealing with floods in riverine cities.

3 FROM FLOOD DEFENCE TO FLOOD ADAPTATION

Perhaps a first step towards a more inclusive approach to floods is to fully recognize them as hybrids, complex phenomena in which natural and human processes converge. Obviously, fluvial floods are not only the result of river overflows, but mostly of the confrontation of its water with human interventions (including on the river system itself), valuable assets and ultimately human life. Therefore, in spite of being included among 'natural' (hydrometeorological) risks, floods are increasingly driven by human actions, which have the power to impact equally on the hazard, exposure and vulnerability risk components. In this vein, people cannot be taken merely as fragile passive elements to be protected but also as active co-producers of flood events.

Defensive approaches to flood management, which generally aim to control either the hazard or the exposure (or both – see Fig. 1), have largely overlooked the existing feedback loops between hydrological and social processes that lead to flood events [8]. Consequently, despite centuries of flood prevention practices, flood risk as a whole has been steadily increasing. At the same time, climate change has been challenging such defensive approaches [3], which have mostly been built and developed under the paradigm of certainty. In this sense, some flood management initiatives have started to put more emphasis on precautionary measures targeting either the exposure or the vulnerability risk components, envisaged as a no-regret, more robust approach to face climate change uncertainties [3]. Accordingly, it seems that the human approach to floods has started to change: instead of endeavouring to prevent the unpreventable, the focus has been shifting towards the acceptance of flood processes, in a 'living with floods' perspective that nevertheless keeps people's safety as its prime goal [2].

While in defensive approaches floods are normally taken solely as a threat, a technical problem to be solved, 'living with floods' as an adaptive perspective entails also dealing with floods through a more positive stance. Adaptation refers thus to proactive changes in the predominant mindset, to include the recognition and exploitation of possible societal opportunities that may arise from the acceptance of flood processes. At first sight, it is not really easy to grasp any human advantage deriving from urban floods; yet, regular and minor tangible flood experiences reminding that absolute prevention is unattainable can for example be of a great value in order to plan for possible emergency situations [3]. On the other hand, these quasi disasters can have an intangible effect of repositioning cities and humankind as regards the cyclical flows of life and the unforeseen, which is totally fine-tuned with climate change prospects.

An adaptive attitude thus incurs the active anticipation of hazardous events, comprising the control of the potential disaster situations (not of the hazard itself), and even the tolerance of

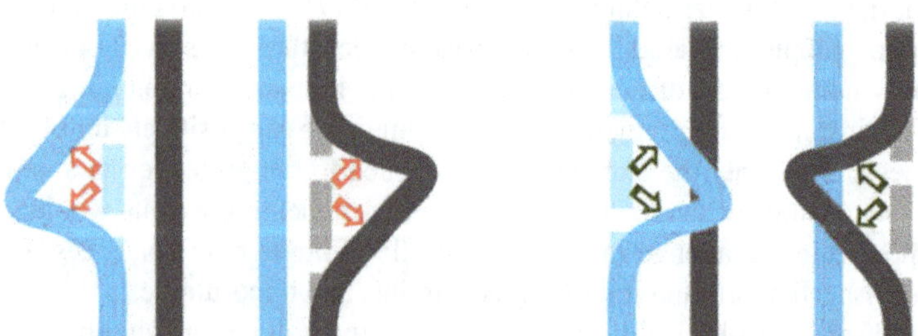

Figure 1: From defensive (left) to adaptive (right) strategies to floods.

some disturbances from time to time. Although a controlling perspective is still present – and must always be, whenever people's life is at stake –, it is not river processes that are subject to it, but a more complex entity that integrates water and human dynamics. Adaptation certainly requires a thorough understanding of the flooding conditions at two distinct and complementary scales. At a lower scale, it is important to grasp the hydrologic behaviour of the watershed as a whole (in particular the rainfall-runoff relationship), which continually interplays with human-driven impacts such as land-use changes, soil erosion, or the introduction of retention ponds or dams. At a higher scale, hydraulic determining factors (topography, soil permeability, underground flows, surface roughness, physical barriers and flood defence functioning) translate the fluvial flow into the floodable area at the floodplain level, being more directly influenced by urban project interventions.

Nonetheless, adaptation should not be seen as a one-size-fits-all strategy that will finally free cities from nuisances related to flood risk [2]; its limitations lie exactly in the very hybrid trait of floods. For example, rapid-onset riverine floods surely pose additional challenges to 'living with floods', as the water depth, velocity and sediment load might imply too dangerous conditions that should, of course, be kept away from people's experience. Also, since this acceptance perspective goes in the opposite direction of the prevalent flood defence strategies, strong cultural barriers and organizational inertia should not be overlooked. Such social constraints may indeed be as important as river dynamics themselves to embrace an adaptive track.

4 SPATIAL DESIGN AS A MEANS TO FOSTER A FLOOD ADAPTIVE APPROACH

As a human-natural process that is permanently being constructed, fluvial floods could be regarded as an 'unconsciously designed' condition. Yet, design as an activity largely refers to the mental plan or the deliberate conception of artefacts (in terms of both form and substance) that aim to make human life and activities easier; meanwhile, the term design covers both the process of reaching such artefacts (design as a means) and the final output itself. The core of design has been to tackle complex issues through the adoption of solution-focused strategies, which may indeed challenge the problems initially posed. From a spatial design point of view, floods should be dealt with differently in regard to other site constraints such as noise or ground pollution. While the latter may be taken only as technical issues calling for solutions, floods potentially have positive aspects to be environmentally, socially and aesthetically explored. At the same time, it is noteworthy that when traditional spatial design restrictions are dealt with through an approach that is not strictly technical and reduced to efficiency, the ensuing design response tends to be more sensitive and innovative, adding value and meaning to such spaces. For example, London's future Garden Bridge was designed by Thomas Heatherwick and Arup not only as a crossing over the Thames River but also as an urban landmark and a linear garden in which biodiversity will be allowed to flourish [9].

Considering that it is possible to design 'in a unitary mode, beyond simple dichotomies of nature and culture', Prominski proposes the concept of '*and*scape' to refer to the state or the condition of unity, which takes into account the overall dynamic relationships between living and non-living organisms and their environment, all of them encompassed within our cultural-natural world [10]. The notion of '*and*scape' thus seems to go hand-in-hand with the emerging field of socio-hydrology [11], focused on the intertwining character and co-evolution of social and hydrological processes. Although being primarily conceived to address and communicate 'the synthetical, integrative character of landscape architectural design' [10], '*and*scape' can be a suitable guiding concept to all design situations in which nature and culture come together.

Starting from such conceptual framework, we propose to explore a 'design *and* floods' stance, examining the possibility of a reflexive relationship between these two processes, in which they would feed one another. Therefore, 'design *and* floods' contrasts with the dominant 'design *against* floods' mindset, usually geared towards the development of defensive structures that create barriers and even lower the sense of urbanity. It also aims to expand the 'design *for* flood risk' alternative [5], which still put floods (and keeps the negative term *risk*) in an unnecessary central position in relationship to other urban issues and does not really shed light on floods' human dimensions. In fact, both 'design *against* floods' and 'design *for* flood risk' can be equated to problem-solving approaches that somehow understate the design remit of uncovering new issues and raising new meanings to usual situations. We thus argue that 'design *and* floods' should embrace such a holistic and not strictly technical endeavour, through a positive adaptive stance. In this sense, besides acknowledging flood as a hybrid, it requires to fully recognize urban spaces as places, that is, territories with meaning to human life. Hence safety and urbanity (place-making) should be at the core of 'design *and* floods', which implies the challenge of strengthening visual and physical links between cities and urban rivers, and even providing access to water whenever possible [5].

5 'DESIGN *AND* FLOODS': THREE URBAN PROJECTS

To explore possible synergies of 'design *and* floods', our research has focused on flood-prone urban regeneration projects (and not on flood management projects that, due to their location and scope, happen to have multiple functions and present an urban character). The three studied cases were intentionally selected to cover diverse situations (see Fig. 2), and comprise:

- an intra-urban riverine green park along both riverbanks of the Mondego River, in Coimbra, Portugal (already partially implemented);
- a flood-compatible intra-urban neighbourhood nearby the Garonne River, in Bordeaux, France (detailed design is still ongoing); and
- an intra-urban flood defence infrastructure bordering the Scheldt River, in Antwerp, Belgium (implementation is ongoing).

It was only in the 1980s, when a system of dams and river training works were finalized, that the Mondego River lost its highly fluctuating nature, which had up to then secured large unoccupied intra-city parcels. A new stable landscape, with a permanent storage pond in the centre of Coimbra, was then the centrepiece of a regional park project, spanning both

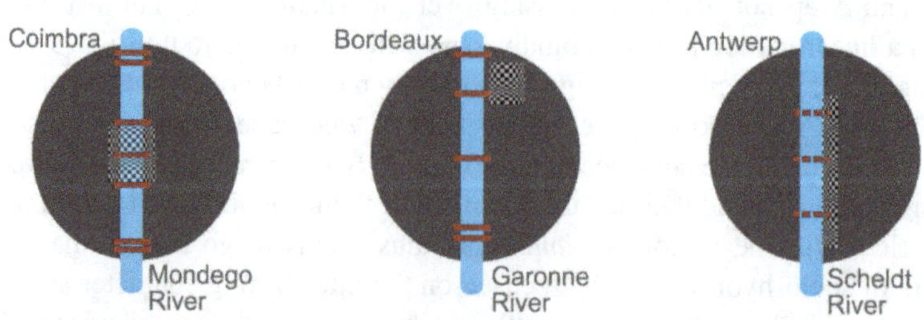

Figure 2: Spatial characterisation diagram of the three studied interventions.

riverbanks. An international competition in the mid-1990s, won by the MVCC agency, has initiated a long-term process of urban riverfront creation. The competition brief overlooked the susceptibility of the area to floods, probably due to the newly implemented regulation works and to the fact that the envisaged project was a green urban park, mostly with lawn areas and some ancillary buildings. The idea of floodplain was indeed taken as obsolete: the site was then referred to as 'the old floodplain of the Mondego River' [12]. Yet, in 2001, when a first section of the park was being implemented, an 'unexpected' flood event revealed that such perception was biased. This event has triggered structural and programmatic changes in the design output, while the attainment of a fixed safety level was the main recommendation; the previous strategy of fostering the lowest exposure was kept. A singular green urban riverfront was subsequently achieved, marked by its close proximity to the river, and has since been intensively used; yet, although the river is allowed to (re)occupy the prairie during its occasional overflows, water was overall treated as a static element of the urban scenery.

Since the 1990s, Bordeaux has been trying to reconquer the Garonne riverbanks, a process that started on the left bank and has more recently expanded to the right one. The construction of a new bridge linking two major urban redevelopment zones in the North brought to light the strengths of Brazza, a typical declining industrial urban fringe, rightly in the centre of Greater Bordeaux. From 2010 on, Brazza has been subject to a sequence of planning phases, with flood proneness being recognized by the city authorities as one of its main constraints, besides ground pollution. In contrast to most of Bordeaux's riverbanks, this area is not protected by a dyke; therefore, adapting to floods has been taken as a driving concept behind all the design proposals, the latest one elaborated by a team led by the YTAA agency. Anticipating upcoming (stricter) flood regulations (the current ones are considered outdated), two main concepts were adopted to adapt the new neighbourhood to floods: minimal alteration of the local topography, and hydraulic transparency of and between buildings. Such approach aims at the same time to protect the site's future population and not to engender greater flood risk to adjacent urban areas. In terms of urban design, this is reflected in three wide green strips perpendicular to the river, which allow nature (including the river overflows) to enter into the quarter; two main building typologies (on stilts and with floodable undercrofts) also favour such encounters with minimal damages. These simple and passive solutions were driven by the logic that 'water must be seen' within the neighbourhood during floods, as stressed by the involved lead engineer [13].

The 2005 update of the Sigma flood protection plan for the whole Scheldt estuary, with its foreseen physical barriers putting under threat the remaining contact between city and river, called for a reinterpretation of Antwerp's wasted land along the right bank. A restricted international competition was jointly promoted in 2006 by the city of Antwerp and Waterwegen en Zeekanaal NV (the Flemish waterway administration agency) to gather insights on how to equate urban character (the riverfront's sense of place) and flood defence, while the existing quay wall should be treated as an urban heritage. Considering that behind every problem 'lies an opportunity to prepare a project, and the bigger and more complex the problem is, the more extraordinary the opportunities are for finding unusual solutions' [14], the PROAP-led winning team put forward a flexible proposal for the reclamation of the riverfront, with some guiding principles instead of a defined master plan. Spatially adapted to the river's dynamics, this proposal tackles multi-objectives at once and is, by nature, eventually adaptable to ever-changing societal needs. An open design process has led to an output that is a composite defensive structure – a civil element, consisting of dykes, floodwalls and movable devices –, which nevertheless does not constitute a physical or visual barrier. On the contrary, it functions as well as a civic urban element [15], a riverine promenade through which the relationships between city and river are renegotiated, while the moods of the river can be further experienced (for instance, increasing the accessibility to the water).

6 FROM 'DESIGN *AND* FLOODS' TO 'DESIGN *WITH* FLOODS'

These three cases exemplify a 'design *and* floods' perspective, even if the intensity of this interaction is quite different in each of them. Of course, their overall context and flood processes are not comparable (fluvial floods in Bordeaux and Antwerp are also influenced by tides and storm surges), nor are the related design outputs. Yet, the local stance towards floods may be taken as the key feature that somehow connects them, being thus a significant aspect of pursuing 'design *and* floods' as an integrated approach. Considering that it underlies the design context, process and product of these urban projects, the stance towards floods was then explored, among others, through the contents of the design briefs, the profiles of key involved actors and the actual proposed layouts. As already stressed, the very traits of the flood risk also shape such stance.

Regarding the three competition briefs, while floods were rather underestimated in the case of the green park around the Mondego, they were in contrast highlighted as an important site constraint for redeveloping the Brazza sector; for the Scheldt quays, floods were even the *raison d'être* of the ambitious local regeneration project. The fact that this initiative has been partaken by a waterway agency potentially anticipates an integrated planning process. As a consequence, in this case floods could be taken as an opportunity to envision the quays as a decompressing space between city and river [14], in which both could temporarily and plentifully manifest themselves. In the Coimbra case, the susceptibility of the riverbanks to floods was also perceived as the opportunity that had guaranteed such a large undeveloped parcel within the city centre; yet, such opportunity was only partially capitalized by the design of the green park.

On the other hand, floods were not exactly viewed as an opportunity in Bordeaux, but pragmatically taken as a condition that had to be duly accounted for to make feasible the revamping of that area, as formally required by existing flood regulations. In fact, this is the only studied case in which permanent urban activities (including housing) are planned in a floodable zone, while in the other two mostly temporary uses are envisaged, in recognition of the transient nature of riverine urban spaces. In any case, the proactive attitude towards floods shown by the local authorities in Bordeaux results from both the vast intra-urban floodable area under regeneration (the scope of the challenge has been an important driver for flood adaptation) and the gradual development of in-house expertise on the subject (a major asset that is also possible to identify within the Brazza design team). Actually, two complementary factors were fundamental for raising awareness to floods: the overall impacts of the storm Xynthia (2010) on the French Atlantic coast and, more locally, the results of an archaeogeographical study for the Garonne right bank (2010), which recalled the strong role that water had held in configuring such territory [16].

All three cases had an explicit problem to solve involving flood management and urban development; as such, the three ensuing design outputs present simple layouts that withstand flood events, responding as well to other issues such as implementation costs or operationalization. Yet, in Antwerp and Bordeaux, the problem-solving approach was clearly geared towards a more complex and process-oriented perspective, as witnessed by the openness of the design practice: the final outputs were not fully anticipated in both cases, being allowed to evolve along with societal needs. For example, the elaboration of design proposals for the Brazza quarter has been supported by several rounds of iterative investigations of spatial alternatives coupled with flood modelling, gathering diverse categories of spatial designers (engineers, urbanists, architects and landscape architects). Such dynamic design processes could have been encouraged by the fact that these two cities already experience on a daily basis river variations due to sea tides, which keep people's awareness of riverfronts as an unstable (and possibly vibrant) urban setting. Conversely, in Coimbra, the Mondego's seasonal variations – which in the past were more intense and perceived as undesirable – were drastically pacified after regulation works, leading to the fading of the pre-existing flood risk culture [17]. The new apparent 'stable' condition of the river has likely induced an underestimation of flood-related human and natural processes, namely the unmanaged silting problem, which is presently the main factor behind more intense damages within the green park and the nearby areas traditionally exposed to floods.

Although arising from a coupled flood management and urban development demand, 'design *and* floods' may not correspond to an actual intertwining of these two processes, which can be perceived and performed as parallel ones, as shown by the Coimbra case. As a result, since the accomplishment of the green park, flood events (even minor ones) are locally misinterpreted as a failure of both the management of the dams' system and the park's design. Conversely, in a deeper 'design *and* floods' perspective, experiencing minor and manageable floods is taken as a mechanism for preserving the much needed flood risk culture [3]. In fact, spatial design is not a self-referencing tool: the goal of bringing together design and floods must be intentionally pursued by local authorities, territorial players and all stakeholders involved in the redevelopment of a flood-prone urban area. For example, in the cases where such interlinkages were explicitly required by the commissioners since the design brief, 'design *and* floods' has indeed extrapolated the problem-solving approach to reach a more thorough adaptive stance that fosters the sense of urbanity, enriching thus the urban realm and experience with floods. When the design task was performed as a dynamic endeavour that consciously integrated flood processes themselves as a compositional element, 'design *and* floods' eventually evolved into a more intense 'design *with* floods', in which these two processes somehow transcend their respective precincts to synergistically merge as one comprehensive process.

7 DEGREES OF FLOOD ADAPTATION BY DESIGN

The three analysed cases have clearly shown that 'design *and* floods' can never be a straightforward stance, being completely dependent on the existing human-natural context behind each particular urban project (i.e., on people, place and processes). In any case, 'design *and* floods' remains a search for an acceptable and sustainable compromise between safety and urbanity. In this sense, we have carried out an extrapolation exercise, based on the lessons that could be learned from the cases, as a means to propose a tentative scale to qualify the different degrees of flood adaptation by design. Such a scale does not intend to express various levels of urbanity and safety by design, but the growing intensity of the interactions between these two design components (see Fig. 3):

- **Tolerating floods**: the most basic level of 'design *and* floods', with no clear signs of interaction. In the Coimbra case, despite the high level of urbanity attained, not much room was left for experiencing river dynamics; instead, a sense of resignation seems to have led to the tolerance of water overflow within the new park area.

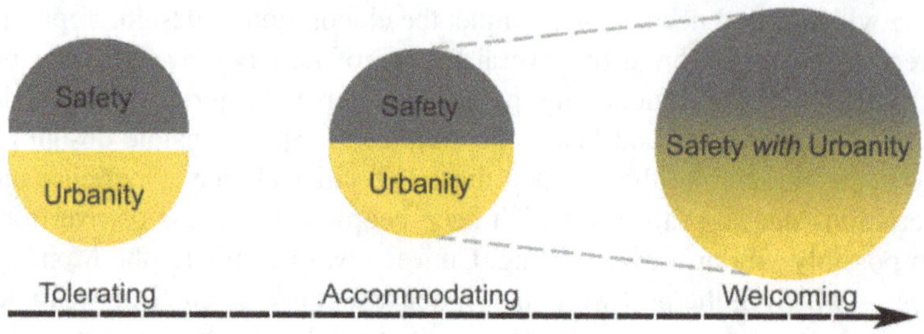

Figure 3: Different degrees of flood adaptation by design – from coexistence (left) to integration (centre) to synergy (right).

- **Accommodating floods**: from this point on, we can distinguish a 'design *with* floods' approach, with visible interactions between the two dimensions. This is best illustrated in the Bordeaux case, through its pragmatic pursuit of hydraulic transparency (preserving thus the space needed for the occasional Garonne overflows).
- **Welcoming floods**: the most challenging level, given spatial and cultural restrictions, yet in which '*and*scape' can be exercised at its best. A synergetic stance is thus clearly portrayed. In the Antwerp case, even if a defensive device was the intended design output, the welcoming perspective could unexpectedly be attained, thanks to the fostered close and intense relationships between design and floods.

In any of these adaptation levels, damage and disturbance are not impeded from happening, although they should be kept to a minimum; yet, at the higher end of the scale (welcoming floods), overall benefit is also obtained: for instance, disturbance introduces a new temporary wet landscape, resulting in a more diversified and dynamic urban realm. According to the specificities of each case, tolerating, accommodating and welcoming floods can nevertheless coexist in the same urban project; for example, in Brazza, having urban agriculture plots placed under the buildings on stilts could be understood as a welcoming floods sign, since they can somehow benefit from the nutrients sporadically brought by the river. Also, welcoming floods cannot be directly taken as a desired approach, since it may be unfeasible in some circumstances. In any case, this is probably the most difficult level for spatial designers, since welcome implies a double design request that anticipates both dry and wet scenarios. Such spatial flexibility, which must be the rule as a means to face uncertain futures, is only attainable through a truly interdisciplinary design practice.

8 FINAL CONSIDERATIONS

'Design *with* floods' ideally means futureproofing riverine cities vis-à-vis the climate change prospects, requiring thus blurring barriers, not only with respect to human and natural processes, but also disciplinary fields. In fact, spatial design can only be successful in its task if carried out as a negotiation platform in which all the concerned stakeholders act consciously as 'flood designers'. This is certainly not yet the mainstream; but just as other environmental issues that are today taken for granted have required decades to be internalized into planning practice, time has now come for the gradual improvement of riverine urban projects through the positive incorporation of floods [13, 16]. The analysed cases indicated that the promotion of flood risk culture is at the same time a condition for and a consequence of well-designed flood-adapted urban projects. A proactive flood risk culture, acting as a catalyst of involved stakeholders (including spatial designers), can indeed promote the proper understanding of the flood phenomenon and the existing urban vulnerabilities, as well as higher degrees of flood adaptation, leading to safer, more liveable and more meaningful urban places.

ACKNOWLEDGEMENTS

This study was partially funded by the Portuguese Foundation for Science and Technology, through a PhD grant. The authors are thankful to all designers and territorial players who generously provided information to this research.

REFERENCES

[1] European Union, Directive 2007/60/EC of the European Parliament and of the Council of 23 October 2007 on the assessment and management of flood risks.

[2] Centre Européen de Prévention du Risque d'Inondation (CEPRI), *Comment Saisir Les Opérations De Renouvellement urbain pour réduire la vulnérabilité des territoires inondables face au risque d'inondation?* Available at http://www.cepri.net/tl_files/ Guides CEPRI/CEPRI rapport principe amenagt.pdf, 2015.

[3] Jha, A., Bloch, R. & Lamond, J., *Cities and Flooding: a Guide to Integrated Urban Flood Risk Management for the 21st Century*, World Bank: Washington DC, 2012. http://dx.doi.org/10.1596/978-0-8213-8866-2

[4] Viganò, P., Extreme cities and bad places. *International Journal of Disaster Risk Science*, **3**(1), pp. 3–10, 2012. http://dx.doi.org/10.1007/s13753-012-0002-6

[5] Royal Institute of British Architects (RIBA), *Designing for Flood Risk*, RIBA: London, 2009.

[6] Rossano, F. & Hobeica, L., Design as negotiation platform: new deals and spatial adaptation in flood-prone areas. *WIT Transactions on Ecology and the Environment*, **184**, pp. 287–298, 2014. http://dx.doi.org/10.2495/FRIAR140241

[7] Lévi-Strauss, C., *Tristes Tropiques*, Criterion: New York, p. 127, 1961.

[8] Di Baldassarre, G., Kooy, M., Kemerink, J. & Brandimarte, L., Towards understanding the dynamic behaviour of floodplains as human-water systems. *Hydrology and Earth System Sciences*, **17**, pp. 3235–3244, 2013. http://dx.doi.org/10.5194/hess-17-3235-2013

[9] Arup, http://www.arup.com/Projects/Garden_Bridge.aspx, 2015.

[10] Prominski, M., Andscapes: concepts of nature and culture for landscape architecture in the 'Anthropocene'. *Journal of Landscape Architecture*, **9**(1), pp. 6–19, 2014. http://dx.doi.org/10.1080/18626033.2014.898819

[11] Sivapalan, M., Savenije, H. & Blöschl, G., Socio-hydrology: a new science of people and water. *Hydrological Processes*, **26**, pp. 1270–1276, 2012. http://dx.doi.org/10.1002/hyp.8426

[12] Assessoria de Planeamento e Ordenamento do Território da Câmara Municipal de Coimbra. *Plano de Pormenor do Parque Verde do Mondego*, Coimbra [not published], p. 1, 1999.

[13] Dunogier, C., Personal communication, Engineer, head of the infrastructure department at Ingérop's Bordeaux agency, Bordeaux, France, 2014.

[14] Nunes, J., Master plan for Antwerp waterfront. *Portus*, **15**, pp. 68–75, 2008.

[15] De Meulder, B., Personal communication, 5 October 2013, Architect-Urbanist, former partner of WIT Architecten, Louvain, Belgium.
[16] Scheurer, F., Personal communication, Geographer-Urbanist, Brazza project manager at Ville de Bordeaux, Bordeaux, France, 2015.
[17] Gameiro-Fernandes, J.F., Personal communication, Landscape Architect, former coordinator of the Coimbra Polis Programme, Lisbon, Portugal, 2014.

INVESTIGATING THE FLOOD RESPONSIVENESS OF SMALL AND MEDIUM ENTERPRISES USING AGENT-BASED MODELLING AND SIMULATION

G. COATES[1], C. LI[1], N.G. WRIGHT[2] & S. AHILAN[3]
[1]Durham University, United Kingdom.
[2]De Montfort University, United Kingdom.
[3]University of Leeds, United Kingdom.

ABSTRACT

In recent years, flooding has caused major damage and disruption to businesses in the United Kingdom (UK) resulting in significant losses to the economy. The effect of flooding on Small and Medium Enterprises (SMEs) is of particular importance given this size of business represents a major component of the UK's economy. More specifically, according to the UK's Department for Business Innovation and Skills, SMEs account for 99.9% of all private sector businesses and approximately 47% of annual turnover. In contrast to large businesses, SMEs often have limited resources and in most cases lack business continuity plans, which diminishes the effectiveness and efficiency of their response to flooding. This paper presents an overview of an approach taken to agent-based modelling and simulation of manufacturing SMEs experiencing a flood event, and related organisations. In this overview, the sources used to develop the attributes and behaviour of SMEs are highlighted. In addition, the paper presents preliminary results of simulations of the behaviours and actions of manufacturing SMEs in response to the 2007 flood event in the Lower Don Valley region of Sheffield in the UK. Results include key performance metrics of SMEs which provide an indication of the effectiveness and efficiency of the response of the businesses modelled.

Keywords: agent-based modelling and simulation, flooding, small businesses.

1 INTRODUCTION

Flooding is the most common and widespread natural disaster in the UK [1]. In addition to households being affected by flooding, businesses can also suffer and experience significant losses in terms of damage to property and disruption to operations. For example, in terms of losses to UK businesses, the Environment Agency (EA) estimated the financial cost of the 2007 and 2012 floods at approximately £740 million and £600 million respectively [1]. The impact of flooding is arguably more acutely felt by small and medium enterprises (SMEs) due to their lack of resources, planning and organisational resilience in the face of such disruptive events [2–3]. The European Union classifies SMEs according to the number of employees and turnover or balance sheet total [4]. In terms of number of employees, micro-businesses have less than 10, small-sized have greater than or equal to 10 but less than 50, and medium-sized have greater than or equal to 50 but less than 250. In the UK, 99.9% of private sector businesses are SMEs accounting for approximately 60% of employment and 47% of annual turnover [5]. Table 1 presents a breakdown of these figures according to SME size. Furthermore, Table 1 shows that micro-businesses dominate in terms of proportion of SME businesses, employment and turnover.

DOI: 10.2495/SAFE-V6-N3-627-635

Table 1: Breakdown of private sector SMEs in the UK [5].

SMEs	Businesses (%)	Employment (%)	Turnover (%)
Micro (1–9 employees)	95.5	32.7	18.1
Small (10–49 employees)	3.8	15.3	14.6
Medium (50–249 employees)	0.6	12.3	14.5
Total	**99.9**	**60.3**	**47.2**

With its emphasis on SMEs, given the composition of UK business and significance of companies of this size, the UK's Engineering and Physical Sciences Research Council funded SESAME project aims to explore and improve the ways in which these businesses prepare for and respond to flooding thus making them more resilient to these disruptive events [6]. While the SESAME project consists of a multi-disciplinary team of researchers pursuing a number of research goals, this paper focuses on the computational modelling and simulation framework developed to facilitate an investigation into how the behaviours, activities and actions of these SMEs influence the continuity of their business operations during and in the aftermath of flood events.

The remainder of this paper is organised as follows. Section 2 presents a brief synopsis of related work. In Section 3, an overview of the modelling and simulation framework is given. Section 4 presents a preliminary case study related to the flooding of Sheffield's Lower Don Valley. Furthermore, for illustrative purposes, a simulation involving an analysis of SMEs with and without potential behaviours is considered. Finally, conclusions and future work are provided in Section 5.

2 RELATED WORK

Research into modelling the preparedness, response and recovery of SMEs to flooding is scarce. SMEs' responses to flood events, along with recovery post flooding, are complex processes requiring a suitable modelling approach. One approach to modelling such processes is agent-based modelling. This approach has received increasing attention in recent years to facilitate the study of complex systems, and it has been applied to a range of fields such as business and social science [7]. Hitherto, several agent-based models have been developed to simulate either flood risk related to individuals [8, 9] or business aspects of SMEs [10, 11]. As examples, an agent-based model has been developed to understand the process of growth and innovation that evolves in SMEs over time [10], while another ABM has been developed to study supply chains involving SMEs [11]. Despite this research on agent-based models in the context of small businesses, modelling SMEs in the context of disruptive events, in particular flood events, remains largely unexplored.

3 MODELLING AND SIMULATION FRAMEWORK

In order to investigate SMEs' behaviour when faced with flooding, agent-based modelling and simulation (ABMS) using the Repast platform [12], allied with flood modelling, has been employed. Achieving this aim has involved developing: (i) the agent-based model's (ABM's) virtual geographic environment (VGE), which is able to combine Ordnance Survey information with flood model output, to identify the businesses affected by flooding [13]; (ii) agents to model businesses in terms of their attributes, behaviour, actions and interactions in response to flood events [14]; (iii) an agent interaction network to enable simulations to be performed. An overview of the modelling and simulation framework is presented in Fig. 1 and described in further detail in [13].

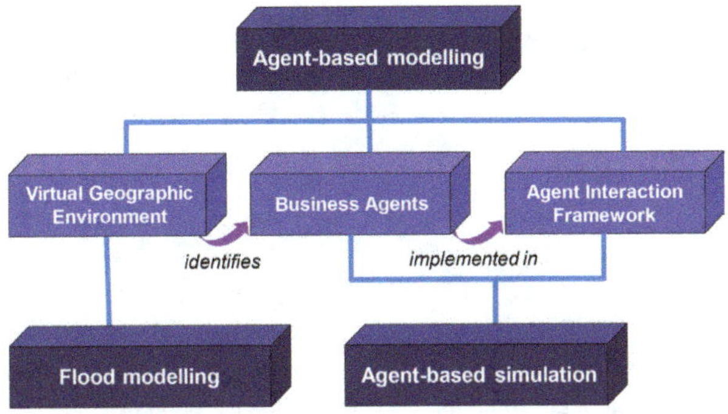

Figure 1: Overview of modelling and simulation framework [13].

4 PRELIMINARY CASE STUDY

In the context of the SESAME project, the Lower Don Valley region of Sheffield has been identified as a suitable case study area due to its high concentration of SMEs from a range of sectors, particularly manufacturing, allied with their experience of significant flooding in 2007. In June 2007, approximately 100 mm of rainfall fell in 24 hours in the Lower Don Valley region having a devastating impact on more than 1,000 businesses including key manufacturing companies with one suffering over £15 million worth of damage [15].

4.1 Flood modelling

A one in 1,000 year flood event has been simulated for the River Don in the Lower Don Valley region of Sheffield, which is similar in scale to the flood event experienced in June 2007. The simulation used a hydrodynamic model to obtain flood depths at half hour intervals over the flood duration. The 1D-2D ISIS-TUFLOW hydrodynamic model of the River Don provided by the EA has been used. The 1D ISIS model represents the Don River channel whereas the 2D TUFLOW model represents the surrounding floodplain and urban areas. In recent years, rainfall run-off boundaries have been improved with the introduction of the Revitalised Flood Hydrograph (ReFH) methodology. One in 1,000 year flood hydrographs were (i) used as inflow units to the hydrodynamic model and (ii) developed in compliance with recent guidelines produced by the EA in which the ratios of the ReFH are multiplied by the statistical 100 year estimates to gain the 1,000 year estimate.

Simulation results for a one in 1,000 year flood event are shown in Fig. 2 at 10 hour time intervals over a 40 hour period. For the case study of the River Don Valley in Sheffield, the dynamic flood inundation data associated with the maps shown in Fig. 2, and at other time intervals, provide the input for the ABM's VGE, which, in turn, provides input for agent-based simulations. That is, in agent-based simulations, the actions and interactions of agents representing flooded SMEs are driven by the depth of water at their respective premises, and adjacent roads, at different times throughout the flood event being modelled. An example of VGE output is shown in Fig. 3a, which is based on the inundation data for Sheffield at t = 20 hours corresponding to Fig. 2b. Although only four maps are shown in Fig. 2, hydrodynamic modelling has provided estimated water depths at regular time intervals, i.e. each half hour, during the simulated flood event.

The VGE is constructed using Geographic Information System (GIS) data from OS MasterMap's *Topography Layer*, *Integrated Transport Network™ Layer* and *Address Layer 2*

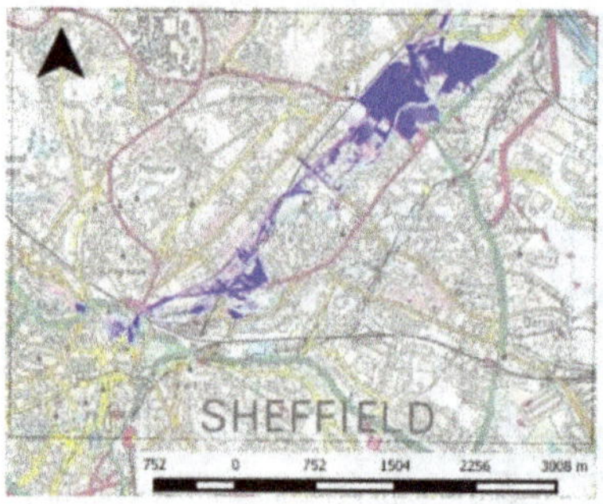

(a) 1000 year flood extent at t = 10 hours

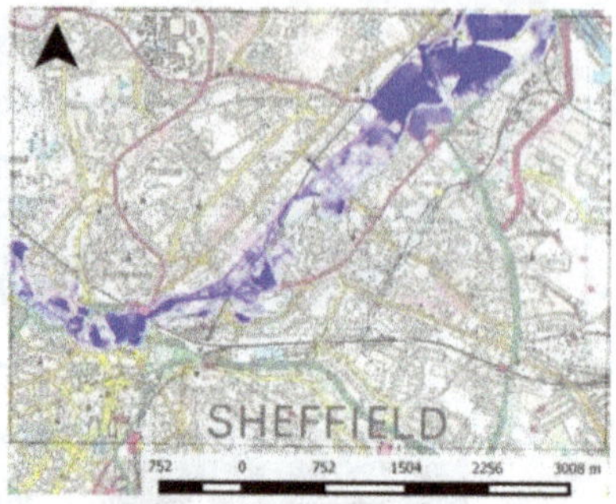

(b) 1000 year flood extent at t = 20 hours

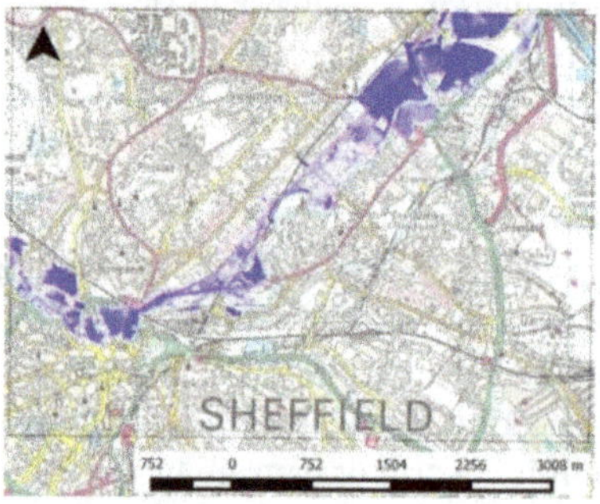

(c) 1000 year flood extent at t = 30 hours

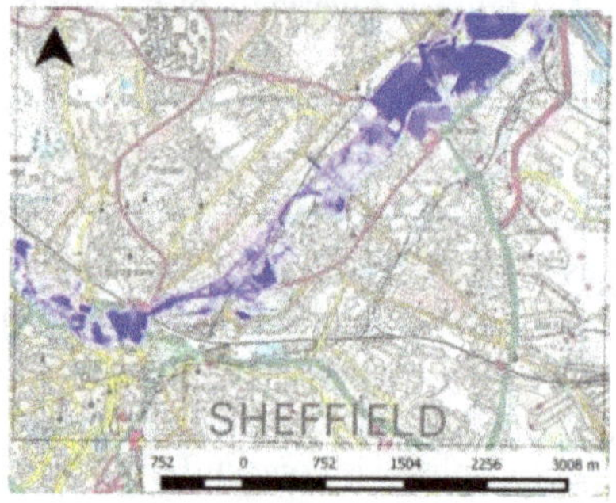

(d) 1000 year flood extent at t = 40 hours

Figure 2: Flood inundation maps for Sheffield's lower don valley.

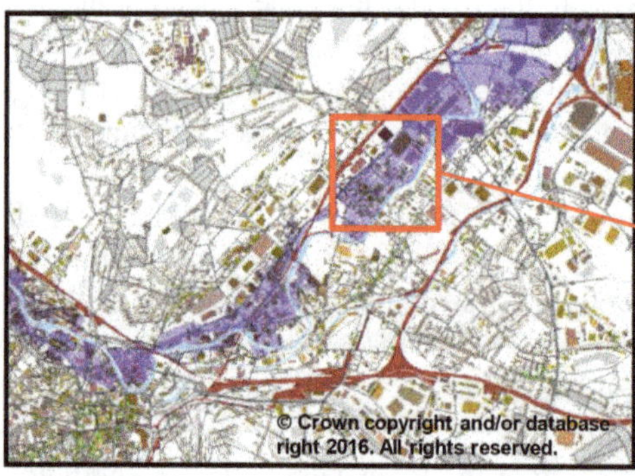

(a)

(b)

Figure 3: Sample VGE output for the Lower Don Valley in Sheffield.

[16]. The *Topography Layer* provides information on individual buildings; the *Integrated Transport Network™ (ITN) Layer* provides information on the road network; the *Address Layer 2* provides information on commercial properties including the precise location and the identification of the associated building in the *Topography Layer* and road link in the *ITN Layer*. In addition to showing the extent of flood water displayed in the VGE for the case study area under consideration, Fig. 3b indicates the location of each business using a

circle symbol, which is colour-coded according to whether their associated building is flooded (red), or their building is not flooded but their associated road link is flooded (orange), or neither their associated building nor road link is flooded (green). Throughout a simulated flood event, the status of each business modelled varies according to the dynamic inundation data.

The ABM's VGE software classifies businesses and organisations according to the National Land Use Database (NLUD) code, which consists of 41 groups, within the *Address Layer 2* of OS MasterMap. An analysis of Sheffield's Lower Don Valley region has been performed using the VGE software to identify the types of businesses in this area. The VGE identified 5,570 business/organisations with retail (U091 for shops, U092 for financial and professional services) along with industry and business (U101 for manufacturing, U102 for offices) dominating. This analysis provides input for the ABM in terms of identifying the precise geographic location of each SME along with their type and, based on inundation data for the particular flood scenario considered, which of these businesses experiences flooding and to what extent in terms of depth of water and length of time before the water recedes.

4.2 Agent-based modelling

In the context of our work, agent-based modelling has involved capturing and representing the attributes, actions, interactions and dynamic behaviours of SMEs, and related organisations, such that simulation can be used to assess ways in which SMEs can improve their response to flood events. With a focus on SMEs from the manufacturing sector, agents have been developed to model small businesses in terms of their behaviour and attributes relevant to responding to and recovering from flood events. Simultaneously, an agent interaction network has been developed containing agents representing other related organisations such as suppliers, customers and service companies.

SME agents' attributes and behaviours have been derived from a number of sources: (a) information extracted from transcripts of semi-structured interviews with SMEs at risk of flooding and/or which have experienced flooding; (b) discussions with colleagues who conducted the aforementioned interviews to ensure the attributes and behaviours, and the duration range of behaviours along with the relationships between them, being used to model SMEs are appropriate and relevant; (c) EA literature providing advice for businesses preparing for flooding; (d) literature detailing business continuity management (BCM) systems' requirements (ISO223301). In this work, SME behaviours extracted from interview transcripts are referred to as 'existing behaviours' whereas those found in EA and BCM literature are referred to as 'potential behaviours'. This distinction is made as it provides the basis for a comparative analysis in our modelling environment in terms of what SMEs did when flooded in the past and what they may do in the future.

In addition to modelling the attributes and behaviours of SMEs, simulations of the behaviours of business agents during and in the short-term aftermath of a flood event has required the development of other agents to model organisations such as suppliers, customers, mutual aid partners and service companies. Furthermore, relationships between SMEs and these organisations have been modelled since some will influence the response and recovery of an SME when a flood event occurs. For example, the existence of a pre-agreed priority relationship between an SME and a service company may lead to preferential treatment thus expediting recovery and resumption of normal operations. Similarly, the existence of a pre-existing mutual aid agreement between SMEs may enable the flood-affected SME to resume some degree of production operations, in the case of manufacturing SMEs, thus enabling customer demands to be met to a greater degree than if such an arrangement was not in place.

4.3 Agent-based simulation

Simulating SMEs during and in the short-term aftermath of flood events provides a means of enabling what-if scenarios/experiments to be performed to investigate the effect of different behaviours on operational response and business continuity. In terms of agent-based simulations, we are currently in the process of designing experiments for simulations which will enable us to assess the effect of different combinations of behaviours on the performance of the small businesses modelled. However, as part of the process of developing SME agents incorporating existing and potential behaviours, preliminary simulations have been conducted to assess the functionality of the associated software. This paper presents some results from those preliminary simulations.

For illustrative purposes, a preliminary simulation involving manufacturing SMEs is considered. Furthermore, the SMEs considered are micro-sized since, as indicated in Section 1, this size of business accounts for the vast majority, i.e. 95.5% of 99.9%, of SMEs in the UK. In response to the flood event (tick 56–137) in Sheffield's Lower Don Valley as described in Section 4.1, Fig. 4 shows the average production capacity level of all manufacturing micro-sized SMEs modelled for 20 repeat simulations: (a) without potential behaviours; (b) with some potential behaviours. Also, in Fig. 4, the simulation time of 1,200 ticks corresponds with 25 days since there are 48 ticks per day, i.e. one tick per half hour.

In terms of what is meant by 'with some potential behaviours', in addition to the SMEs being registered to receive alerts from the EA, they have also raised the level/height at which their machines, raw materials, paper documents and electrical sockets are located to mitigate damage should a flood event occur. Each of these potential behaviours has associated benefits such as being registered to receive alerts from the EA provides an SME with time to take preparatory action in terms of, for example, lifting up IT equipment and taking measures to stop the flood water entering premises before employees are evacuated. Raising the levels of machines, raw materials and electrical sockets lessens the damage to them, whereas raising

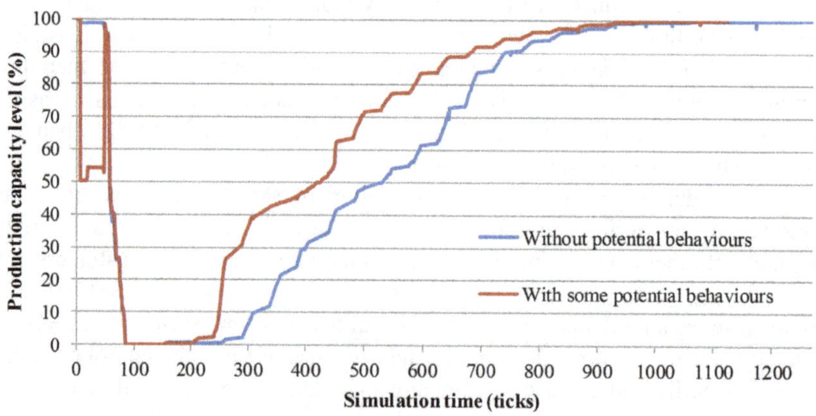

Figure 4: Simulation output for all manufacturing micro-sized SMEs.

paper documents enables an SME to contact service companies, customers and suppliers more quickly than if these documents were damaged. In Fig. 4, it can be seen that manufacturing SMEs with the potential behaviours mentioned, are able to achieve a greater production capacity level earlier than those SMEs which do not have these behaviours. Also, in relation to Fig. 4, the production capacity level, PC_{level}, at each time interval, i.e. each tick signifying half an hour, is determined using the following equation,

$$PC_{level} = R^{e}_{level} \times R^{m}_{level} \tag{1}$$

given there is a power supply to the premises and there is sufficient raw materials available for this level of production, and where R^{e}_{level} is ratio of an SME's available employees to total employees, and R^{m}_{level} is the ratio of an SME's available machines to total machines. To note, at each time interval, one unit of raw materials is consumed given (i) at least this amount is available, (ii) all employees are available and utilised, and (iii) all machines are available and utilised.

In the same simulation associated with the data shown in Fig. 4, the average production capacity level of a single manufacturing micro-sized SME for 20 repeat simulations is shown in Fig. 5.

In a typical simulation for this SME without the aforementioned potential behaviours, once the flood water approached the SME's premises, employees evacuated before water entered the property reaching a height of 0,913 metres. During the flood, power was lost and a proportion of the SME's machines were damaged and raw materials ruined due to the water depth reached. Employees returned to the premises only when the flood water had receded and thus recovery operations could begin. However, due to the flood event, with some machines inoperable and raw materials damaged, the SMEs' ability to manufacture products, once the flood water had receded, was degraded. Furthermore, during the immediate recovery period, employees returning to the premises carried out activities such as cleaning and salvaging raw materials, contacting the insurance company and electrical service company. After an electrician attended to restore power to the premises and the insurance company organised a cleaning contract company to clean-up the premises, the SME carried out other activities including cataloguing damage, repairing machinery, cleaning-up the storage area for raw materials, ordering raw materials, and contacting suppliers and customers. Manufacturing products only resumed at the beginning of the third day after the flood (~ tick 241) once employees were available for production,

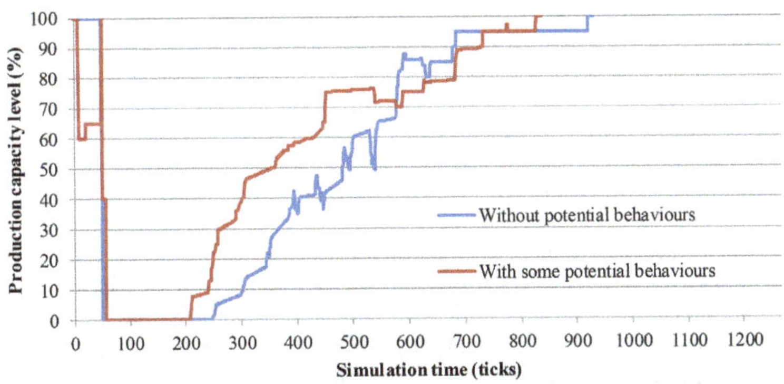

Figure 5: Simulation output for a single manufacturing micro-sized SME.

machines were repaired and raw materials available. In contrast, while similar activities were undertaken in a typical simulation for this SME with the aforementioned potential behaviours, the company was able to resume production during the second day after the flood (~ tick 207).

5 CONCLUSIONS AND FUTURE WORK

Flooding continues to cause businesses to suffer significant losses in terms of damage to property and disruption to operations. Given the significance of SMEs to the economy, research aimed at understanding how these businesses behave and how they may make changes to their behavior to better respond and recover from flood events may provide an insight to improving their business continuity.

In the strand of the SESAME project reported in this paper, research has focused on modelling existing, or typical, behaviours of SMEs in response to flooding and potential, or enhanced, behaviours aimed at improving their response. Based on data obtained from the field, an agent-based model has been developed to enable the impact of potential behaviours to be assessed. A preliminary case study focussing on manufacturing SMEs experiencing a one in 1,000 year flood event in the Lower Don Valley region of Sheffield has indicated the benefit of registering for EA alerts and raising machines, raw materials, paper documents and electrical sockets in terms of expediting a return to production operations. Future work will focus on the design of experiments to simulate SMEs during and in the aftermath of flood events to allow the investigation of the impacts of the range of potential behaviours on operational response and business continuity. Consequently, an assessment will be made as to which combination of behaviours may be most suitable for SMEs to implement in order to improve business continuity.

ACKNOWLEDGEMENTS

The authors gratefully acknowledge the funding provided be the UK's EPSRC under grant EP/K012770/1. The data used in the preliminary case study presented in this paper is unavailable due to it including data from OS MasterMap's *Address Layer 2* obtained under a research agreement.

REFERENCES

[1] http://www.environment-agency.gov.uk

[2] McGuinness, M. & Johnson, N., Responding to changing paradigms of risk: managing flood risk and enhancing organisational resilience. In *Proceeding of the Association of Geographical Societies in Europe (EUGEO) 2013 Congress, 'Changing Geographies and Geographies of Change'*, Rome, Italy, 2013.

[3] Reuter, C., Towards efficient security: business continuity management in small and medium enterprises. *International Journal of Information Systems for Crisis Response and Management*, **7**(3), pp. 69–79, 2015.

[4] http://ec.europa.eu/growth/smes/business-friendly-environment/sme-definition/index_en.htm

[5] Rhodes, C., Business statistics, briefing paper 06152, house of commons library, 2015.

[6] Coates, G., Hawe, G.I., McGuinness, M., Wright, N.G., Guan, D., Harries, T. & McEwen, L., A framework for organisational operational response and strategic decision making for long term flood preparedness in urban areas'. In *Proceeding of the 3rd International Conference on Disaster Management*, A Coruña, Spain, 2013.

[7] Heath, B., Hill, R. & Ciarallo, F., A survey of agent-based modeling practices (January 1998 to July 2008). *Journal of Artificial Societies and Social Simulation,* **12**(4), p. 9, 2009.

[8] Dawson, R.J., Peppe, R. & Wang, M., An agent-based model for risk-based flood incident management. *Natural Hazards,* **59**, pp. 167–189, 2011.

[9] Liu, Y., Okada, N., Shen, D. & Li, S., Agent-based flood evacuation simulation of life-threatening conditions using vitae system model. *Journal of Natural Disaster Science,* **31**, pp. 69–77, 2009.

[10] Blackmore, K., Bossomaier, T., Jarratt, D. & Nesbitt, K., Intelligent agent framework for modelling the evolution of small and medium sized enterprises. In *Proceeding of the 8th Australian and New Zealand Conference on Intelligent Information Systems*, Sydney, Australia, 2003.

[11] Tounsi, J., Habchi, G., Boissière, J. & Azaiez, S., A multi-agent knowledge model for SMEs mechatronic supply chains. *Journal of Intelligent Manufacturing,* **23**, pp. 2647–2665, 2012.

[12] http://repast.sourceforge.net/

[13] Coates, G., Hawe, G.I., Wright, N.G. & Ahilan, S., Agent-based modelling and inundation prediction to enable the identification of businesses affected by flooding'. *In Proceeding of the 4th International Conference on Flood Recovery, Innovation and Response*, Poznan, Poland, 2014.

[14] Li, C. & Coates, G., Design and development of an agent-based model for business operations faced with flood disruption. *International Journal of Design & Nature and Ecodynamics*, **11**(2), pp. 97–106, 2016.

[15] Environment Agency, Review of 2007 summer floods, December 2007.

[16] http://www.ordnancesurvey.co.uk

ANALYSIS OF INSURANCE COMPANIES OPERATING IN SLOVAKIA IN TERMS OF FLOOD RISK INSURANCE

M. MADAJOVÁ & Ľ. SOLÍN
Institute of Geography, Slovak Academy of Sciences, Bratislava, Slovakia.

ABSTRACT
Analysis of the trends and spatial variability of flood insurance coverage in Slovakia pointed to the decreasing interest of people in property insurance against flood risk. There are several causes for this. One of them possibly lies in the conditions of insurance companies concerning flood risk insurance, which among other things also specify limits and exclusions from insurance. The majority of the insured are not aware of these facts at the time of signing the contract and the rejection or reduction of payments for flood damage is then a source of frustration. The paper analyses flood insurance conditions in all insurance companies in Slovakia, which offer building and contents insurance in terms of five aspects: (1) definition of concepts connected with flood hazard, (2) inclusion of flood risk into insurance modules of companies, (3) scope of property which is the object of the insurance, (4) exclusions and limits to flood risk insurance, and (5) the amounts of the premiums and claims.
Keywords: exclusions, flood, flood hazard insurance, limits, premium.

1 INTRODUCTION

The strategy for the mitigation of negative consequences caused by floods is one of important components of integrated flood risk management [1]. It is the Government and insurance companies that play the principal role in this sense [2–8]. In many countries, there is no strict legal duty for governments to provide assistance for flood damage. In the Slovak Republic (SR), the legislative system does not impose on the government the obligation of participating in the compensation of damage caused by flooding to people and firms, but it may provide, especially in case of massive damage caused by flooding, a non-recurring financial contribution to affected persons. Therefore, private insurance companies play a more important role in the strategy of mitigation of negative consequences caused by floods. There are 14 insurance companies active in Slovakia [9] that insure buildings and household content in the sphere of non-life insurance. Their market share is shown in Table 1.

Analysis of the trends and spatial variability of flood risk insurance in Slovakia [11] based on data provided by the Allianz–Slovak Insurance Company for the period 2002–2011 points to a comparatively distinct decreasing interest of people in property insurance against floods. For instance, as of the end of 2011, only 54.4% of insurance policies were active against the situation in 2002. This decrease can be attributable to several causes. Madajová *et al.* [12] point to, for instance, the socio-economic status of the population and the general discontent of the insured with the compensation amounts. The essential reason is also the refusal of insurers to provide any compensation. This leads to frustration and distrust that the insurance company will not compensate the real cost of damage liquidation. These statements were also confirmed by empirical findings obtained by questionnaire survey, which was concerned four

DOI: 10.2495/SAFE-V6-N3-636-647

Table 1: Market share of Slovak insurance companies in non-life insurance.

Insurance companies	Market share (%) on 31/12/2013
Aegon	x
Allianz	34.87
Astra	x
Axa	1.59
ČSOB	2.89
Ergo	0.2
Generali	10.05
Groupama	0.38
Komunálna poisťovňa	6.45
Kooperatíva	26.73
Poštová banka	0.12
Union	3.32
Uniqa	8.01
Wüstenrot	2.35

x: not specified (insurance companies not active in non-life insurance in 2013)
Source: Slovak Insurance Association [10]

aspects of flood risk: (i) flood hazard and flood damage, (ii) individual flood protection, (iii) flood hazard awareness, (iv) economic and social vulnerability. The questionnaire survey was conducted during June – July 2015 in six municipalities (3,082 households) located in the upstream basins of Myjava river.

We consider that the discontent of the population with the insurance claim is due to the fact that people do not thoroughly read the general and specific insurance conditions (ICs), which contain the limiting factors of insurance claims and their amounts. The general ICs of companies operating in SR are as a rule similar, but the insurance coverage and compensation of damage caused by floods depends on the specific conditions concerning the risk of flooding, which are different in each company. The majority of insurance companies in the period between 2010 and 2014 (after the extreme flood events of 2010) updated their ICs for property insurance.

The aim of this article is to analyse the ICs of companies operating in the SR insurance market in terms of building and household content insurance against risk of flooding. The ICs are available online on the websites of individual insurance companies. All insurance companies were also approached individually to give us additional information about the flood risk insurance (for examle, whether they use the flood maps or they use own database about the flood insurance events, what other criteria are applied when deciding on property insurance against the risk of flooding, how they differentiated the insurance premium). The analysis is focused on five basic aspects that should be paid attention to at the moment of contracting property insurance, especially if the person arranges the policy in order to insure damage caused by flood hazard. The first aspect is the definition of flooding itself in ICs and its compliance with the definition of flooding in the literature and legislative standards of the SR. The second and third aspects are the inclusion of flood risk into the insurance modules of the company and the scope of property that is the object of insurance. A very important aspect in terms of insurance claims is that of exclusions and limits to flood risk insurance. The amount of the premiums, as well as of the claims in the case of an insured event, also plays a crucial role when deciding to conclude an insurance policy. The final remarks sum up the obtained results.

The results of the analysis are expected to help better orientate the population in ICs concerning flood risk and to provide them with a rational basis for their expectations in the case of damage caused by flooding.

2 DEFINITION OF FLOOD

The definition of the term *flood* in ICs of companies should be based on the interpretation of this concept in the scientific literature or on the definition of flooding in the legislative system of the SR. The agreement of the definition of flooding on the science–legislation–insurance axis reduces the risk of different conflicts that may arise due to vague interpretation of the term *flood*. Before the detailed analysis of the term *flood* in ICs of companies, a brief commentary on the interpretation of this concept in the scientific literature and the legal system of the SR will be provided.

2.1 Scientific literature

Flooding is considered a natural phenomenon caused first of all by climatic factors (rain, snow and ice melt, ice jams). Apart from the main causes of floods, local flood-intensifying factors also play an important role. Ward and Robinson [13] and Smith and Ward [14], for instance, classify them into four groups: (a) stable basin factors – area, shape, slope, aspect, altitude; (b) variable basin factors – climate, geology, soil type, vegetation cover, anthropogenic influences on storage capacity, infiltration, and transmissivity; (c) drainage network factors; (d) channel factors. Therefore, there are various definitions of flooding in the literature. Some of them are limited to the process of overflowing, for example:

> *'A flood is a relatively high flow which overtaxes the natural channel provided for the runoff'* [15].

Other definitions interpret flooding in a broader sense:

> *'A flood is body of water which rises to overflow land which is not normally submerged'* [16].

This means that apart from the overflow, flooding of a territory can also be caused by processes such as flat areas being flooded by heavy rainfall ponding on the surface, sheet overland flow on slopes, or the groundwater table rising above the level of the ground surface, etc., due to which areas outside the floodplains are also flooded.

Floods are normally classified into several types in terms of factors that participate in their origin and the scope of flooding: (i) *river or regional floods*: the natural overflow of a stream due to long-lasting rainfall or snow melt occurring in a number of streams; (ii) *flash floods*: local floods, originally caused by excessively heavy rainfall, and/or intense snow and ice melt. Apart from the overflow of the river bank, there is also overland flow on slopes that threatens structures and objects outside the floodplain zone; (iii) *ice-jam floods*: impoundment of the water table and overflow due to ice in the river channel; (iv) *inner floods*: flooding of a territory by surface waters that did not soak into the soil and that had no natural outflow or the rise of groundwater to the surface; and (v) *floods as a consequence of the disruption of water management objects*: flooding of a territory due to overflow, seepage, or dam/dam wall failure. Different types of floods fall under the common term of flood hazard.

2.2 Legislative system of the SR

The legislative system of the SR defined the term flood in Act No. 71/2015 [17], but in time when the majority of insurance companies defined their ICs, that is, in the years 2010–2014, Act No. 7/2010 [18] was in force. This Act (§ 2, clause 1) defined a flood as the temporary inundation of a territory, which is not usually covered by water. A flood occurs when:

a. The water table of a stream temporarily significantly increases and there is the hazard of overflow or water has already overflowed;
b. Natural runoff of precipitation or snow melt waters into the recipient is temporarily blocked and inner waters flood a territory. Inner waters are those that occur in the territory protected by dams or counter-flood lines, especially waters that cannot flow out in a natural way because of increased water table in a recipient; water brought by intensive precipitation or snow melt in the territory with no option of runoff via a stream;
c. There is a hazard of overflow of a stream or water has already overflowed due to ice drift, ice-jam, or other obstructions in channel, bridges, and locks or in the inundated area;
d. The area is flooded due to intensive rainfall or accumulation of water from snow melt;
e. The protected area (area protected by a water management structure) is flooded due to increased groundwater table surfacing, which causes a long-lasting high water level in a stream;
f. There is hazard of overflow or the stream has already overflowed due to a crash or defect in a water management structure.

2.3 Insurance conditions (ICs) of insurance companies

Flood hazard is expressed in the ICs of all companies in two terms: flood (F) and inundation (I).

The term *flood* appears in three variants:

- F is flooding of the insured place by water that has naturally overflowed the banks of water streams/water reservoirs or broken these banks. In the case of a watercourse enclosed by dikes, the bank means the body of the dike (*Allianz, Astra, ČSOB, Groupama, Union, Wüstenrot*).
- F means flooding of territories by water that has overflowed the banks of streams or water reservoirs or which has broken the banks and dikes or the flooding has been caused by an abrupt and accidental diminishment of the discharge profile of the stream (*Aegon, Generali, Ergo, Komunálna poisťovňa, Kooperativa, Uniqa*);
- F is a temporary distinct increase of the water table in a stream or other surface waters when water floods the area beyond the stream channel and causes damage (*Axa, Poštová banka*); this also applies in the cases when water cannot outflow in a natural way from a certain area due to the perturbed stability of a stream (*Poštová banka*).

Varied interpretations of the term *inundation* also appear in the conditions of insurance companies:

- I is the formation of a continuous water area, which is either standing or flowing for some time over the insured place (*Aegon, Generali, Komunálna poisťovňa, Kooperativa*);

- I is a phenomenon caused by natural effects results in the formation of a continuous water area, which is either standing or flowing for some time over the insured place (*Allianz, Astra, Grupama, Union, Wüstenrot*). *Union* quotes that the facts of where the water is from, how it penetrated the insured place, and how it caused the damage (rainfall, snow melt, backwater of groundwater level, and the like) are not decisive.
- I is the formation of a continuous water area that is standing or flowing for some time without natural outflow at the insured place (*Poštová banka*);
- I is the formation of a continuous standing or flowing water area on the ground at the insured place where there is no natural outflow and that has been caused by natural effects (*Axa*)/atmospheric precipitation (*ČSOB*)/atmospheric precipitation, snow, and/or ice melt (*Ergo*).
- I is the formation of a continuous water area in the insured place that was caused by increased surface waters (standing or flowing) of a channel or by atmospheric precipitation (*Uniqa*). This definition is remarkable for its ambiguity because the phrase 'increase of surface water of its channel' is already incorporated in the definition of *flood*.

The reason why the term inundation was introduced into ICs of companies is not altogether clear. It is probably the result of the fairly limiting definition of flood, which only takes into account the flooding of the riverside zone by an outflow of a stream and dam failure. Presumably, insurance companies try to encompass inundation of areas outside the riverside zones.

We can conclude that from the matter-of-fact point of view definitions of terms F and I in ICs do not completely correspond to the interpretation of these terms in the scientific literature or in the legislative system. ICs are first of all unclear in their consideration of overland flow caused by sheetwash flooding on slopes. This is an important phenomenon occurring in hill and foothill areas. Furthermore, no insurance company recognises flooding of a territory caused by the controlled release of water from dams.

3 INCLUSION OF FLOOD HAZARD INTO THE INSURANCE MODULES

Only one insurance company in the SR (*Axa*) offers insurance for damage caused by F and I individually through supplementary insurance to standard property insurance, and at the same time it does not offer insurance for F and I risk in any other product. All other companies provide insurance packets with different content (number of covered risks) and amounts of insurance coverage for the given risks. Three companies (*Uniqa* and partially *Poštová banka* and *Wüstenrot*) offer only one module, that is, complete property insurance covering all risks in one product. The remaining companies offer several modules of insurance coverage, which differ in the scope of included insurance risks, the price of insurance, and the limits of the claim amounts. With the exception of *Allianz* and *Union*, all other companies include the F and I risks in their basic property insurance packets.

Such a business model of insurance is normally applied when the demand for insurance is in a positive relationship with the risk of loss (*adverse selection*, [16]). Larger risk generates greater inclination to buy insurance, and the insurer is not able to effectively reflect this correlation in the price of the insurance. Particularly in the case of natural events, it is more economically effective if these risks are part of broader insurance. If the clients want to insure their property against flooding, they have to accept the fact that their insurance premium also includes risks that they do not need/want to insure.

4 SCOPE OF PROPERTY INSURANCE

Property insurance is divided into two basic categories: building insurance and household insurance. In general, the object of building insurance is a building (family house or flat) where damage to fixed (outer and inner) inseparable building elements and coupling connections is reimbursed. The object of household insurance covers the content of a household in a family house, flat, or holiday cottage/house, which serves the operation of the households or meets the needs of the insured or the insured household members. It is possible to contract separate building insurance and separate household insurance.

Particularly if a primary reason for insurance is damage caused by flood hazard, insurance needs to focus also on the protection of property situated outside the building itself. It is important to enquire whether and under what conditions the other buildings situated on the plot belonging to the property, the plot itself (garden), crops/harvest, etc. are the objects of insurance. The majority of insurance companies insure outbuildings (buildings not intended for living in built on solid foundations or firmly attached to the terrain, such as summer houses, garages, farm buildings, and the like). However, plots, crops, greenhouses, and plastic greenhouses are not generally objects of insurance. *Aegon* is the only insurer that offers insurance of crops grown for private consumption and agricultural produce after harvest within the basic insurance packet as part of household contents. Likewise, one insurance company (*Axa*) offers the insurance of growths, garden architecture, and greenhouses in its top variant (within building insurance). Three insurance companies (*Union*, *Allianz*, and *Groupama*) offer the option of additional insurance for gardens, plants, and trees, including the harvest, for their basic insurance. A special category is that of yard animals and pets, which are either not included in the property insurance at all (*Astra*, *Ergo*, *Poštová banka*, *Union*, and *Uniqa*), or they are insured under the basic household insurance but always with a fixed claim amount (remaining companies).

Concerning the insurance coverage, in the case of insured event caused by F and I the cost of the following plays and an important role: tidying/cleaning and drying of insured things; removal/disposal of the remaining insured things that have been impaired or damaged by insured event; alternative accommodation if the building/household has become inhabitable; storage of insured things in hired premises after the insured event and measures expended to divert/mitigate the imminent insured event.

Before signing the policy in addition to familiarity with scope of insurance, it is necessary to require information about all specific conditions concerning the exclusions and limits of insurance coverage.

5 EXCLUSIONS AND LIMITS OF FLOOD AND INUNDATION RISK INSURANCE

Situations when the insurer applies exclusion (property is not insured against F and I risk) or imposes limits to insurance claims (property is insured but the amount of insurance claim is limited) are generally quoted in specific ICs of each company. The following part of the study will be devoted to exclusions from F and I risk insurance. The question of limits will be treated along with the insurance claims in the next chapter.

Exclusions of insurance coverage in Slovak insurance companies are defined:

- On the basis of flood frequency at the insured place: the insurer will not compensate damage caused by F if F occurred at the insured place within 10 years before the policy was contracted (*Aegon*, *Groupama*, and partially *ČSOB*);
- On the basis of N-year maximum discharge, that is, discharge repeated on average once

every N-years. Insurance does not apply to the damage caused by flooding which is repeated on average less than every 10 years (*Allianz, Astra, Union*);

- Both exclusions can be applied simultaneously and the insurance does not apply to damage caused by F with a return period on average less of than 20 years and for the insured place where a flood occurred in the last 10 years at least twice before the time of insurance (*Axa, Ergo*);
- Based on their own definition of the risk area in combination with some of the above-mentioned exclusions: *ČSOB* does not insure in areas where a place of insurance is situated within 200 metres of a riverbank and its elevation does not surpass 3 m from the bank. It also does not insure in places which have been hit by F or I within 10 years before conclusion of the policy. *Poštová banka* requires information concerning the distance of the family house from the nearest stream (more or less than 10 m) and whether the building was hit by F or increasing ground water in 5 years prior to the conclusion of the policy. *Generali* uses flood maps with the statistical probability of F and I risk. If the system assesses that the place of insurance falls into a zone which has been hit by flooding four or more times in the last 10 years, the building/household cannot be insured against F and I risk.

The insurance companies *Komunálna poisťovňa, Kooperatíva, Uniqa*, and *Wüstenrot* did not apply any exclusions in 2015; they insured all areas regardless of flood hazard (some of them with no limits: *Wüstenrot* and *Uniqa*, other than some limits to insurance coverage). This is the reason why these companies together with *Poštová banka* and *Generali* do not quote any exclusions in their ICs either. In the case of the two companies mentioned last, all relevant information on flood insurance is discovered before the conclusion of contracts and the insured is so notified of the exclusion of insurance coverage at the outset.

However, it is not a rule to inform clients about exclusions of flood risk insurance by insurers who have such exclusions in their ICs. The companies that declare that they inform their clients about the clauses in their ICs based on which they are not obliged to provide an insurance claim or in which claims are limited are *Aegon, Axa, Groupama* (by word of mouth), and *Union* (via questionnaire). It should be pointed out here that companies do not pay special attention to the F and I risk, also because this risk is normally part of a packet along with other risks. It is therefore the client's obligation to become thoroughly familiar with the ICs concerning the individual risks. Empirical evidence shows, however, that the policyholders do not read ICs related to insurance restrictions in detail upon signing the contract, and study them only when the insurance event has occurred.

An insurer may also refuse the overall claim/reduce the insurance coverage if the insured fails to fulfil the obligations connected to flood hazard insurance. Some companies impose special obligations concerning protection of property against inundation or flooding. For instance, in places located in the underground floors of buildings, the insured is obliged to have the insured items stored at least 12 cm (*ERGO*), 14 cm (*Kooperativa*), or 15 cm (*Generali*) above the lowest level of the basement floor. Each company also quotes in its general conditions that the policyholder is obliged to prevent the insurance event and to carry out reasonable measures against it or measures that mitigate their scope if the event occurs.

6 INSURANCE PREMIUM AND CLAIMS

One of the key factors that plays an essential role in deciding on property insurance is its price – the premium (the amount that customers pay for insurance on their property) and the amount of insurance claim after an insured event.

6.1 Amounts of premiums

The insurance premium, in general, is established depending on the insured value, which should correspond to the value of the insured property (building or household content) at the time the contract is concluded. The insured value for a building is established either by:

- the insurer as the recommended insured value established by calculation based on the parameters of the building;
- the insured, at their own liability or by adaptation of the amount set by the insurer (while it can be grossed up without limits, reductions are only limited by *Allianz*, that is, up to 20% of the amount established by the insurer);
- based on an expert testimony, budget, contract, or other document requested by the insurer.

In the case of household content, it is the insured that decides on the insured value and this should correspond to the new value of the group and subgroup of household items at the time the policy is signed. *Allianz*, as the only insurance company, has the minimum insured value for movable assets established at €6,640.

If the insurer sets the insured value of a building, its calculation is based on the product of the area in square metres and the internally established coefficient. In the case of family house insurance, the useful (built-up) area is considered, and in the case of flat and household insurance, it is the floor area. Some companies, however, take into account a broader spectrum of factors. For example, apart from the built-up area, *Uniqa* also uses the realisation of the building (basement, number of storeys, habitable attic) or the territorial factor (capital, regional centre and selected bigger towns, other communes). In *Poštová banka,* it is the year of final building approval or the year of refurbishment of the family house. *Allianz* calculates the insured value for a family house taking into account the built-up area, the year of the construction/last refurbishment, the materials used, the number of above-ground storeys, the existence of a basement, and the type of roofing. If a price is being established for a house older than 30 years, 20% is deducted from the price calculated from other parameters. If the object of insurance is a flat, the above-mentioned companies not only take into account the floor area but also the territorial factor, and *Allianz* in addition also considers its construction (standard).

Based on the analysis of ICs and on the individual consultations with each insurance company, we can say that the majority of insurance companies in SR practically do not apply the differentiation of premiums based on the level of flood hazard in a given locality. On the one hand, this is because the given locality is assessed by the insurer as a risky one and directly excludes it or because the property insurance against the flood hazard is not concluded separately, but is part of the broader insurance package. Only *Axa* and *Generali* differentiate the insurance premium depending on the risk zone (*Axa* applies the system of coefficients based on the flood event frequency in the place of insurance. *Generali* makes use of flood maps, which display the statistical probability of flood hazard).

The insurance premium is also modified by what is referred to as a deductible, which is represented by an amount in euros agreed in the policy when the insured participates in insurance claim. Some companies automatically negotiate insurance without a deductible (*Allianz*, *Uniqa*, *Komunálna poisťovňa*, *Kooperatíva*, *ČSOB* only with the top packet), some of them do it with an obligatory deductible (*Aegon*, ČSOB, *Ergo*, *Generali*, *Groupama*, *Poštová banka*, *Wüstenrot*), although the insured may choose from several options (*Astra*, *Axa*, *Komunálna poisťovňa*, *Kooperatíva*, *Union*). *Aegon* and *Generali* also offer zero deductible, albeit with a 5% and 20% surcharge added to the insurance, respectively. In the case of an obligatory deductible, some companies offer certain benefits; for instance, *Ergo* applies a deductible reduced to the amount of €38 if there was no insurance event during a year, and after two- and three-year periods without claims the amount decreases to €25 and €13, respectively. No deductible is required in the case there has been no insurance event in 4 years. *Union* provides a discount in the case of a deductible of €100 and €150.

The insured value is the basis for the calculation of premiums. It should respond to the new value of the insured items, and it is the top limit of an insurance claim as a rule for all insurance events in one insurance year. The insured value does not only influence the amount of a premium, but also the amount of an insurance claim.

6.2 Insurance claims

As indicated in the introduction, a distinctly lower compensation of damage caused by flooding reimbursed by an insurance company compared to the actually expended means necessary for the elimination of damage is the most frequent reason for the insured's discontent. In general, the insurance company satisfies the claim up to the insured value quoted in the contract, but always depending on the limits of the insurance claim established for the individual insurance risks and/or insured things according to the chosen insurance module.

In the case of building insurance, the insurance claim limit for property damaged by F or I is set either as the:

- insured value, that is, 100% of the cost is reimbursed (*Aegon, Allianz, Axa, Ergo, Groupama, Komunálna poisťovňa, Kooperatíva, Poštová banka, Astra* only in the top packet, *Generali* only if the building is situated in a minimum F and I risk zone, and *Uniqa* only in the case of flats);
- some percentage of insured value (*Astra, ČSOB, Uniqa* in the case of building insurance). *Generali*'s principle is that if the building is situated in an increased F and I risk zone then there is a limit to the claim amount lower than the insured value. *Komunálna poisťovňa* and *Kooperatíva* may cut the insurance claim by 80% in the case the insurance event was caused by F or I which occurred more frequently that once in 5 years prior to the date of the event;
- maximum claim limit as it is established and quoted in the contract as a fixed amount (ČSOB, *Union, Wüstenrot*).

In the case of household insurance, there are more specified items and their reimbursement is defined as some percentage of insured value for household content or a fixed amount in euros is set. Some companies even quote in their conditions a table for all groups of household things. Particular limits to claims are separately set for what is referred to as special movable property (money, jewels, artwork, etc.) or other specific things and equipments (animals, in some companies food articles, tools, sports equipment, electronics, optical equipments).

Moreover, the cuts in insurance claims also occur due to:

- the depreciation of property. If the depreciation or devaluation exceeds a certain percentage or age, the claim amount is cut proportionally. This means that the company does not compensate in the 'new' but in 'time' prices or a limit to the new price of the building/thing is set.

Time prices are also paid if the things are kept below the terrain level (*ČSOB*), or for 'special' things such as someone else's things (*Axa, ČSOB*), building components, carpets, garments, and shoes (*ČSOB*), small mechanisms (*ČSOB, Komunálna poisťovňa, Kooperatíva*), damaged outbuildings (*Komunálna poisťovňa, Kooperatíva*), accessories of buildings with a depreciation of 5% per year deduction (*Komunálna poisťovňa, Kooperatíva*), vehicles, trailers, boats in a garage, agricultural machines, and instruments kept inside buildings (*Uniqa*);

- the declaration of other value (underinsurance), a status of the property other than real, or incorrect data in the insurance contract. If at the time of the insurance event the insured value is lower than the real value of the insured item, it is underinsurance, and the insurer may reduce the amount of the claim proportionally of the insured value to the real value of the insured item.

7 FINAL REMARKS AND CONCLUSION

If people were sufficiently aware of what the object of insurance is and for what risks the insurance is negotiated when buying insurance, many misunderstandings would be avoided. However, they rather concentrate on the lowest possible premiums and ignore the scope of the insurance coverage, believing that the conclusion of an insurance contract means insurance of any damage caused to their property. Regarding the flood risk insurance, familiarity with the ICs is crucial and necessary. The insured will thus clearly avoid possible frustration in time of claim when it can happen that the building/household is insured against flood hazard but the damage is not compensated either not at all or in a limited rate due to exclusions and limits to insurance. It is therefore important to:

1. Be thoroughly familiar with the object of insurance, the exclusions, and the limits of the insurance claim at the time of concluding the contract.
   It is important to have what is and what is not the object of insurance and the situations and cases (risks) for which the insurance is concluded explained, as well as what the specific situations and limits of the insurance claim are.
2. To establish the correct value of the insured items.
   It should not be too low (underestimated), or out of the date if the policy was concluded many years ago. It is recommended that the value be set in the way that would enable the client to procure alternative accommodation in the case of an unexpected insurance event that causes total damage. The client should check on the contract condition regularly once every few years and always revalue the contract when the building/household has been refurbished or renewed.
3. Choose an appropriate method of damage repair.
   Experience shows that the amount of a claim depends on the method of damage repair. The insurance companies offer options on how to do it. The insured may repair damage by themselves (on a budget) or hire a company for the job. Especially if the case is small damage, the clients may choose the first option. They receive money from the insurance company earlier, albeit usually a smaller amount compared to the second option. As explained by an employee of one unnamed insurance company: '*People who decide to remove damage by themselves or not to repair it at all receive less money, because the insurance company only pays up to the time price. If they hire a contractor and the damaged building has been insured to a new value, and present an invoice to the insurer, the insurance company reimburses it including the price of the work.*' According to him, people believe that the first option is more profitable. They estimate the damage, for ex-

ample at €3,000, and plan to repair the damage for half that amount, expecting to obtain the other half from the insurer. Of course the reality is different. Clients are advised not to try to gain any surplus through the claim.

ACKNOWLEDGEMENTS

This article is one of the outputs of the 2/0038/15 project, financially supported by the Scientific Grant Agency of the Ministry of Education, Science, Research and Sport of the Slovak Republic and the Slovak Academy of Sciences (VEGA).

REFERENCES

[1] APFM (Associated Programme on Flood Management), *Integrated Flood Management*, APFM technical document no. 1, 2nd edn., World Meteorological Organization and Global Water Partnership: Geneva, 2004.

[2] Lamond, J.E., Proverbs, D.G. & Hammond, F.N., Accessibility of flood risk insurance in the UK: confusion, competition and complacency. *Journal of Risk Research*, **12**(6), pp. 825–841, 2009.
http://dx.doi.org/10.1080/13669870902768614

[3] Thieken, A.H., Petrow, T., Kreibich, H. & Merz, B., Insurability and mitigation of flood losses in private households in Germany. *Risk Analysis*, **26**(2), pp. 383–395, 2006.
http://dx.doi.org/10.1111/j.1539-6924.2006.00741.x

[4] Thieken, A.H., Kreibich, H., Müller, M. & Merz, B., Coping with floods: preparedness, response and recovery of flood-affected residents in Germany in 2002. *Hydrological Sciences Journal*, **52**(5), pp. 1016–1037, 2007.
http://dx.doi.org/10.1623/hysj.52.5.1016

[5] Siegrist, M. & Gutscher, H., Natural hazards and motivation for mitigation behavior: people cannot predict the affect evoked by a severe flood. *Risk Analysis*, **28**(3), pp. 771–778, 2008.
http://dx.doi.org/10.1111/j.1539-6924.2008.01049.x

[6] Holub, M. & Fuchs, S., Mitigating mountain hazards in Austria: Legislation, risk transfer, and awareness building. *Natural Hazards and Earth System Sciences*, **9**(2), pp. 523–537, 2009.
http://dx.doi.org/10.5194/nhess-9-523-2009

[7] Johnstone, W.M. & Lence, B.J., Assessing the value of mitigation strategies in reducing the impacts of rapid-onset, catastrophic floods. *Journal of Flood Risk Management*, **2**(3), pp. 209–221, 2009.
http://dx.doi.org/10.1111/j.1753-318X.2009.01035.x

[8] Michel-Kerjan, E. & Kunreuther, H., Redesigning flood insurance. *Science*, **333**(6041), pp. 408–409, 2011.
http://dx.doi.org/10.1126/science.1202616

[9] Slovak Insurance Association. Insurance products on the Slovak insurance market, 2015. Online [in Slovak], avaliable at https://www.slaspo.sk/tmp/asset_cache/link/0000048299/RS%20SLASPO%202013%20-%20kapitola%20D%20-%20SJ.pdf

[10] Slovak Insurance Association, Annual Report 2013. Chapter D: Slovak Insurance Market in 2013, 2014. Online [in Slovak], available at https://www.slaspo.sk/tmp/asset_cache/link/0000048267/Poistne%20produkty%20na%20SK%20poistnom%20trhu%20k%203103%202014%20na%20web%20stranku.pdf

[11] Solín, Ľ., Madajová, M. & Skubinčan, P., Mitigating flood consequences: analysis of private flood insurance in the Slovakia. *Journal of Flood Risk Management*, DOI: 10.1111/jfr3.12191, 2015.
http://dx.doi.org/10.1111/jfr3.12191

[12] Madajová, M., Solín, Ľ. & Michálek, A., Vývoj a priestorová variabilita poistenia obyvateľstva proti riziku povodne na Slovensku v období 2002 – 2011. *Ekonomický časopis*, **63**(2), pp. 167–187, 2015.

[13] Ward, R.C. & Robinson, M., *Principles of Hydrology*, 3rd edn., McGraw-Hill Book Company: London, 1990.

[14] Smith, K. & Ward, R., *Floods: Physical Processes and Human Impacts*, John Wiley & Sons: Chichester, 1998.

[15] Chow, V.T., Hydrologic studies of floods in the United States. *International Association of Scientific Hydrological Publications*, **42**, pp. 134–170, 1956.

[16] Ward, R.C., *Floods: A Geographical Perspective*, Macmillan: London, 1978.

[17] Act No. 71/2015, Amendment to the Act Flood Protection No. 7/2010, Slovakia [in Slovak].

[18] Act No. 7/2010, Act on Flood Protection No. 7/2010, Slovakia [in Slovak].

THE STEVENS FLOOD ADVISORY SYSTEM: OPERATIONAL H³E FLOOD FORECASTS FOR THE GREATER NEW YORK / NEW JERSEY METROPOLITAN REGION

N. GEORGAS, A. BLUMBERG, T. HERRINGTON, T. WAKEMAN, F. SALEH, D. RUNNELS, A. JORDI, K. YING, L. YIN, V. RAMASWAMY, A. YAKUBOVSKIY, O. LOPEZ, J. MCNALLY, J. SCHULTE & Y. WANG
Davidson Laboratory, Stevens Institute of Technology, USA.

ABSTRACT

This paper presents the automation, website interface, and verification of the Stevens Flood Advisory System (SFAS, http://stevens.edu/SFAS). The fully-automated, ensemble-based flood advisory system dynamically integrates real-time observations and river and coastal flood models forced by an ensemble of meteorological models at various scales to produce and serve street scale flood forecasts over urban terrain. SFAS is applied to the Greater NY/NJ Metropolitan region, and is used routinely by multiple forecast offices and departments within the US National Weather Service (NWS), regional and municipal Offices of Emergency Management, as well as the general public. Every six hours, the underlying H^{3}E (Hydrologic–Hydraulic–Hydrodynamic Ensemble) modelling framework, prepares, runs, data-assimilates, and integrates results from 375 dynamic model simulations to produce actionable, probabilistic ensemble forecasts of upland and coastal (storm surge) flooding conditions with an 81-h forecast horizon. Meteorological forcing to the H^{3}E models is provided by 125 weather model ensemble members as well as deterministic weather models from major weather agencies (NCEP, ECMWF, CMC) and academia. The state-of-the-art SFAS, a replacement of the well-known, but deterministic, Storm Surge Warning System (SSWS) that was highlighted during Hurricanes Irene and Sandy and more recently extratropical cyclone Jonas, has been operational since the end of 2015.

Keywords: early warning systems, emergency management, ensemble forecast, flood advisory system, integrated flood forecasts, river flooding, storm surge.

1 INTRODUCTION

In the United States of America alone, annualized inland hydrologic flood losses over the 30-year period ending at the end of water year (WY) 2014, stood at $7.96 billion in damages, and 82 fatalities, per year [1]. These hydrologic losses are adjusted for inflation but do not include losses from coastal storm tide flooding due to Extratropical and/or Tropical Cyclones (ETCs and TCs, respectively). Storm tide flooding can be defined as coastal flooding caused by a storm's surge pushing ocean waters to rise above local astronomical tide levels, with the phase of the local astronomical tide having important effects on total water levels above ground [2].

New York Harbor (NYH) and its connected tidal waterways lie at the apex of the New York Bight, and are surrounded by over 21 million people in cities such as New York, Newark, Jersey City and Hoboken. The metropolitan region with the highest risk for storm tide flood-

DOI: 10.2495/SAFE-V6-N3-648-662

ing is the New York City Core Based Statistical Area (CSA), which includes New York City (NYC), Long Island and a portion of New Jersey, with a total 685,152 homes at risk and a reconstruction-cost-value of over $244 Billion [3]. As with Hurricane Sandy in 2012, TCs can cause significant coastal storm tide flooding in the Eastern United States and the New York / New Jersey Metropolitan Region. In New York counties alone, aggregated storm tide losses from Hurricane Sandy were estimated to total $23B of over $50B country-wide [4, 5]. New Jersey was the second coastal state worst hit. Historically, many of the historical records for storm tide flooding in the region have been caused by cold-season ETCs, mid-latitude cyclones locally called Nor'Easters [6]. On January 23 2016 for example, Nor'Easter Jonas caused severe tidal flooding in southern New Jersey: Jonas' storm tide broke the local records previously set by Sandy or the historic December 1992 Nor'Easter, while dumping near 30 inches of snow in the northern part of the State and New York City; an area that did not experience major coastal flooding during the event [7].

Importantly, TCs are not only a coastal flood hazard but an important inland flooding agent in the Eastern United States [8]. As recently as between October 1–5 2015, locally extreme rainfall totals exceeded 20-inches in several eastern US states resulting from the convergence of a powerful low pressure system / frontal boundary and copious moisture from Hurricane Joaquin in the Atlantic [9]. A year before Hurricane Sandy, Hurricane Irene and Tropical Storm Lee in the summer of 2011 brought severe inland flooding into the NY State region and broke records for both storm surge and river stage flooding [10, 11].

To characterize deaths related to Sandy, the US Centers for Disease Control and Prevention (CDC) analyzed data on 117 hurricane-related deaths captured by American Red Cross mortality tracking during October 28–November 30, 2012. 53 of these deaths occurred in New York and 34 in New Jersey, with drowning being the most common cause of death related to Sandy, and 45% of drowning deaths occurring in flooded homes in NYC's Evacuation Zone A. The CDC report states: 'Drowning is a leading cause of hurricane death but is preventable with advance warning systems and evacuation plans' [12].

Seventy-eight percent of emergencies are weather related [13, 14], with flooding now being the second leading cause of death behind heat-related incidents according to the CDC. Driving a car into flood waters is the leading cause of flood-related deaths, making the 'Turn Around, Don't Drown' message of the US National Weather Service (NWS) one of the most ubiquitous and recognizable signs in the US. NWS is working hard to implement its Weather-Ready Nation plan with a goal to minimize preventable losses in life and property from flooding. To effectively minimize flood risks, flood forecasting and warning systems must provide precise flood information to emergency managers and decision-makers. Producing accurate flood forecasts requires numerical models that properly take into account the important physical processes that affect a storm tide or river stage and consequently the flooded areas [11, 15–17]. An accessible interface that is quick to understand and easy to use in decision making is then needed to effectively communicate these forecasts to the public.

Stevens Institute of Technology has built and operates an automated Storm Surge Warning System (SSWS) for New Jersey and New York since 2002 [18]. The system extended its capacity over the years with support from the New York Harbor Observing and Prediction System (NYHOPS) and its models [19, 20] and was highlighted during Hurricane Sandy as being preferred by the NYC Office of Emergency Management for its ease of use [2, 21]. With significant funding through the Port Authority of New York and New Jersey (the PANYNJ), the system has recently been updated and renamed the Stevens Flood Advisory System (SFAS, http://stevens.edu/SFAS). In this paper, the design of the latest generation

SFAS and its website interface is presented, followed by a discussion on the system's verification and improvements upon other current SSWSs in the US, and closing with a discussion of ongoing research and development.

2 DESIGN OF THE STEVENS FLOOD ADVISORY SYSTEM

The system diagram shown in Fig. 1 abstracts the components of SFAS. Real-time water level observations (coastal storm tide and river stage) and an ensemble of water level forecasts produced by operational computer models at stations in and around the NY/NJ Greater Metropolitan area are acquired and stored in relational databases, registered to equivalent vertical datums, and then compared against critical flood levels to interpret whether a station is presently flooding or whether it is expected to flood in the future over the next 81 h. It also indicates how severe the flooding may be. The data are then visualized on the SFAS website. For tidal stations, storm surge is also computed as the deviation between the observed or forecast water level from the astronomical tide prediction for each station.

To provide common context for the severity of occurring or anticipated flooding, we use the high water level terminology of the NWS for the critical flood levels [22]: 'Near' flooding (equivalent to 'Action stage' for inland river stations) and 'Minor', 'Moderate', and 'Major' flood levels. These levels are set and provided through regional NWS Weather Forecast Offices (PHI, OKX, ALY, and GYX) for accuracy and consistency [23]. Station locations displayed on the main SFAS map (Fig. 2) start blinking at Near (Action) flood stage. Minor Flooding represents minimal or no property damage but possibly some public threat (e.g. inundation of roads). Moderate Flooding signifies some inundation of structures and roads near the coastline or stream and possible evacuations of people and/or transfer of property to higher elevations. Major Flooding indicates extensive inundation of structures and roads and significant evacuations of people and/or transfer of property to higher elevations. An SFAS user has the ability to register for automated flood alerts for the stations he/she is most interested in by providing his e-mail and information on the website; an automated e-mail will be sent when water level at a registered station is forecast by the modeling ensemble to rise above Minor Flood stage.

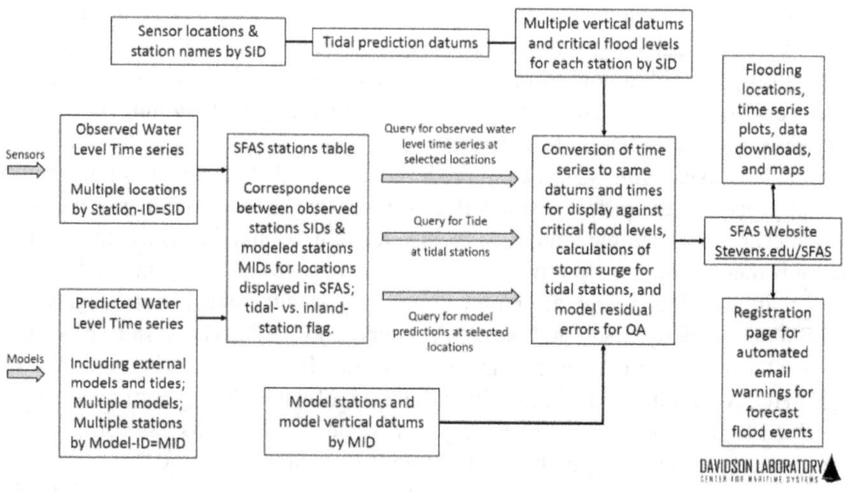

Figure 1: Stevens flood advisory system diagram.

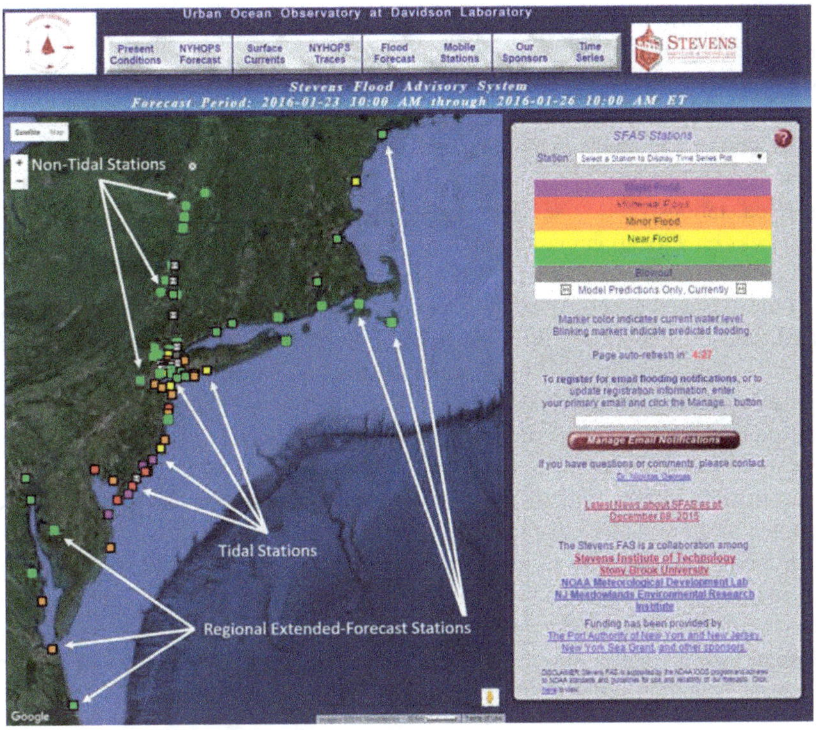

Figure 2: Screenshot taken from the opening page of the SFAS website (http://stevens.edu/SFAS), at 10 am EST (UTC-5) January 23 2016, during Nor'Easter Jonas, showing Major (purple) and Moderate (red) flooding occurring in the Delaware Bay and southern New Jersey coastal regions, with Minor flooding (orange) occurring in New York Harbor and western Long Island. All stations with forecast water levels above Near-flood stage are blinking and outlined with thicker black lines. Arrows point to examples of inland (non-tidal) flood station locations (with circles), and tidal station locations (squares), either within the NY/NJ greater metropolitan area that the NYHOPS model covers, or regional extended-forecast stations forecast by SNAP.

2.1 SFAS Website

Figure 2 shows the opening page of the SFAS website (http://stevens.edu/SFAS). This page is also accessible from the NYHOPS website (http://stevens.edu/NYHOPS) by clicking on the top banner's 'Flood Forecast' button. White arrows are included on Fig. 2 to highlight different types of stations shown on the map: inland stations (rivers, using circles), and tidal stations (ocean coast, estuarine, or tidal rivers, using squares). The latter station are further broken down to indicate those stations for which flood advisories are based on the NYHOPS model for the NY/NJ Greater Metropolitan area or by the SNAP extended regional model (see section 2.2). The user can see if a station is presently experiencing flooding by comparing its marker colour to the colour key found on the right hand side of the webpage. Note that the grey colour is used if a 'blowout tide' condition is presently occurring, a blowout tide being a 'negative storm surge' condition where the waters have dropped too low and safe vessel passage through navigable waters is a concern.

If a user would like to receive email notifications of predicted flooding, he/she can enter their primary email address in the control panel to the right, and click the 'Manage Email Notifications' button. On the page that appears, users may select the stations for which they would like to receive notifications, optionally, provide an alternate email, or deregister previously registered stations.

If a user clicks on a location marker, an 'info window' will appear collectively noting the current observed water level, the most severe forecast level in the forecast period, and the date and time of the expected event. By default, an ensemble average of forecast predictions created at the Davidson Laboratory of Stevens Institute of Technology is used in SFAS. To display a time series plot (Fig. 3), which includes 81 or more hours of forecast water levels, and observed data as available, users should click on the 'Show Time Series Plot' in the info window or use the Station drop down in the control panel to select a station.

On the top-left panel of the SFAS-station time series webpage (Fig. 3), the user can shift between stations through the 'Station' drop-down menu. The best Stevens ensemble forecast for the selected location will be displayed by default. Where available, a user may select to display a different forecast time series from a number of forecast models described in the following section by selecting them from the 'Forecast:' drop down box, which includes three models from the NWS (Fig. 3). The numerical values for the selected forecast time series can be downloaded by clicking on the 'Extended download' button at the bottom of the control box. For consistency, time series of observed (where available) and forecast water levels are

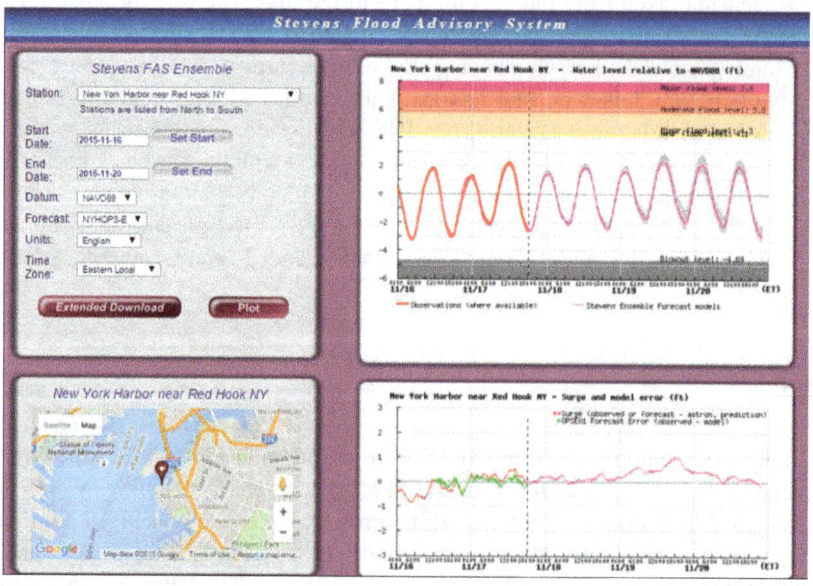

Figure 3: SFAS-station, water level time series page. The selected model prediction (in this case, the NYHOPS Ensemble Forecast, NYHOPS-E; pink line) appears on the top right panel, for a new tidal station (New York Harbor near Red Hook, NY), including the 5th to 95th percentile forecast uncertainty (grey area), along with past and real-time observations (red colour).

shown and can be downloaded referenced to a 0 level being the North American Vertical Datum of 1988 (NAVD88). The user can, however, select among multiple vertical Datums, Units, etc., depending on familiarity or need, as well as look or download past data and predictions using the control box options shown on Fig. 3. The map on the bottom left of the webpage shows the selected station location in anticipation of predicted overland flood depth maps that will be placed there in the future, while the bottom right plot shows the model's deviation from observed data (green line) together with the observed (red) or model-forecast (pink) storm surge above astronomically-predicted tide levels (for tidal stations only).

2.2 Supporting infrastructure: Observations and Models

The Davidson Laboratory has been installing 11 dual (for redundancy) environmental monitoring stations (Aquatrack 5002 & Greenspan Analytical EC250 C/T sensors and power cells) near critical infrastructure sites, such as airports and marine terminals, through a collaboration with the PANYNJ. This environmental data and collected water levels are sent to the Stevens Institute databases in real-time through cellular connections from the station data loggers every 6 minutes. Real-time data from partner institutions and agencies (see Acknowledgements section) are retrieved through the internet. Such regional collaborations have helped greatly in expanding the scope of the system.

Regarding water level forecasts at SFAS stations, most water level forecast data are generated at the Davidson Laboratory. The overall research objective is to predict inundation occurring from a storm surge|rainfall event at the street level scale. The Storm Surge flood forecasts, and automated e-mail flood advisories that previous SSWS users were familiar with are now of improved quality because they are ensemble-model based. This is accomplished by running many numerical hydrodynamic and hydrologic models on the in-house Pharos Hyperscale Computing Facility, a 1,320-core Hewlett-Packard supercomputer built on sheltered high-ground based on HP Proliant microserver architecture, Mellanox FDR InfiniBand network and a 2.2 PB Seagate online storage solution, dual head nodes, Uninterrupted Power Supplies, Cooling Towers, and a Caterpillar generator for storm resiliency.

Figure 4 shows the domains and scope of the presently operational Davidson Laboratory models. Three sets of 125 linked coastal and inland flood models with a forecast horizon of at least 81 h, are reinitialized every 6 h and are based on different atmospheric model predictions of surface meteorological factors, such as near-surface winds, barometric pressure, and rainfall. Meteorological forcing for the 125 members of the 'Stevens Ensemble' is provided through output fields from the ECMWF ensemble and ECMWF High-Resolution [24], SREF Ensemble [25], GEFS Ensemble [26], GFS, including its experimental version GFS-e [27], NAM [28], Rutgers-WRF [29], and the Canadian Meteorological Centre's Ensemble (CMC) [30] models (Fig. 4).

The computational grid of the three-dimensional NYHOPS built using the sECOM hydrodynamic model, consists of 147×452 nodes and its horizontal resolution ranges from approximately 7.5 km at the open ocean boundary to less than 50 m in NY/NJ Harbor [19, 20]. The grid has 10 vertical layers that stretch from the water surface to the local bottom relief. The NYHOPS grid is nested in (derives open ocean boundary conditions from) the regional-scale Stevens Northwest Atlantic Prediction (SNAP) model domain, which is also based on the sECOM code but using a 5-km constant resolution grid [31]. River stages at SFAS stations are forecast by the Stevens HYDRO model, based on the HEC-HMS code [16]. River discharges from HYDRO are coupled into the NYHOPS model. This operational

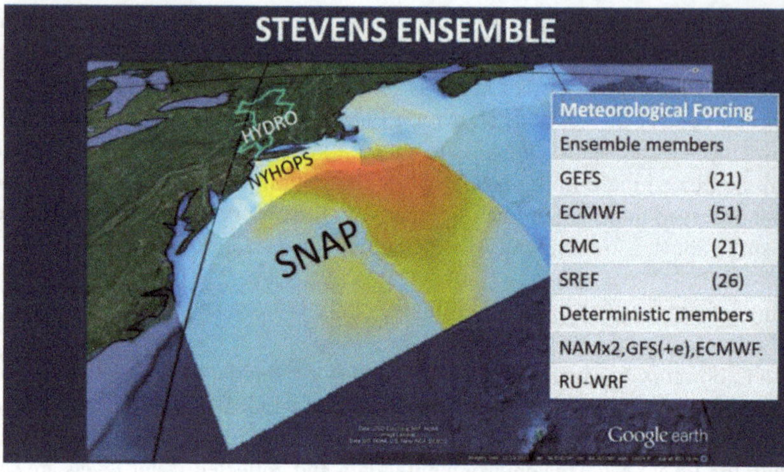

Figure 4: sECOM-NYHOPS three dimensional 125-member hydrodynamic ensemble linked to offshore Stevens Western North Atlantic Prediction (sECOM-SNAP) model and inland Stevens Hydrologic (HMS-HYDRO) ensembles.

framework offers an improvement over AHPS in this particular region [16, 32]. The AHPS streamflow forecasts are at 6-h time intervals using one weather control member for input and with a lead time that is less than 60 h in this region [33]. A new ensemble AHPS is being evaluated presently by the NWS.

The NYHOPS model ensemble members span 81 h from present. Every 6 h, each of 125 NYHOPS ocean model forecasts is run based on a different meteorological forecast. These 125 different water level predictions provide uncertainty estimates and are processed to create the consensus weighted ensemble average prediction that the automated e-mail advisories are based on, called 'NYHOPS-E.' Ensemble-averaged water level predictions are displayed in the SFAS time series and are enveloped by the uncertainty around that ensemble (5th to 95th percentile) prediction to better depict a range of possible futures (Fig. 3). Selecting the most probable of these possible futures, the one that is most likely to be experienced over the next 3–4 days, is the focus of ongoing research in ensemble forecasting.

An extended storm surge forecast ensemble (SNAP-Ex) that spans at least 105 h from present is also available at some regional stations based on 76 of 125 total members of the SNAP regional ensemble that predict further than 81 h out (51 ECMWF, ECMWF-HR, 21 GEFS, GFS, GEFS-e, and RU-WRF). 17 new regional stations, from Maine to North Carolina, including cities like Portland, Boston, Baltimore, and Annapolis, were added to SFAS, with predictions based on that extended regional ensemble.

3 COMPARISONS OF ENSEMBLE WATER LEVEL PREDICTIONS TO EXISTING DETERMINISTIC SYSTEMS

The NWS has developed two deterministic models for ETCs [34]: The Extratropical Storm Surge model (ETSS), based on the SLOSH numerical model, and the Extra Tropical Storm and Tide Operational Forecast System (ESTOFS) based on the 2D ADCIRC model. For Extra-Tropical storms, both models use a nominal 2.5 km, are forced by GFS winds, and are initialized every six hours.

Gridded water level results from these models at specific stations are debiased against observations by an anomaly-correction procedure used by NOAA's Meteorological Development Lab (MDL), and subsequently posted on MDL's Extra-Tropical Surge Guidance website. From there, debiased water levels from the two models at SFAS stations are downloaded each hour, stored at the Stevens databases along with other models, and displayed in SFAS as NOAA guidance together with the Stevens NYHOPS-E, SNAP, and SNAP-Ex ensembles. Di Liberto *et al.* [15] compared NYHOPS water level predictions to ETSS for 75 available days during the 2007-08 and 2008-09 cold season to assess the benefit of creating a multi-model storm surge ensemble using NYHOPS, ETSS, and a Stony Brook University ADCIRC-based ensemble. Forecasters from the NWS OFSs who use SFAS routinely and regularly note on the benefit of having these different guidance sets for the same stations and valid times displayed on one website, SFAS. They routinely do their own multi-model ensemble averaging to devise their coastal flooding forecasts. In the process of verifying that ensemble, Di Liberto *et al.* [15] found that ETSS was significantly biased, but after an imposed debiasing produced quite accurate results; yet, it was still somewhat inferior to a deterministic NYHOPS model based on NAM winds.

Here, as part of the new ensemble verification, we expanded and updated that work by comparing the accuracy of the water level forecasts that debiased ETSS and now ESTOFS provided for 130 days mostly in the cool season of 2015-2016 to that of the 3 new three-dimensional model ensembles (the NYHOPS-E, SNAP, and SNAP-Ex ensembles created at the Davidson Laboratory based on the sECOM model). The method used was the following:

a. Extract the predictions at several stations from the five systems collected hourly at Stevens Institute databases, four times a day, at 0 z, 6 z, 12 z, and 18 z each day beginning on December 30 2015 at 0 z, real time. The process was automated and continued henceforth each day. The dataset considered here ends on April 23 2016 at 18 z, real time, some 135 days later. This real-time operation disallows any posterior data or forecast corrections, but ensures that the downloaded model predictions are truly as they were stored and displayed at SFAS at the time of the scheduled downloads, across all 5 models considered.
b. The forecasts from each model were then binned in six hourly increments as the 0–6 h, 6–12 h, 12–18 h forecast-bin for each model, etc., to the end of each model's forecast horizon.
c. The data were used to compare the models performance at each station, primarily using the same metrics used at [15]: The Mean Error (ME), and the Root-Mean-Square-Error (RMSE). For ease of presentation, model results on Fig. 5 are aggregated over several stations in four station categories, somewhat arbitrarily, and where available from each model, corresponding to SFAS stations on Fig. 2:
   i. Coastal NYHOPS stations: Ocean City Inlet, MD; Lewes, DE; Cape May, NJ; Atlantic City, NJ; Montauk, NY; Newport, RI.
   ii. Estuarine NYHOPS stations: Reedy Point, DE; Sandy Hook, NJ; The Battery, NY; Kings Point, NY; New Haven, CT; Bridgeport, CT; New London, CT.
   iii. SNAP-South: Annapolis, MD; Cambridge, MD; Lewisetta, VA; Baltimore, MD; Chesapeake Bay Bridge Tunnel, VA; Duck, NC.
   iv. SNAP-North: Woods Hole, MA; Nantucket Island, MA; Boston, MA; Fort Point, NH; and Portland, ME.

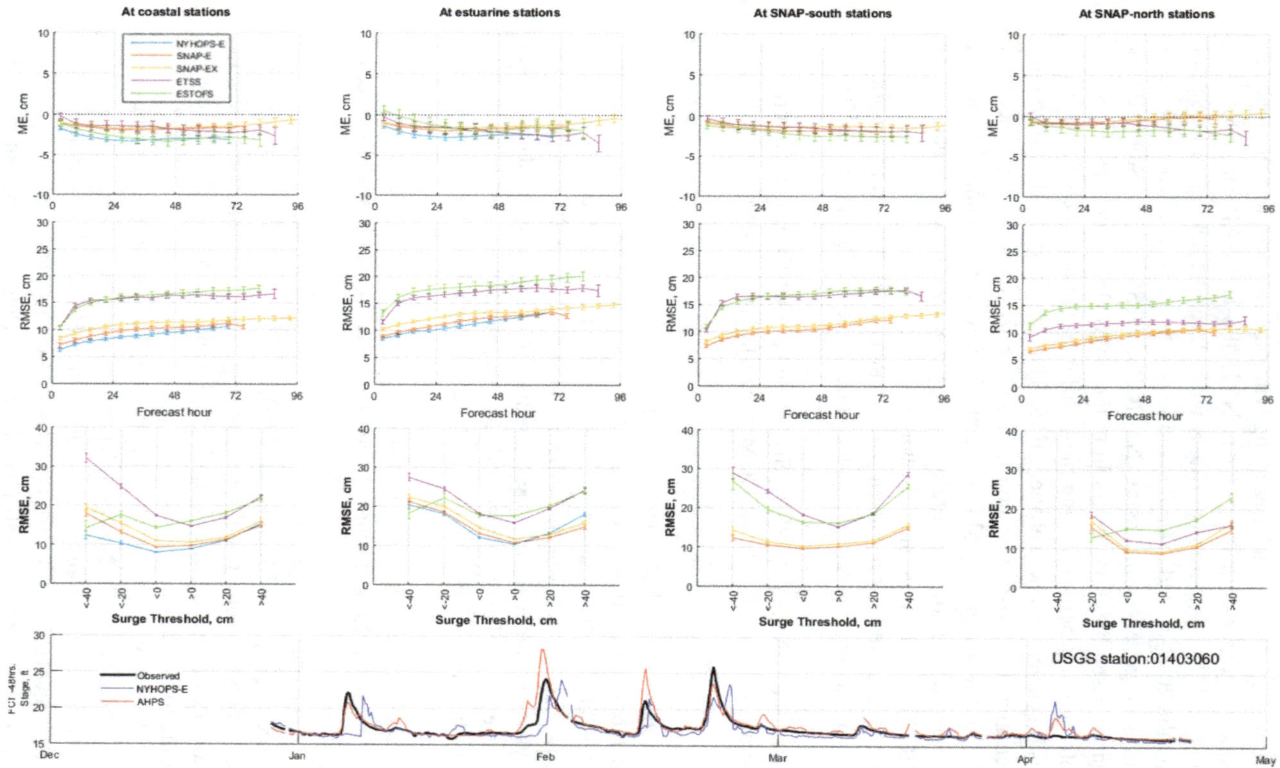

Figure 5: (Previous page). Top three panels: Multi-model comparison for storm surge forecasts against observations across several stations operated by the National Ocean Service (NOS) and included in SFAS. Top row: ME by forecast lead hour. Second row: RMSE by forecast lead hour. Third row: RMSE by surge threshold, similar to [15], across all 0–72 h forecasts. Bottom row: Comparison between the derived HYDRO hydrologic ensemble stage ('NYHOPS-E') time series and the AHPS deterministic model stage against observations collected by USGS at an inland river station: 01403060, Raritan River below Calco Dam at Bound Brook NJ. Stages were derived from 6 h-averaged NYHOPS-E discharge data, through the same rating curves, to match AHPS's six hourly outputs. Only the 48 h forecast lead time time-series is shown.

d. 90% confidence intervals around the calculated means were generated using Matlab's boot-strapping by random replacement that was repeated 1000 times for each interval shown. If the confidence intervals of two particular samples did not overlap, they were considered different at the 90% level.
e. River stages between AHPS and HYDRO were treated similarly, except that the latest rating curve obtained from USGS per station was used to convert river discharge, which is what the hydrologic models predict, to river stage, and is what is used in SFAS for comparison to flood stages. Also, the AHPS forecast horizon was short, near 60 h.

The top three panels of Fig. 5 clearly show that all five models produce reasonable results, with the NYHOPS-E storm surge predictions continuing to have overall the least amount of error residuals at all stations considered. ME's are small, on the order of 2–3 cm for all models and regions, a significant improvement for the NOAA ETSS model compared to 5 years ago. The NYHOPS-E across-station forecast-wide RMSE is ~ 10.5 cm during this 135-day period that included three Nor'Easter storms, including ETC Jonas, which was a record-setting storm for some locations. The NYHOPS-E RMSE is about 50% less than that of the NOAA models (ETSS and ESTOFS). All models have higher errors at estuarine stations compared to coastal stations. Model errors grow with forecast lead time, with ETSS and ESTOFS experiencing a very quick increase in RMSE shortly after forecast start, and then following the gradients of the Stevens models; the net effect being that the Stevens models maintain comparable performance to the NOAA models initial 12 h performance for a much longer lead, up to at least 96 h in the SNAP-Ex case. The outlier is the ETSS model at SNAP-North stations, where that model's error does not appear to grow significantly with lead time. Interestingly, the differences between the higher resolution NYHOPS model performance from the lower resolution SNAP model, both based on the sECOM code, but with SNAP having less comprehensive forcing that NYHOPS, are barely significant at the 90% level. They are about the same magnitude different as SNAP to itself using less Ensemble models (SNAP vs. SNAP-Ex), with the leaner SNAP-Ex ensemble having an increase in error compared to the full 125-member SNAP for similar lead times. As expected, RMSE grows with positive or negative surge threshold on Fig. 5, third row from top. The exception is the constancy or increase of skill for the ESTOFS model at low surges in three of the four regions. Comparing ESTOFS and ETSS alone, no consistently better model emerges for the period considered here.

For river stage, the example in Fig. 5 is largely representative of what happens in other stations too. In the case of inland stations, overall for the period considered, NOAA AHPS did significantly better than the NYHOPS HYDRO ensemble in predicting the observed river stage, by as much as 50% in upper Hudson watersheds (not shown). Resolution and spread of rainfall appear to be key here. In the case of the Stevens inland hydrology ensemble, NYHOPS-E is presently the median of all precipitation-driven hydrologic predictions. Given the high spread in the precipitation data and the way it propagates in the hydrologic model this appears to be a limiting approach; research is ongoing on other ensembling techniques. The AHPS performance in RMSE is of the same order of magnitude as the tidal NOAA models (not shown). The example shown in Fig. 5 is for the 48 h lead forecast time. The error growth with lead time appears to be quasi-linear to lead time, with the two models being roughly equivalent at small lead times, and growing toward the forecast horizon.

4 PERFORMANCE COMPARISONS FOR INDIVIDUAL MEMBERS IN THE NYHOPS ENSEMBLE TO EACH OTHER, AND THE ENSEMBLE MEAN

Figure 6 compares the performance of individual members of the 125-member NYHOPS ensemble to each other and to NYHOPS-E ensemble mean for coastal stations, using the Grand mean RMSE (GM-RMSE) over all runs in the verification period as a performance metric. Figure 6 clearly shows that the NYHOPS-E weighted ensemble used in SFAS has indeed a lower GM-RMSE than any of the 125 NYHOPS members that comprise it. This is both true at The Battery NOS station, but also when all 13 stations are considered (Fig. 6).

As expected, NYHOPS members forced by deterministic members of the same meteorological models show better performance in simulating storm tide than NYHOPS members forced by the lower resolution meteorological ensemble members of the same model (GFS deterministic vs. GEFS ensemble, for instance). Comparing across meteorological ensembles, the

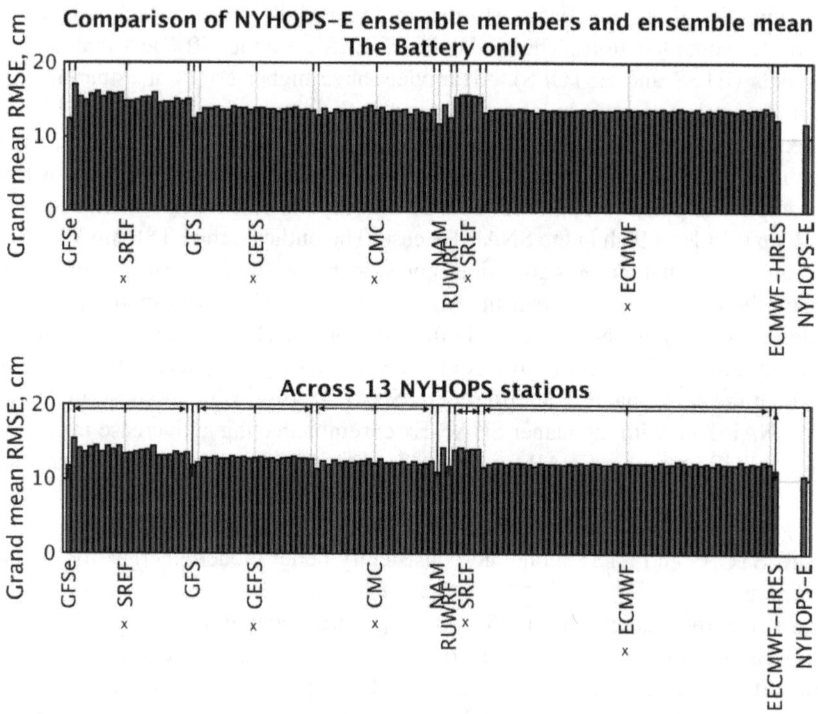

Figure 6: Comparison of 125 NYHOPS Ensemble Members to each other and to the NYHOPS-E weighted ensemble mean (right-most bar): The bar charts depict Grand Mean RMSE across all forecast times, four forecast cycles a day, and 135 days in the verification period at The Battery, NY (top panel), or at all 13 NYHOPS stations considered together (bottom panel). NYHOPS members forced with deterministic high-resolution meteorological models are highlighted with black vertical lines on the bottom panel. Perturbed meteorological ensembles of a single model Horizontal arrows span NYHOPS members forced by a perturbed ensemble of a single meteorological model, highlighted with an x below its name.

SREF-forced ensemble members perform quantitatively worse than average, even though the meteorological fields from SREF have a relatively high 16 km resolution. Also note the big discrepancy in Grand Mean RMSE between individual SREF members. This appears to be a sign of an over-dispersed SREF ensemble; the model is run with two different physical packages and strong negative and positive perturbations. This finding does not appear to necessarily mean that the SREF ensemble is not a good ensemble to use to force an ensemble for storm surge, however. For example, the SNAP-Ex model that does not use SREF, nor the CMC, was slightly inferior to the SNAP ensemble that did (Fig. 5).

On the other end of the spectrum, Fig. 6 shows that the NYHOPS members forced with the ECMWF ensemble have slightly lower GM-RMSE than the ones forced with the CMC and GEFS ensembles, and also that ECMWF is not over-dispersed because the standard deviation in GM-RMSE among its members is the lowest compared to that of other ensembles (its bar-chart profile is flatter). CMC and GEFS also appear to be similarly-dispersed and quite similar in performance here. Further, and as expected, careful consideration of each ensemble internal structure reveals that control members tend to have the smaller GM-RMSE compared to an ensemble's perturbed members.

5 CONCLUSIONS AND FUTURE WORK

The redesigned, and now ensemble-based SFAS, was presented and its flood-predicting performance was compared to other operational flood prediction systems. The work shows that the new ensemble water level forecasts in SFAS provide improved predictions at tidal stations, thus improving the assessment of coastal flood risks when compared with deterministic water level forecasts, and leading the way for similar changes occurring at the NWS level. Section 4 showed that the weighted ensemble average has improved performance (significantly lower storm surge RMSE) at coastal stations compared to even some of the best members of the NYHOPS-E ensemble. Uncertainties in weather inputs may result in false warnings and missed flooding events, reducing the potential to effectively mitigate flood damage, and need to be accounted for through ensemble techniques [35].

The research presented also indicates the value of using increased resolution meteorological forecasts in such an ensemble: with the exception of SREF, the NYHOPS ensemble members that were forced with the higher resolution versions of each meteorological model, as well as the control members of the lower resolution ensembles produced results with the lowest RMSE in forecast water level. Statistical downscaling of meteorological ensembles to their high-resolution equivalents might increase accuracy further, and such downscaling methods should be investigated in the future.

Stevens SFAS forecasts were very accurate during winter storm Jonas, predicting no significant coastal flooding effects at Port Authority facilities well in advance of the storm, unlike the significant flooding predicted for South Jersey (https://twitter.com/PortAuthOEM/status/690897301905227776). SFAS has become part of the NWS information flow: The web site has been entrained into the NWS pages (http://www.weather.gov/phi/tides; Clicking on the New York Harbor map, the link takes a user to the Stevens web site). The NYHOPS ensemble tracked *Jonas* tidal water levels at the 85 SFAS stations for the most part within the 5%–95% uncertainty envelope displayed in the SFAS website. It is of concern that at most locations the observations were skewed high, closer to the 95% of the forecast ensemble envelope for this Nor'easter Storm, as this may be an indication of the weighted ensemble process over-smoothing peaks. There was also large uncertainty around the river water level ensemble prediction, primarily due to the large uncertainties associated with precipitation forecasts. Alternative methods to minimize that uncertainty and produce a more accurate ensemble are currently under active research and development with some promising theoretical results. Work needs to continue on developing better ensemble and debiasing techniques for tidal and inland stations and improving downscaling of the lower-resolution members of the ensemble.

With regard to SFAS website improvements, although inundation maps are now created with bath-tubbing techniques (http://hudson.dl.stevens-tech.edu/njdemo/), dynamic street-scale resolution sECOM models (as in [32]) for the PANYNJ facilities is now a focus activity. For the SFAS region as a whole, the hydrodynamic resolution deficit at the NJ Back Bays became obvious during Nor'Easter Jonas, with observations at a few sites in the Intracoastal Waterway (Sea Isle City NJ, Stone Harbor NJ, and even Ocean City Inlet MD) coming out of the 5%–95% Stevens Ensemble predictions at times.

ACKNOWLEDGMENTS

This work was partially funded by a research task agreement entered between the Trustees of the Stevens Institute of Technology and the PANYNJ, effective August 19, 2014. New York Sea Grant, New Jersey Sea Grant, Connecticut Sea Grant, the Long Island Sound Study, the NJ Coastal Protection Technical Assistance Service, the Borough of Manasquan, NJ, the Monmouth County Office of Emergency Management, the NJ Meadowlands Research Institute of the New Jersey Sports and Exposition Authority, the New Jersey Office of Emergency Management, the Hudson River Environmental Conditions Observing System, the United States Geologic Survey, and the NOAA Integrated Ocean Observing System program and the Mid-Atlantic Regional Association for Coastal Observing Systems (MARACOOS) are acknowledged for their contributions and support to the SFAS. We are thankful for the academic numerical models contributed to SFAS by Rutgers University and Stony Brook University, as well as for the operational models contributed by the US National Centers for Environmental Prediction, the Canadian Meteorological Centre, the European Centre for Medium-Range Weather Forecasts, and the NOAA Meteorological Development Laboratory. Finally, the authors are grateful to Ms. Elizabeth Lamb for help editing this document.

REFERENCES

[1] NWS, United States Flood Loss Report – Water Year 2014, available at: http://www.nws.noaa.gov/hic/summaries/WY2014.pdf (accessed 01 Febuary 2016), 2014.

[2] Georgas, N., Orton, P., Blumberg, A., Cohen, L., Zarrilli, D. & Yin, L., The impact of tidal phase on hurricane sandy's flooding around New York City and Long Island Sound. *Journal of Extreme Events*, **1**(01), p. 1450006, 2014.

[3] Botts, H., Du, W., Jeffery, T., Kolk, S., Pennycook, Z. & Suhr, L., CoreLogic Storm Surge Report 2015, Irvine, CA. p. 44, 2015.

[4] Schubert, C.E., Busciolano, R., Hearn Jr, P.P., Rahav, A.N., Behrens, R., Finkelstein, J., Monti Jr, J. & Simonson, A.E., *Analysis of Storm-Tide Impacts from Hurricane Sandy in New York*, US Geological Survey, 2015.

[5] Blake, E.S., Kimberlain, T.B., Berg, R.J., Cangialosi, J. & Beven II, J.L., Tropical cyclone report: Hurricane Sandy, National Hurricane Center, **12**, pp. 1–10, 2013.

[6] Anderson, M.L., Chen, Z.Q., Kavvas, M.L. & Feldman, A., Coupling HEC-HMS with atmospheric models for prediction of watershed runoff. *Journal of Hydrologic Engineering*, **7**(4), pp. 312–318, 2002.
http://dx.doi.org/10.1061/(ASCE)1084-0699(2002)7:4(312)

[7] USGS, January 22–24, 2016 *Summary of Coastal Flooding in New Jersey Caused by the Nor'easter,* available at: http://nj.usgs.gov/hazards/flood/flood1601/ 2016.

[8] Villarini, G. & Goska, R., Smith, J.A. & Vecchi, G.A., North atlantic tropical cyclones and U.S. flooding. *Bulletin of the American Meteorological Society*, **95**(9), pp. 1381–1388, 2014.
http://dx.doi.org/10.1175/BAMS-D-13-00060.1

[9] Berg, R., National Hurricane Center Tropical Cyclone Report Hurricane Joaquin (AL112015) 28 September – 7 October 2015, available at: http://www.nhc.noaa.gov/data/tcr/AL112015_Joaquin.pdf. 2016.

[10] Lumia, R., Firda, G.D. & Smith, T.L., Floods of 2011 in New York: U.S. geological survey scientific investigations report 2014–5058, p. 236, avaialble at: http://dx.doi.org/10.3133/sir20145058. 2014.

[11] Orton, P., Georgas, N., Blumberg, A. & Pullen, J., Detailed modeling of recent severe storm tides in estuaries of the New York City region. *Journal of Geophysical Research: Oceans*, **117**(C9), 2012.
http://dx.doi.org/10.1029/2012JC008220

[12] Noe, R.M., Casey-Lockyer, M., Mertzlufft, C., Heick, R., Yard, E. & Wolkin, A., Deaths associated with hurricane sandy. *Morbidity and Mortality Weekly Report (MMWR) Centers for Disease Control and Prevention*, **62**(20), pp. 393–397, 2013, available at: http://www.cdc.gov/mmwr/preview/mmwrhtml/mm6220a1.html. 2013.

[13] Weaver, J., Harkabus, L.C., Braun, J., Miller, S., Cox, R., Griffith, J. & Mazur, R.J., An overview of a demographic study of united states emergency managers. *Bulletin of the American Meteorological Society*, **95**(2), pp. 199–203, 2014.
http://dx.doi.org/10.1175/BAMS-D-12-00183.1

[14] Hoss, F. & Fischbeck, P., Increasing the value of uncertain weather and river forecasts for emergency managers. *Bulletin of the American Meteorological Society*, **97**(1), pp. 85–97, 2016.
http://dx.doi.org/10.1175/BAMS-D-13-00275.1

[15] Di Liberto, T., Colle, B.A., Georgas, N., Blumberg, A.F. & Taylor, A.A., Verification of a multimodel storm surge ensemble around New York City and Long Island for the cool season. *Weather and Forecasting*, **26**(6), pp. 922–939, 2011.
http://dx.doi.org/10.1175/WAF-D-10-05055.1

[16] Saleh, F., Ramaswamy, V., Georgas, N., Blumberg, A. & Pullen, J., A Retrospective streamflow ensemble forecast for an extreme hydrologic event: a case study of Hurricane Irene and on the Hudson River basin. *Hydrology and Earth System Sciences*, **2016**, pp. 1–29, 2016.
http://dx.doi.org/10.5194/hess-20-2649-2016

[17] Marsooli, R., Orton, P.M., Mellor, G., Blumberg, A.F., Georgas, N.A., Coupled Wave-Circulation Model for Numerical Simulation of Storm Tides and Waves, Ocean Modeling, Submitted.

[18] http://www.weather.gov/media/phi/veteransday2009_pres.pdf.

[19] Georgas, N. & Blumberg, A., Establishing confidence in marine forecast systems: the design and skill assessment of the New York Harbor observation and prediction system, version 3 (NYHOPS v3). Estuarine and Coastal Modeling, 660-685. 2010.

[20] Georgas, N., Blumberg, A.F., Bruno, M.S. & Runnels, D.R., Marine forecasting for the New york urban waters and harbor approaches: the design and automation of NYHOPS. In 3rd International Conference on Experiments/Process/System Modelling/Simulation & Optimization. Demos T. Tsahalis, Ph. D., University of Patras, Greece, 2009.

[21] NWS, Service Assessment: *Hurricane/Post-Tropical Cyclone Sandy. October 22–29, 2012*, Silver Spring, MD, USA: National Oceanic and Atmospheric Administration National Weather Service. 2013.

[22] NWS, *National Weather Service Manual 10-950,* DECEMBER 4, 2012, Operations and Services Hydrologic Services Program, NWSPD 10-9, available at: http://www.nws.noaa.gov/directives/. 2012.

[23] http://hudson.dl.stevens-tech.edu/SFAS/SFAS_Latest_News.pdf.

[24] Molteni, F., Buizza, R., Palmer, T.N. & Petroliagis, T., The ECMWF ensemble prediction system: Methodology and validation. *Quarterly Journal of the Royal Meteorological Society*, **122**(529), pp. 73–119, 1996.
http://dx.doi.org/10.1002/qj.49712252905

[25] http://www.spc.noaa.gov/exper/sref/

[26] https://www.ncdc.noaa.gov/data-access/model-data/model-datasets/global-ensemble-forecast-system-gefs.

[27] http://www.ncdc.noaa.gov/data-access/model-data/model-datasets/global-forcast-system-gfs.

[28] http://www.ncdc.noaa.gov/data-access/model-data/model-datasets/north-american-mesoscale-forecast-system-nam.

[29] http://rucool.marine.rutgers.edu/COOL-Data/ruwrf-model-output.

[30] https://weather.gc.ca/ensemble/index_e.html.

[31] Blumberg, A.F., Georgas, N., Yin, L., Herrington, T.O. & Orton, P.M., Street-scale modeling of storm surge inundation along the New Jersey Hudson river waterfront. *Journal of Atmospheric and Oceanic Technology*, **32**(8), pp. 1486–1497, 2015.

[32] McEnery, J., Ingram, J., Duan, Q., Adams, T. & Anderson, L., NOAA'S advanced hydrologic prediction service: building pathways for better science in water forecasting. *Bulletin of the American Meteorological Society*, **86**(3), pp. 375–385, 2005.
http://dx.doi.org/10.1175/BAMS-86-3-375

[33] Adams, T., Verification of the NOAA/NWS MMEFS Operational Hydrologic Ensemble Forecasting System in the Ohio River Valley. *World Environmental and Water Resources Congress 2015 at Floods, Droughts, and Ecosystems*, pp. 2688–2700, 2015.

[34] http://www.stormsurge.noaa.gov/models_obs_modeling.html.

[35] https://ams.confex.com/ams/96Annual/webprogram/Paper279348.html.

ESTIMATING PEAK FLOW-DISCHARGE DURING EXTREME RAINFALL EVENTS FOR THE GAO-PING RIVER, TAIWAN

J.-C.CHEN[1], C.YANG[2] & J.CHING[2]
[1]Department of Environmental & Hazards-Resistant Design, Huafan University, Taiwan.
[2]Department of Civil Engineering, National Taiwan University, Taiwan.

ABSTRACT

Typhoon Morakot struck central and southern Taiwan on August 8, 2009. The high intensity and accumulation of rainfall induced floods, landslides and debris flows. During Typhoon Morakot, the flood level was beyond the design limit of many river embankments in southern Taiwan. It is, therefore, desirable to estimate the peak flow-discharge for similar extreme rainfall events in future. In this study, the destructive flood in the Gao-ping river in southern Taiwan, caused by Typhoon Morakot, was selected as an area for case analysis. First, hydrological and geomorphological data for historical typhoons together with rainfall data were collected and analyzed. Next, the rational method, combined with a number of equations, was proposed to estimate peak flow-discharges in the river produced by extreme rainfall events. The estimated peak flow-discharges of historical events were then compared with the observation data acquired from stream gauge stations and previous studies. The results showed that the proposed method is able to estimate the flow-discharge in the Gao-ping river with reasonable accuracy if the flow rate of interest is greater than 5000 cms or the return period of interest is more than 100 years. It could, therefore, be used to determine the peak flow-discharge for extreme rainfall events.
Keywords:extreme rainfall events, peak flow-discharge, rational method, Typhoon Morakot.

1 INTRODUCTION

The peak flow-discharge is an important variable when designing flood mitigation structures such as culverts, flumes, bridges, revetments, and dams. The unit hydrograph and rational methods are common ways to determine the peak discharge in hydrological analysis. Generally, the unit hydrograph method is used when long-term hydrological data for a river is available. The rational method, an empirical equation proposed by Kuichling [1], was widely used to estimate peak flows from urban and small rural un-gauged catchments [2]. The main advantage of the rational method is that there is ample experience in its application over many years of its use [3]. It also adopts a simple concept and can be computed without the use of computers. However, much of the literature indicates that the rational method is appropriate only for small catchments and the definition of a 'small catchment' is not consistent among practitioners. For example, the catchment area (A) limit is 0.08 km^2 in New York [4, 5], 25 km^2 in rural Australia [6], and 10 km^2 in Taiwan [7]. The accuracy of the peak discharge determined by the rational method could be uncertain when A exceeds its limit. Many streams or rivers in mountainous areas usually lack stream gauge stations or have no flow-rate data record. Even if stream gauge stations were installed, these stations have a high possibility of being destroyed by massive floods during extreme events. The flooding induced by the extreme

DOI: 10.2495/SAFE-V6-N3-663-673

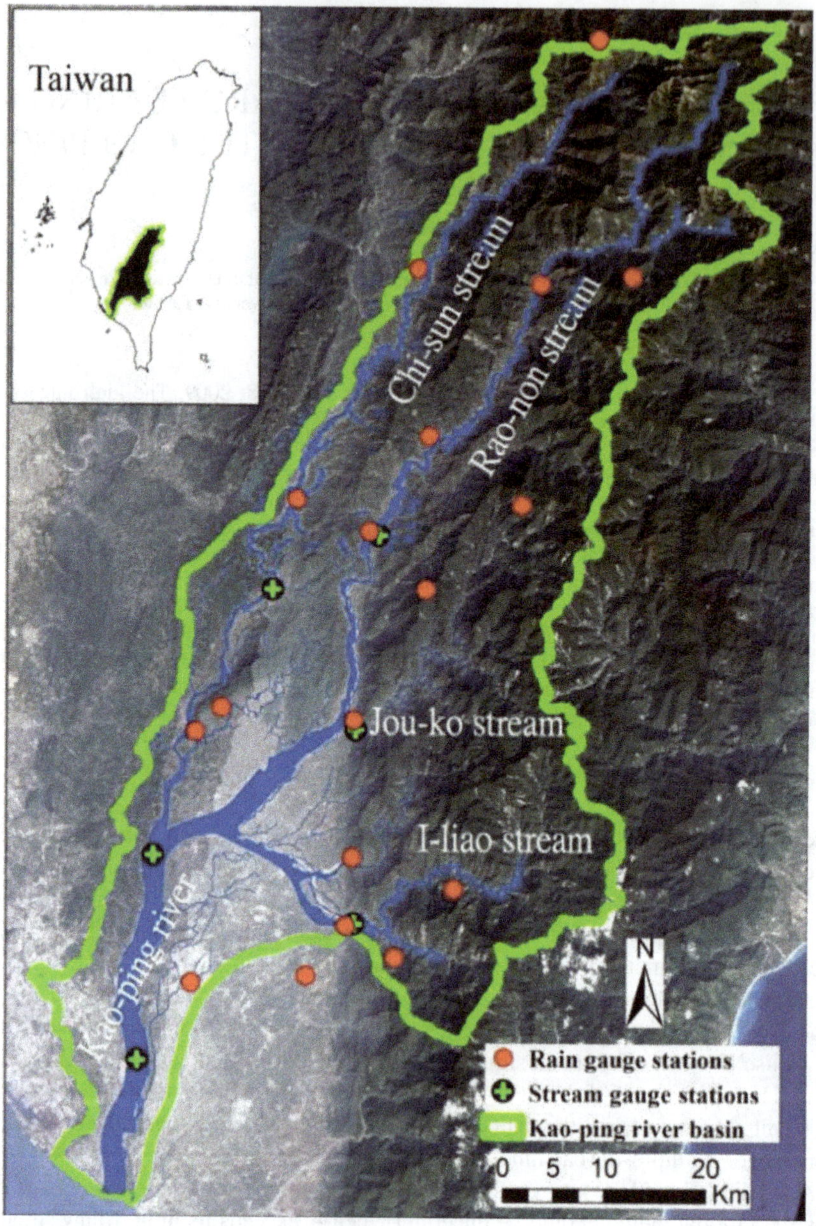

Figure 1: Map showing 18 rain gauge stations and 6 stream gauge stations in the study area of the Gao-ping river catchment.

rainfall during Typhoon Morakot is an example of a huge flood that took place in southern Taiwan in 2009. During this extreme rainfall event, most stream gauge stations were destroyed or were unable to operate normally. Due to the fact that the peak flow discharge is an important hydrologic parameter in hazard prevention and hydraulic engineering design, especially for extreme rainfall events, the rational method was used to examine its reasonability in the analysis of peak flow discharge. The Gao-ping river, located in southern Taiwan, was selected as the study area. The peak flow discharge in the river produced by historical rainfall events were estimated by the rational method and they were compared with observation data. The estimated peak flow-discharge was also compared with previous studies.

2 STUDY AREA

The study area is located in the Kao-ping river basin of southern Taiwan. The river has a catchment area of 3342 km^2 and includes four streams: the Rao-non stream, Jou-ko stream, I-liao stream, and Chi-sun stream. There are 18 rain gauge stations and 6 stream gauge stations in the basin. The six stream gauge stations are with names: Da-Jing bridge (DJB), Shin-Fa bridge (SFB), Sun-Di-Men (SDM), Li-Lin bridge (LLB), Sun-Lin bridge (SLB), and Wan-Da bridge (WDB), as shown in Fig. 1. The hourly rainfall and flow discharge data taken from the rain and stream gauge stations during the Typhoon events that occurred from 2004 to 2009 were collected and used to analyze the rainfall intensity and peak flow discharge.

3 METHOD

3.1 Equations

3.1.1 Rational method

The rational method has been used for over 150 years and remains the most widely used method for estimating peak flows from urban and small rural un-gauged catchments[2]. It relates peak flow discharge, Q(m^3/s), to the catchment area, A (ha), rainfall intensity, I (mm/h), and runoff coefficient, C. The equation has the form:

$$Q = CIA/360 \tag{1}$$

This equation is also generally used to determine the peak discharge for water flow in mountainous gullies [7, 8] or the watershed where there is a lack of hydrological observation data. This is because it is a simple equation and all the parameters involved can be easily obtained. The method is based on the assumptions that the rainfall intensity and storm duration is uniform over the study area, that the storm duration must be equal to the time of concentration of the catchment, and that the runoff coefficient must be constant during the storm [9].

The runoff coefficient C_j is a dimensionless parameter that represents the percentage of rainfall appearing as runoff. C includes interception, infiltration, evaporation, depression storage, and groundwater flow in the hydrological cycle and it depends on soil type, land use, degree of imperviousness, watershed slope, surface roughness, antecedent moisture condition, and so on. A suggested range of runoff coefficients are available in many papers and handbooks. The average runoff coefficients for various surface types are commonly used. The average runoff coefficient C for a catchment was determined in this paper by the following equation:

$$C = \frac{1}{m}\sum_{j=1}^{m} C_j \tag{2}$$

in which, the catchment area was divided into m number of units, following the land use and landform. The runoff coefficient for each unit C was determined by referring to the handbooks of SWCB [7] for Taiwan.

3.1.2 Average rainfall intensity

The rainfall intensity is the volume of rain that has fallen per unit of time. The rainfall intensity varies with time during a given storm for different geographical regions and also for different locations within these regions, resulting in different rainfall distributions [3, 10]. The reciprocal-distance-squared method [11] was used to determine the average rainfall for a catchment. Since this method is simple and can directly reflect the distance weighting, it has been widely used in determining un-gauged or regional average rainfalls [12]. The average rainfall intensity I estimated by the reciprocal-distance-squared method can be expressed as follows:

$$I = \sum_{i=1}^{N} w_i I_i \quad (3)$$

in which

$$w_i = \left(\frac{1}{d_i^2}\right) \Big/ \left(\sum_{i=1}^{N} \frac{1}{d_i^2}\right), \quad (4)$$

where N is the number of gauge stations; I_i is the rainfall intensity at the ith gauge station; w_i is the weighting factor corresponding to the ith gauge station; d_i is the distance from the ith gauge station to the center of the chosen catchment.

3.1.3 I-D curve and Horner formula

The analysis of I_i in this paper is divided into two parts. Part one focuses on rainfall events and the value of I_i is determined from the relationship between the maximum rainfall intensity and its duration (the I-D curve). Part 2 focuses on the design rainfall and I_i is determined from the Horner formula [13]. The rainfall intensity can be read from the I-D curveand from the Horner formula at the chosen catchment, for a duration equal to the time of concentration. To obtain the design rainfall intensity at each rain gauge station, thedesign rainfall formula for the study area was collected. The rainfall-duration-frequency curve at each rain gauge station was collected to determine the design rainfall intensity and the rainfall intensity for a return period. The design rainfall intensity in this paper was expressed by a form of the Horner formula:

$$I_i = a/(D+b)^c, \quad (5)$$

where I_i is the rainfall intensity (mm/h) at a gauge station; D is rainfall duration (min); a, b, and c are empirical coefficients that depend on the rainfall characteristics of the area and vary with the return period of rainfall. The rainfall intensity at various return periods, such as 2, 5, 10, 25, 50, 100, and 200 years, were determined by the above equation according to the coefficients a, b, and c, developed by WRA[14].

3.1.4 Time of concentration

The time of concentration is defined as the travel time for the runoff to move from the most hydraulically remote point of the contributing catchment area to the point where peak flow is estimated. The time of concentration, tc, is equal to the sum of the overflow time, t1 (the travel time from the most remote point to the channel inlet), and the channel flow time, t2 (the travel time in channels to the catchment outlet). It can be determined using empirical formulas, such as Kirpich's equation, Kerby's equation, or the kinematic wave equation, and also from hydrographs [15] and technical handbooks [7]. The empirical equation for tc (in min) used in Taiwan [7] can be expressed as:

$$tc = l/(60v) + 5L(L/H)^{0.6}/6, \quad (6)$$

in which, l (m) is the overflow length, v is the overflow velocity (generally ranging from 0.3 to 0.6 m/s), L (km) is the stream length, and H (km) is the elevation difference across the stream length.

3.2 Procedure

3.2.1 Select a catchment area form a point of interest and calculate the concentration time

First, we have to select a catchment area from a point of interest and calculate the concentration time. The point is taken from a stream gauge station that has recorded flow-discharge observation data. For example, the catchment area of the Jou-Ko stream at the DJB station is 374 km^2. The catchment area covered 13 rain gauge stations. The concentration time, tc, at the DJB station is 327 min, using eqn(6).

3.2.2 Determine the rainfallintensity for a rain gauge station

A relationship was developed between the maximum rainfall intensity and duration (the I-D curve) at each rain gauge station for a rainfall event. For example, the I-D curve for one of the rain gauge stations, the Mei-Long (ML) station in the Jou-Ko stream catchment, during Typhoon Conson, is shown in Fig. 2. The maximum rainfall intensity I_i is 5.27 mm/h, taken from the I-D curve at tc = 327 min. Following the same procedure as above, the I_i value for

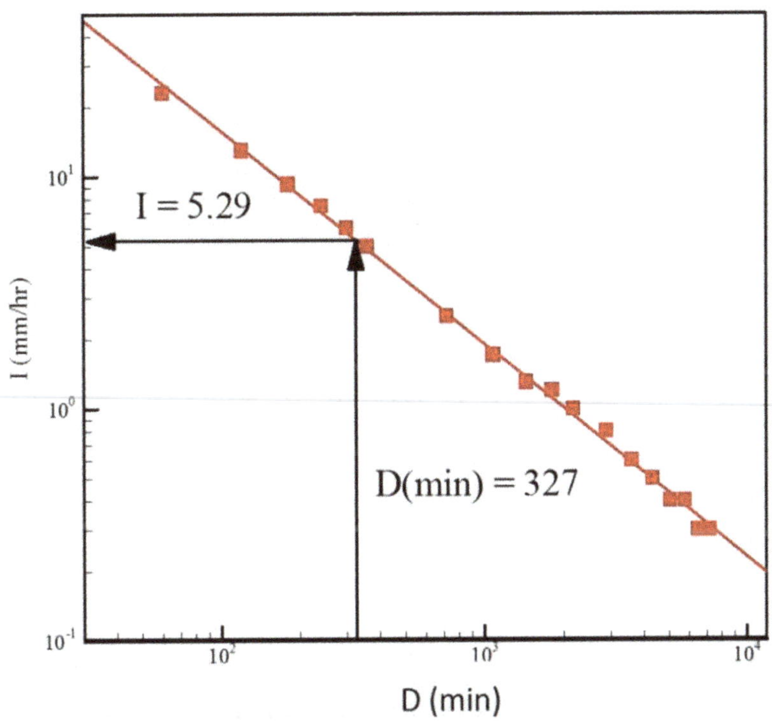

Figure 2: Relationship between the maximum rainfall intensity I and the duration D in the case of the Mei-Long rain gauge station during Typhoon Conson.

the other 12 rain gauge stations in the catchment of the Jou-Ko stream can be determined. In order to obtain the design rainfall intensity at each rain gauge station, the Horner formula eqn(5) for the study area was obtained to determine the rainfall intensity for a return period.

3.2.3 Determine the average rainfallintensity in a catchment

The weighting factor w_i for an individual rain gauge station can be obtained using eqn(4) when the central point of the watershed is determined. In the example of the Jou-Ko stream catchment during Typhoon Conson, there are 13 gauge stations and the average rainfall intensity I can be evaluated from eqn(3). The resulting I is 4.21 mm/h.

3.2.4 Calculate the peak discharge

Finally, the peak discharge can be obtained using the rational equation (eqn1). The resulting Q is 299 cms, obtained with I = 4.21 mm/h, C=0.68, and A= 37,600 ha for the Jou-Ko stream watershed during Typhoon Conson.

4 RESULTS

4.1 Comparison with measured data

The measured flow discharges collected from six stream gauge stations [16], Da-Jing bridge (DJB), Sin-Fa bridge (SFB), Sun-Di-Men (SDM), Sun-Lin bridge (SLB), Li-Lin bridge (LLB), and Wan-Da bridge (WDB), during 15 typhoon events that occurred from 2004 to 2008 are shown in Fig. 3. These measured flow discharges are then used as a comparison with

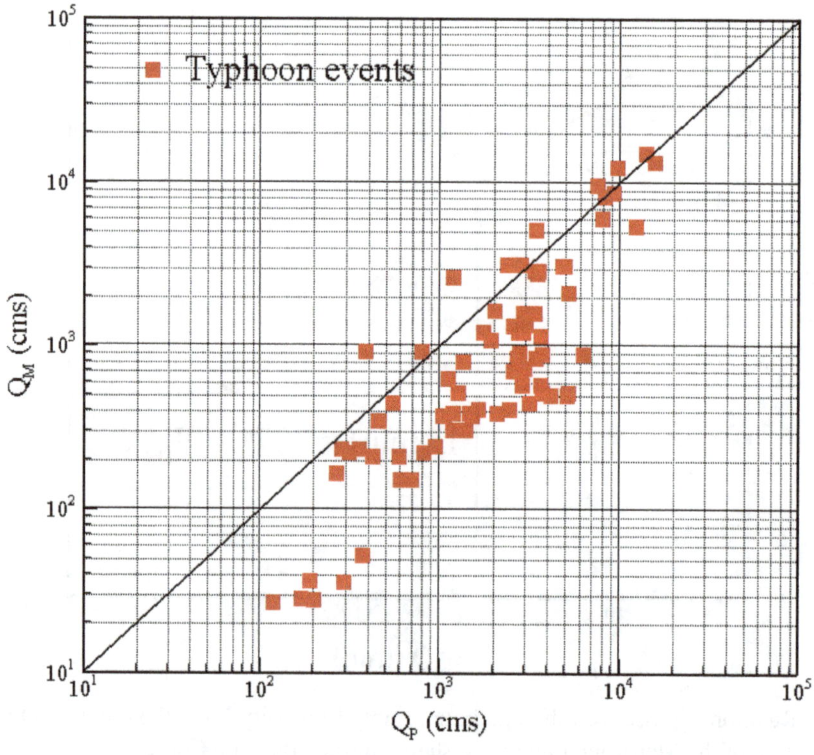

Figure 3:Measured flow discharge Q_M and the estimated flow discharge Q_P

the flow discharge estimated by the method proposed in this study. Figure 3 shows that most of the estimated discharges, Q_P, are greater than the measured discharge, Q_M, that is, $Q_M<Q_P$, when $Q_M< 5000$ cms. At higher discharge, $Q_M>5000$ cms, Q_P is close to Q_M.

4.2 Comparison with previous studies on discharge design

The design peak discharges at return periods of 2, 10, and 100 years were determined by the method proposed in this study for the Kao-ping river basin. This result was then compared with the official result determined using the dimensionless unit hydrograph [14, 17], as shown in Table 1 and Fig. 4. For some data obtained for discharges smaller than 5000 cms or with return periods T <100 years, the estimated discharge produced by this study (Q_P) is greater than the official WRA estimated discharge (Qc). For higher design discharges, exceeding 5000 cms, or return periods T of more than 100 years, Q_P is close to Q_C. Here, we also find that the catchment areas in Table 1 are much greater than those from previous researchers [4–7].

4.3 Estimation of peak flow discharge during Typhoon Morakot

Typhoon Morakot brought extreme rainfall and caused many hazards, such as landslides, debris flows, and floods, in southern and central Taiwan [18]. For such extreme rainfall, most flow discharge stations could not work and, as a result, there was no recorded data for the Gao-ping river in southern Taiwan. Due to the fact that the actual flow discharge

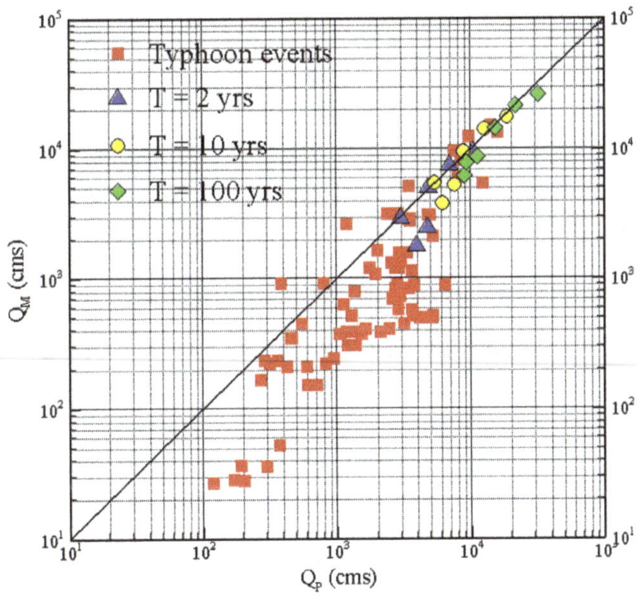

Note: Q_M = measured flow discharge in historical typhoon events; Q_P = estimated flow discharge in this study; Q_C= estimated official flow discharge

Figure 4: Flow discharge estimated from the method proposed in this study (Q_P) compared with that measured at flow discharge stations during various typhoon events (Q_M), and that determined from previous studies (Q_C).

Table 1: Estimation of peak discharge at different control points.

Location		Q (cms) at T = 2 years		Q (cms) at T = 10 years	Q (cms) at T = 100 years			
Catch-ment	Control point	Area (km^2)	WRA	Present study	WRA	Present study	WRA	Present study
RNS	LKB	2014	7310	7004	14,000	12,663	21,100	21,830
RNS	RNS - ILS junction	821	4910	4898	9370	8837	14,200	15,233
ILS	SDM	423	1750	3988	3780	6170	6150	8931
ILS	ILR outlet	1398	2440	4761	5180	7597	8600	11,256
CSS	CSS-RNS junction	819	2890	3008	5380	5395	7780	9297
GPR	KPR main stream	2886	9190	10,457	17,300	18,786	25,900	32,374

Note: RNS:Rao-non stream; ILS: I-liao stream; CSS: Chi-sun stream; GPR: Gao-ping river.

is important in hazard prevention and hydraulic engineering design, especially for extreme rainfall events, researchers evaluated the peak discharge using a hydrological model. For example, the peak flow discharge at some flow discharge stations (SDM, SFB, DJB) in the Gao-ping river were determined using the unit hydrograph method [16, 19]. The estimated peak flow discharge from previous studies [16, 19] are compared with the peak flow discharge estimated by the method proposed in this study, as shown in Fig. 5. Figure 5 shows that the estimated flow discharge in this study (Q_P) agrees with those estimated in previous studies (Q_C). The methods used in previous studies differ from the present study. The design discharges from previous studies were all determined by the unit hydrograph method, with different periods of rainfall and discharge data. The method proposed in this study can easily determine the discharge without a long-term flow rate record. Many streams or rivers in mountainous areas generally have a lack of stream gauging stations or have no flow-rate data record. This means that the method proposed in this study maybe used to estimate the design discharge due to extreme rainfall for an area with a lack of historical rainfall or discharge data. However, this study only focuses on the Gao-ping river area and on typhoon events; therefore, future studies are needed for other areas and rainfall conditions.

5 CONCLUSIONS

The peak discharges for a rainfall event and various return periods were determined using the rational method combined with some equations that were proposed in this paper. The estimated peak discharge was compared with measured data and those from previous studies.

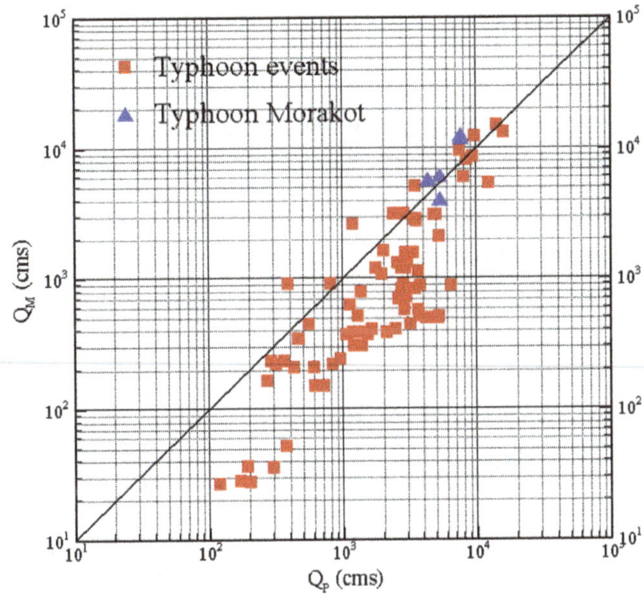

Note: $Q_{M:}$ = measured flow discharge in historical typhoon events; $Q_{P:}$ = estimated flow discharge in Typhoon Morakot in this study; Q_C = estimated official flow discharge during Typhoon Morakot

Figure 5:Flow discharge estimated from the method proposed in this study (Q_P) compared with those measured at flow discharge stations (Q_M) and from previous studies (Q_C) during Typhoon Morakot.

Results showed that the estimated peak discharge is greater than the measured data and those of previous studies for low flow-discharges, Q<5000 cms, or for return periods of T<100 years. The estimated peak discharge is in agreement, however, with the measured data and those of previous studies at high flow discharge, where Q>5000 cms, or with T>100 years. The peak discharges in the Gao-ping river that were brought on by Typhoon Morakot, the extreme rainfall event discussed in this study, were determined by the rational method, and these were found to be close to the results of previous works using the unit hydrograph method. The rational method could be used to determine the peak discharge brought on by extreme rainfalls, or rainfall with T>100 years. The main advantage of the rational method is that it is a simple concept, computed without the use of computers, and it is available for streams without stream gauge stations. However, this study only focuses on the Gao-ping river area and on typhoon events; therefore, future studies are needed for other areas and rainfall conditions.

REFERENCES

[1] Kuichling, E. & Hering, R., *TheRelation Between the Rainfall and the Discharge of Sewers in Populous District*, American Society of Civil Engineers: New York, 20, pp. 1–56, 1889.

[2] Watts, L.F. & Hawke, R.M., The effects of urbanization on hydrologic response: a study of two coastal catchments. *Journal of Hydrology*, **42**(2), pp. 125–143, 2003.

[3] OPW, *Flood StudiesUpdate Programme:Flood Estimation in Small and Urbanized Catchments*, Hydrology and Coastal Section, Office of Public Works (OPW), 2012.

[4] NYS, *Standards and Specifications for Erosion and Sediment Control*, New York State(NYS),2004.

[5] Debo, T.N. & Reese, A., *Municipal Storm Water Management*, p. 209, 1995.

[6] QG, *Queensland Urban Drainage Manual*, Vol. 1, 2nd ed., Queensland Government (QG), 2007.

[7] SWCB, *Technical Handbook of Soil and Water Conservation*, Soil and Water Conservation Bureau (SWCB), Council of Agriculture, Executive Yuan, Taiwan, 2003.(in Chinese)

[8] Chen, J.C., Jan, C.D.&Lee, M.S., Reliability analysis of design discharge for mountainous gully flow.*Journal of Hydraulic Research*, **46**(6), pp. 835–838, 2008. http://dx.doi.org/10.3826/jhr.2008.3173

[9] Hays, D.C.& Young, R.L., Comparison of peak discharge and runoff characteristicestimates from the rational method to field observations for small basins in central Virginia. *USGS Scientific Investigations Report,* 2005–5254, 2006.

[10] Nyman, D., *Hydrology Handbook forConservation Commissions*, MassachusettsDepartment ofEnvironmental Protection, 2002.

[11] Chow, V.T., Maidment, D.R.&Mays, L.W., *Applied Hydrology*. McGraw-Hill: Singapore, 1988.

[12] Chen, J.C., Jan, C.D. &Huang, W.S., Characteristics of rainfall triggering of debris flows in the chenyulanwatershed, Taiwan.*Natural Hazards and Earth System Sciences,* **13**, pp. 1015–1023, 2013.
http://dx.doi.org/10.5194/nhess-13-1015-2013

[13] Horner, W.&Flynt, F., Relation between rainfall and runoff from small urban areas. *Trans.ASCE,* **101**,pp. 104–183, 1936.

[14] WRA, *Technique of Hydrological Observation and Criteria of Hydrological Analysis*, Water Resources Agency (WRA), Ministry of Economic Affairs, Taiwan, 2011(in Chinese)
[15] Chin, D.A.,*Water-Resources Engineering*, Prentice-Hall, 2000.
[16] WRA, *Planning on Gao-ping river treatment*, Lab. of Hydraulic planning, Water Resources Agency (WRA), Ministry of Economic Affairs, Taiwan, 2008(in Chinese)
[17] WRA, *Analysis on extreme rainfall and flow discharge in Typhoon Morakot*, Water Resources Agency (WRA), Ministry of Economic Affairs, Taiwan, 2009(in Chinese)
[18] Chen, J.C. & Chuang, M.R., Discharge of landslide-induced debris flows: case studies of Typhoon Morakot in Southern Taiwan.*Natural Hazards and Earth System Sciences*, **14**, pp. 1719–1730, 2014.
http://dx.doi.org/10.5194/nhess-14-1719-2014
[19] SEC, Investigation on the 88 flood, Sinotech Engineering Consultants (SEC), Inc., Taiwan, 2009 (in Chinese)

IN PURSUIT OF RELIABLE FLOOD PREDICTION

B.E. VIEUX & J.E. VIEUX
Vieux & Associates, Inc., Norman Oklahoma, 73072 USA.

ABSTRACT
Flood forecasting necessarily relies on a number of technologies, namely, precipitation measurement at high resolution for large areas; numerical modelling of the hydraulics and hydrology of runoff and routing; and innovative information technology for synthesizing critical information needed for emergency response and risk management. Data sources for precipitation inputs can come from weather radar and automated rain gauges. Regional and urban hydrologic modelling is increasingly based on distributed physics-based hydrologic models that can leverage the high definition rainfall derived from radar and rain gauge measurements. Statistical quality control in real-time is necessary to transform radar-based precipitation estimates into accurate hydrologic model input, which serves as an input to a physics-based distributed runoff model. Geospatial data are necessary for setup and parameterization of such distributed models. Real-time precipitation estimates from radar and rain gauge monitoring are coupled with the gridded representation of the watershed to produce warning notifications at distributed locations throughout urban, peri-urban, and rural/natural watersheds. Quantification of uncertainties and sources of error are examined within the pursuit of successful flood forecasting. The following case studies illustrate distributed hydrologic forecasting across a range of space-time scales. This paper presents forecasting performance, an evaluation of accuracy, and discussion of factors that contribute to uncertainty in flood forecasts and risk reduction.
Keywords: flood forecasting, forecast uncertainty, hydrometeorology, inundation mapping, numerical modelling, reservoir operations.

1 INTRODUCTION

An approach to reducing flood impacts is to minimize the public's exposure by closing road intersections, or warning residents or businesses to take preventative actions. To provide an effective emergency response, responders must know when and where flooding is likely to occur. This article describes the performance of a flood forecasting system during recent extreme flood events. Public flood warnings are the responsibility of the national meteorological agencies. Increasingly, customized forecasting services are being offered to municipalities, flood control districts, or businesses requiring more specificity in terms of location, time-frame, and type of hydrologic information. Historically, only national governmental agencies had the resources and the ability to perform such services. Vieux and Bedient [1] describe a private-sector flood forecasting system that uses radar and rain gauge input to the finite element model of 260 km^2 watershed at 120-m resolution. It was initiated in 1997 and continues today, providing one of the largest medical centers in the world with required notifications of flooding so that flood defenses can be operated to limit entry of floodwaters into the facilities. Other locations where this approach is being applied operationally include organizations such as a flood control district in the front range of Colorado, municipalities in flash-flood

DOI: 10.2495/SAFE-V6-N3-674-684

alley in Texas, and a hydro-electric plant responsible for power generation and flood mitigation upstream and downstream of a dam.

Enabling technology that makes it possible to offer high-resolution flood forecasts on a commercial basis, includes advances in computing power, information dissemination systems, dual-polarized weather radar, tipping bucket rain gauges or satellite remote sensing technology. Distributed measurement of precipitation at high resolution along with telemetered rain and stream gauge measurements are key elements necessary for the provision of customized high-resolution rainfall-runoff modeling. The web-based services depend on access to real-time weather radar data with hosted computer facilities and high speed internet access. The high-resolution numerical solution to rainfall runoff makes it possible to serve urban flood applications, or for managing reservoir inflow. Within a single distributed framework, it is now possible to provide forecast hydrologic information across a range of time-space scales within the same modeling system. The runoff response resulting in reservoir inflows spanning days across thousands of square kilometers, modeled at 500-m resolution and fifteen-minute time steps. To provide locally specific flood warnings throughout an urban and peri-urban environment, smaller scales are modeled across a range of basin sizes, from less than 30 to over 750 sq. km., at 60-m to 500-m resolution.

2 METHODOLOGICAL APPROACH

The modeling approach taken relies on numerical solution of the conservation equations governing runoff and routing through overland and channel components of a gridded drainage network. The physics-based distributed model, V*flo*®, is setup and parameterized using geospatial data representing terrain, soils, land use/cover, and impervious areas where developed surfaces limit infiltration. The model solves the kinematic wave equations by numerical solution of conservation equations, finite element in space and finite difference in time [2].

1. **Model Schema** Geospatial data are used to setup the drainage network and parameters. Starting with a digital elevation model (DEM) for the derivation of slope and flow direction (D8); remotely sensed land use/cover for overland hydraulic roughness; impervious area and soil characteristics for derivation of Green and Ampt infiltration parameters and soil depth; channel geometry and hydraulic characteristics.
2. **Precipitation Inputs** Rainfall and snowmelt may be input on a gridded-basis as maps, or input as uniformly distributed rainfall rates/accumulations. Distributed precipitation can come from NEXRAD radars.
3. **Runoff Routing** The kinematic wave equations are solved on a grid cell basis within a drainage network composed of finite elements. The formulas are adapted to represent overland and channel hydraulics. Floodplain storage effects can be simulated using the Modified Puls method to account for attenuation due to temporary storage. Complex channel hydraulics are represented by: (1) rating curves for stage area/discharge, (2) surveyed channel cross-sections with slope and Manning's roughness, and (3) trapezoidal cross-sections with slope and Manning's roughness.
4. **Soil Moisture and Infiltration** The Green and Ampt equation is used to compute infiltration-rate excess runoff in each grid cell as a function of soil properties and antecedent conditions. Once the soil moisture storage capacity is filled, then saturation excess runoff is computed. Soil moisture depletion is limited by the potential evapotranspiration (ET) rate and available soil moisture.

5. **Inundation Mapping** Distributed modeling of flow and hydraulic stage supports inundation mapping in two timeframes, (1) post-analysis of flooding extent, and (2) during operational modeling showing maximum inundation forecasts. Inundation maps are produced wherever there are channel cells within the drainage network.

Geospatial datasets are used to setup and parameterize the model, as outlined in [2]. Physical characteristics of the soils, land use/cover, and topography are key sources of distributed parameter maps. Model inputs are derived from radar and gauge precipitation measurements. Evapotranspiration is modeled using measured or climatological rates that are subsequently modified in each grid cell to represent actual conditions, *viz.* rainfall and available soil moisture. The resulting model has applicability to urban, and natural, and mixed landuse in watershed ranging in scale from headwater catchments to larger rivers. Reservoirs and hydraulic structure operation can be incorporated in the model to represent larger systems. Refinement of model parameters is accomplished using distributed model calibration procedures developed for the desktop edition, described in [3, 4].

3 RESULTS

The following case studies illustrate principles and application of physics-based distributed hydrologic modeling, which are described in more detail in [2]. These studies illustrate a common theme, namely a customized hydrometeorological system that produces high resolution rainfall derived from radar and rain gauge, and a physics-based distributed model developed using geospatial data, producing hydrologic predictions that span a range of space-time scales.

3.1 Case study I – reservoir inflow forecasting

Operation of a hydroelectric dam can be complicated by many competing objectives, namely business requirements for power generation, respecting downstream environmental releases, as well as, management of lake levels that affect property owners along the impoundment lakeshore. Because of narrow operational constraints, accurate inflow forecasts are paramount. Reservoir inflow is a key component necessary for both power generation, and flood management in upstream impoundment areas, and in downstream river reaches. A customized hydrometeorological network is provided a service, specially configured for the system. It relies on access to the US National Weather Service (NWS), weather radar and rain gauges operated by various federal, state, and local agencies. The forecast system measures current quantitative precipitation estimates (QPE) and ingests numerical weather prediction out to 7-days. Evaluation of sensitivity to model parameters, precipitation, and evapotranspiration forcing is discussed below for two heavy rainfall-runoff events occurring in 2015. Each of these inputs carries uncertainty that contributes to the uncertainty in forecast reservoir inflow, upstream and downstream flooding, and hydroelectric generation.

3.1.1 Study area

The Lake of the Ozarks (LOZ) is a reservoir located in Central Missouri on the Osage River covering 222 km^2 and storing 2.27 km^3. The lake's serpentine shape is apparent in Fig. 1 (dark blue), resulting from impoundment of the Osage River by a dam completed in 1931. Its height is 45 m tall forming the reservoir comprising a main channel, 151 km in length from a flood control reservoir, Harry S. Truman (HST), connecting downstream to the Dam. The elongated shape and varying depth required finite volume cells for routing flow through shallow water within the inundated body of the reservoir. Releases from HST contributes inflow to the Dam along with runoff from an uncontrolled drainage area from the Niangua River (see labeled drainage network) that drains from south to north, connecting with the LOZs as shown. The uncontrolled drainage area below HST consists of 6,241 km^2 (outlined in Fig. 1). The V*flo*® model of the uncontrolled drainage consists of 25,458 grid cells at 500-m resolution, forming a drainage network comprising both channel and overland flow cells.

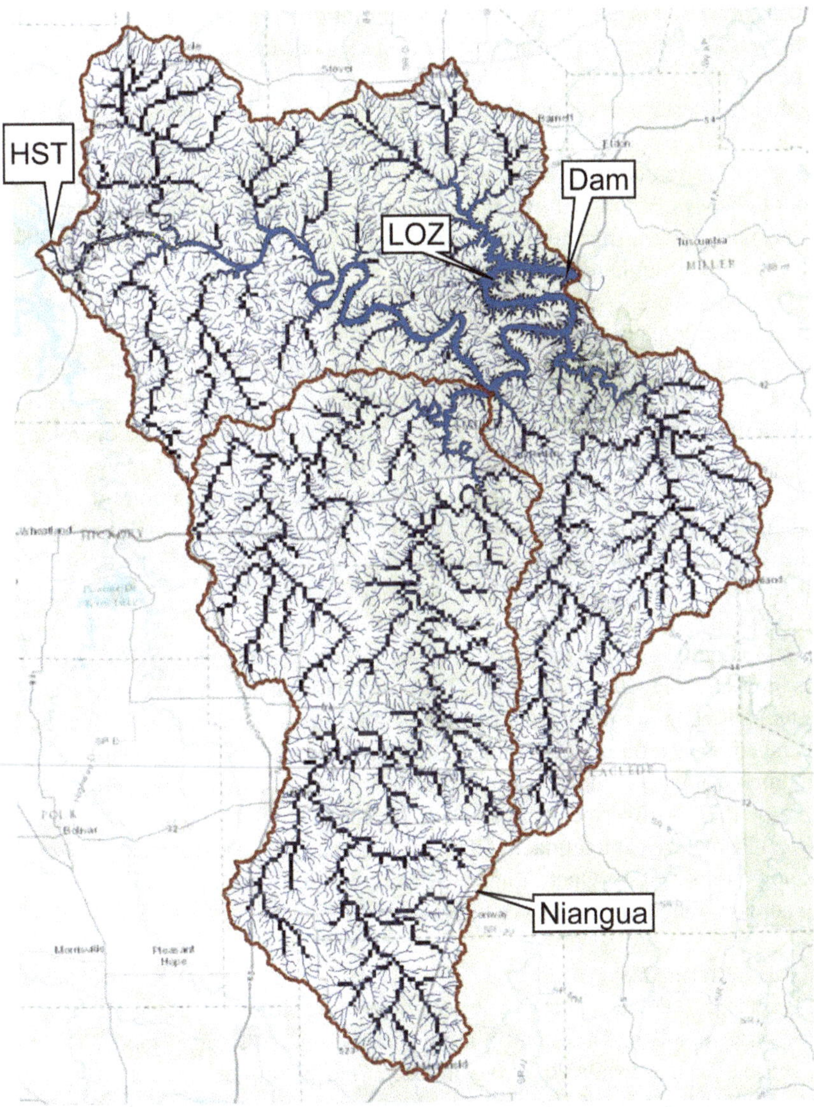

Figure 1: Lake of the Ozarks and tributary drainage area.

3.1.2 Forecast objectives

Forecast services are provided to the dam owner customized for their requirements, relying on a physics-based distributed model of rainfall-runoff, with inputs from current rainfall termed QPE and Quantitative Forecast Precipitation (QPF). Measured evapotranspiration from nine weather stations in and surrounding the basin is used to control soil moisture on a continuous distributed basis across the basin. Besides inflows from uncontrolled drainage areas, lake levels are affected by operational dam releases from the Dam and HST. How much the reservoir level will change in hours, days, or weeks, is important for managing power generation, especially when the lake level rise or downstream discharge impose limits. A continuous physics-based distributed model can be used with advantage to provide knowledge of reservoir inflow from diffuse streams, rivers, combining with discharge from an upstream reservoir. This approach is used to provide reservoir inflow forecasts with input from current and forecast rainfall, uncontrolled watershed runoff, and current or planned releases from the upstream reservoir.

Predicted or forecast reservoir inflow aids in the operation of the dam, provides estimated spill consistent with its operating license, and supports the forecast of lake levels with quantitative precipitation forecasts (QPF) from US NWS, numerical weather prediction at 6-hr intervals out to 7-days. V*flo*® is at the modeling core of the software system that integrates precipitation and evapotranspiration data for making inflow projections, and supporting web-based delivery of hydrologic information.

3.1.3 Hydrometeorological system

Accurate precipitation quantification is necessary for a rainfall-runoff model to yield useful inflow forecasts. Rather than rely on uncertain estimates from the NWS MPE, which is only available from the NWS at hourly and 4 × 4 km, a high resolution product is operated, which is customized to the target watershed. Three radars are bias corrected and merged to produce gauge adjusted radar rainfall (GARR) at 15-minute and 2 × 2 km resolution for model input. Automated failover from primary to secondary radars, and then to gauge-only gridded rainfall occurs in the unlikely event that all three radars become unavailable. The hydrometeorological system integrates reflectivity from three weather radars (KEAX, KLSX, KSGF), which is bias corrected with rain gauge measurements in real-time.

Soil moisture and infiltration rates are simulated in each grid cell using Green and Ampt parameters for a single layer and soil depth. Thus, both Horton and Dunne runoff components are computed, allowing the infiltration rate and saturation excess to operate simultaneously in the watershed. Soil moisture is depleted at hourly rates by potential evapotranspiration (PET) computed from measured humidity, wind speed, solar radiation, pressure, and temperature. Forecast PET is ingested from the US National Digital Forecast Database, while climatological PET serves as a fallback if data feeds of current measurements or forecast PET are unavailable in operations. Operation of the dam requires reasonably accurate continuous simulation across a range of runoff events from small to large.

3.1.4 Forecast performance

In order to create a detailed model of the watershed capable of modeling the often highly variable response to soil moisture and rainfall distributions, a gridded model is setup that leverages geospatial data available across the US. Geospatial data availability means distributed models can incorporate information from maps of land use or vegetative cover, complex terrain, and soils to make the needed predictions of rainfall-runoff. When rainfall derived from weather radar and rain gauges is properly controlled for accuracy, it can be used as a necessary ingredient for operational modeling of the watershed. Model calibration is used to fine-tune the model to be more accurate using historical rainfall data. Review of the model performance from operational review provides insight into how the system can be improved when there is a trend in over- or under-prediction. The information system that manages the rainfall production, modeling, and data/information display and interconnections are tailored to provide the specific requirements for forecasting reservoir inflows. An example of forecast lake level based on current and forecast rainfall is presented in Fig. 2 for a high flow period in July, 2015. The three forecast lines (green, blue, and red) represent lake level response to forecast precipitation at 90, 50, and 10 % exceedance probabilities, respectively. Observed lake level is shown as blue circles to the left of the 'Now' line at 9:01 on July 1, which are used to initiate the forecast in each 6-hr update period.

Continuous simulation with measured PET results in optimization of the model where adjustments to saturated hydraulic conductivity, K_{sat}, and Soil Depth, S_d, help achieve minimization of the Nash Sutcliffe and mean absolute percentage error (MAPE). Model response was tested is for a range of multipliers applied to distributed maps of each parameter, namely, K_{sat} and S_d. For continuous simulation starting January 1, optimal results are achieved with multipliers of $K_{sat} \times 0.3$ and $S_d \times 1.3$. The result is that K_{sat} =1.09 cm/h is reduced to 0.43 cm/h, and S_d increases from 63.5 to 82.6 cm. PET can be formulated by several methods including climatological values or measured atmospheric parameters using the ASCE reference ET method (Walter et al. 2000). In Fig. 3, hydrographs produced with three PET inputs for climatological PET, measured PET with crop coefficients, Kc.

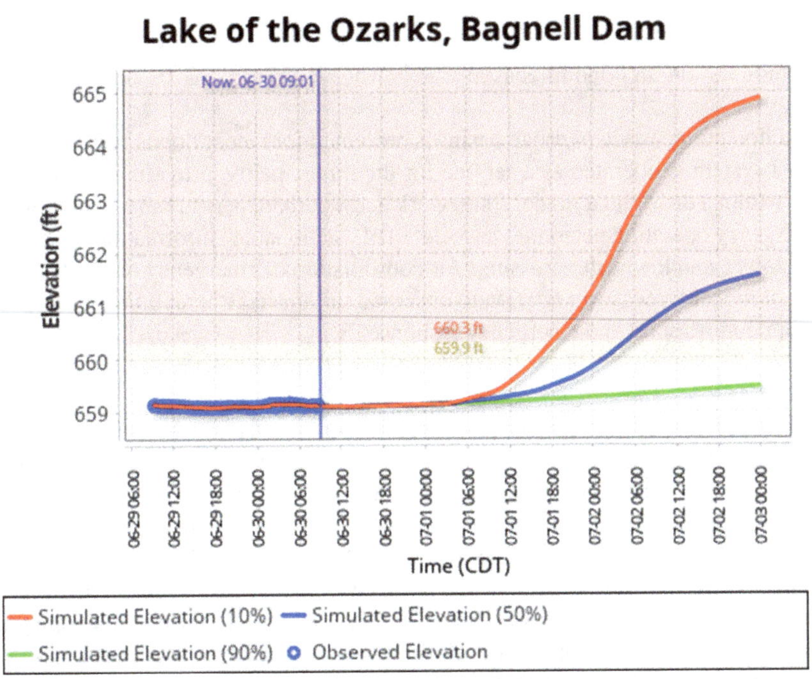

Figure 2: Lake level forecasts (1.0 ft. = 0.3048 m).

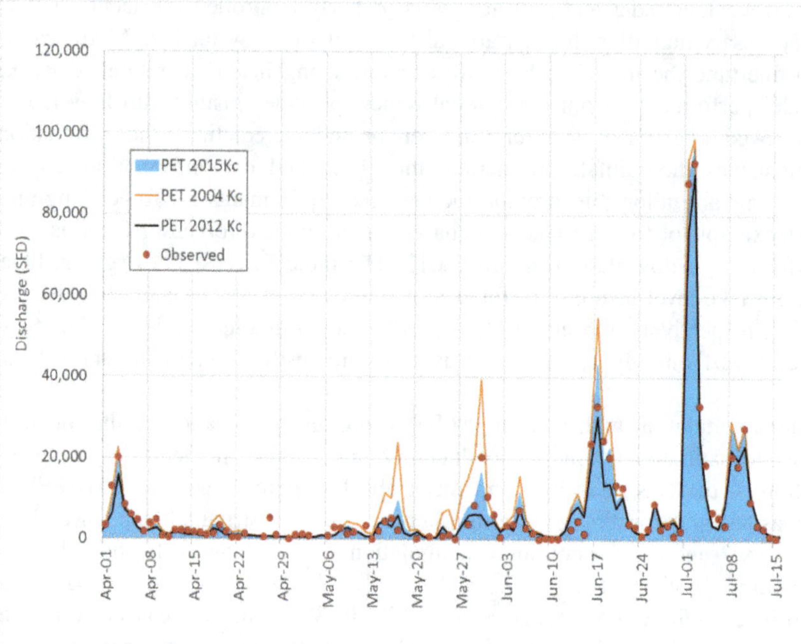

Figure 3: Sensitivity to PET by continuous simulation during. (35.31 SFD = 1 m^3s^{-1}·day).

These simulations are 'hindcasts' used to evaluate PET inputs, and for identification of optimal parameter sets.

3.2 Case study II – urban flood forecasting

Operational flood forecasting of urban basins is presented for recent floods in 2015 and 2016. The forecast system is operated as a service for the municipality, providing flood stage and inundation within a distributed model framework across urban, peri-urban and natural watersheds. Emergency operation personnel have detailed, "actionable" information about flooded buildings, streets, and low water crossings for coordination of emergency response. Analysis and review of model forecasts in relation to observed stream stage reveals that QPFs improve lead time in flood warning and emergency response [3].

3.2.1 Study area

Geographic location near the Gulf of Mexico and terrain typical of the Texas Hill Country results in fast-moving runoff during heavy rainfall, creating dangerous conditions and flash flooding in a growing urban and peri-urban environment. Because of the tremendous runoff potential, central Texas is one of the most flash-flood prone regions in North America, earning the nickname, "Flash Flood Alley." Better rainfall data and flood prediction information available in near real-time allows for a faster, more targeted response to heavy rainfall, which means more time to close roads, reroute traffic, activate high-water signals and deploy emergency personnel before an impending flood. To achieve this goal, both accurate rainfall and model forecasts are provided in a web-based framework to help protect the public from flood hazards posed by stream flooding and twelve high hazard dams.

3.2.2 Forecast objective

Early warning of flooding is an important public safety issue for a city in central Texas. The municipality takes actions to help protect its citizens from food risks. An historical event, motivated the formation of the city's flood early warning system in May (Memorial Day) of 1981 that killed 13 people. In response, the city deployed four telemetered rain gauges. Since then, new technologies were implemented in rainfall monitoring and hydrologic prediction that have proved useful in emergency response to flooding caused by urban stormwater. Rainfall can be highly variable in space and time, making it difficult to characterize accurately over a watershed. Spatially distributed rainfall and physics-based, distributed hydrologic modeling offer high-resolution information for real-time emergency management and response.

Radar rainfall is a powerful tool for hydrologic prediction because it captures the spatial and temporal variability of rainfall as it occurs over watersheds. In recent years, considerable attention has been focused on radar hydrology and its quantitative application in hydraulic and hydrologic modeling, as opposed to weather radar used alone merely as a meteorological indicator of precipitation intensity. Application of radar-derived rainfall in urban drainage modeling is described by [1, 2, 5, 6].

Rainfall monitoring consists of a hydrometeorological system customized to provide real-time model inputs for flood forecasting and inundation mapping. The rainfall used in real-time distributed hydrologic modeling consists of radar at 1-km resolution and rain gauges surrounding the target watersheds. The 4,483 km^2 (1 × 1 km pixels) study area covering the city's watersheds is served by two S-band NEXRAD (WSR-88D) radars. Radar reflectivity is converted to rainfall rate by means of a default Z-R relationship, and then updated continuously through a bias correction in real-time using rain gauge accumulations, with updates every 15 minutes. Through data sharing, gauge data from adjacent networks operated by other governmental entities are incorporated to begin the bias adjustment before rainfall moves into the area.

The capability of radar to provide accurate rainfall estimates over large areas at high resolution has the potential to provide needed rainfall model inputs for inundation forecasts and custom flood alert systems. The WSR-88D radar deployed by US NWS is an S-band radar with a 10-cm wavelength, which can penetrate rainfall with little attenuation out to 230 km in range from the radar [5]. This feature is especially useful for covering large areas where there is a little blockage from terrain or other obstructions. The hydrometeorological system is tailored to the flood forecasting objectives of the municipality. In the online mode, the system ingests radar and rain gauge data and stores in a database with controls for data quality that rejects outliers, corrects for bias, and tracks statistical accuracy. The recent addition of dual polarization to NEXRAD radars results in a QPE product called Digital Precipitation Rate (DPR) is independent of rain gauges. The performance of the radar rainfall versus rain gauge input, and its influence on forecast lead time and accuracy, is described in [6]. Reservoir inflow forecast sensitivity to radar-derived inputs and to soil moisture is evaluated and described in [7].

3.2.3 Forecast inundation

One of the main forecast objectives is a mapping of inundation, with depth rasters showing affected buildings and bridges. The inundation maps are generated on a gridded basis that relies on the definition of the flow direction map extracted from a high-resolution is necessary for mapping stage files produced at the model-grid resolution. A high-resolution flood inundation map from recent flooding is shown in Fig. 4. These maps provide emergency

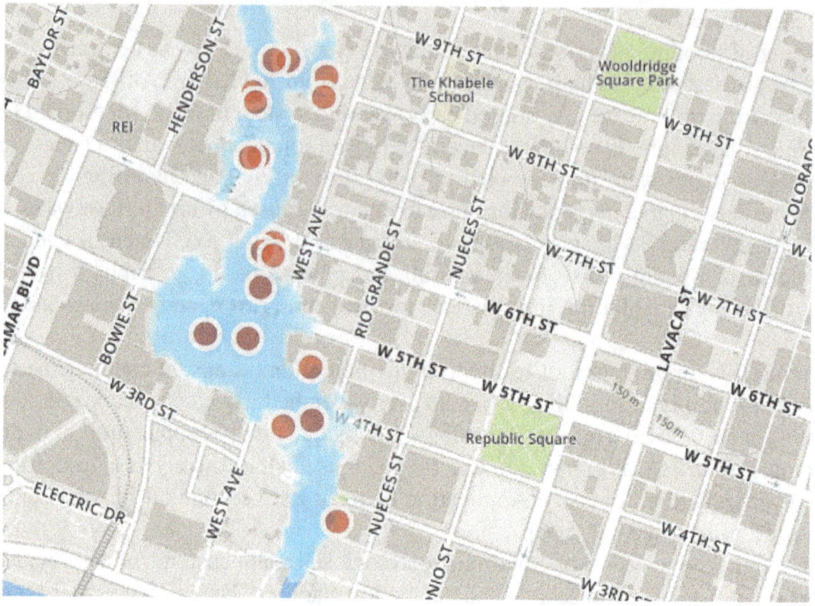

Figure 4: Inundation depth mapping (blue shaded) and flooded structures (red dots).

management personnel with specific locations of flooded structures, and are used for evacuation and barricading of impacted roads.

3.2.4 Forecast uncertainty

An example of a flood forecast hydrograph is shown in Fig. 5, the basin response to copious rainfall amounts, exceeding 355 mm during a 6-hr period (see band of red pixels). Rainfall intensity in some pixels exceeded 235 mm/h in a 15-min period. These broadly distributed rainfall depths and intensities measured by GARR and DPR in real-time, drove the observed and modeled hydrograph peak response under 6 hours. Both GARR and DPR forecast hydrographs are in close agreement with observed stage, to within 30 cm in peak stage.

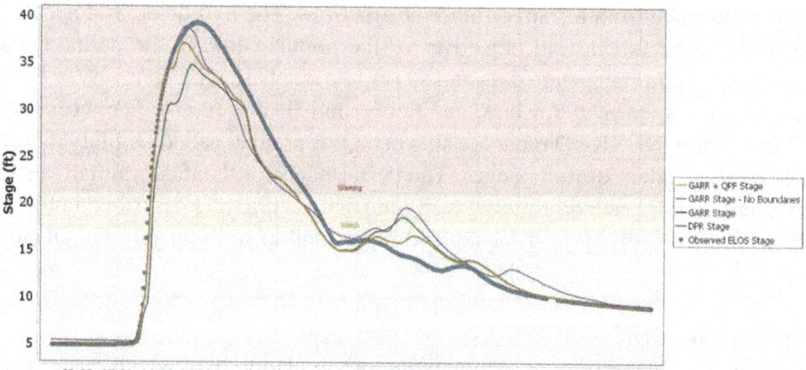

Figure 5: Operational forecasting results during the October 31, 2015 flood in Central Texas. (1 ft. = 0.3048 m).

Because this region experiences intense but highly variable rainfall, significant flooding at multiple locations occurs throughout the urban and peri-urban areas. The QPE is produced at high temporal resolution for input with basins, many of which respond in less than an hour. Real-time operation produces hydrographs that have sufficient lead-time for the emergency response staff to make decisions and take actions during immediate flood threats.

3.3 Forecast uncertainty

Quantification of uncertainties and sources of error are important for understanding forecast reliability in both case studies dealing with reservoir inflow and urban flood forecasting. Sources of error in distributed hydrologic forecasting include an error in current rainfall, QPE, and in forecast rainfall, QPF. As with any modeling approach, the structure of the model can introduce important biases in runoff predictions from a given input. Two runoff mechanisms operate simultaneously throughout the basin, namely infiltration rate excess and saturation excess, thus making accuracy in rainfall depth and rate of paramount importance. Soil moisture can only be interpreted indirectly through modeling of PET from atmospheric measurements, and applied to variable soil profiles across a range of spatial and temporal scales. Continuous simulation using measured PET, as opposed to climatological PET, revealed that model performance could be improved. Recalibration of soil parameters was necessary, once measured evapotranspiration rates were incorporated into the analysis, accomplished using scalar adjustments. Precipitation accuracy is controlled operationally but the hydrometeorological system using rain gauges to adjust radar-based QPE. Accuracy of this product is controlled by the timely ingest of rain gauge data, and the number and distribution of correctly functioning rain gauges across the forecast domain.

4 SUMMARY

In the pursuit of reliable flood forecasting, multiple technologies must be harnessed. Precipitation measurement at high resolution for large areas must be configured to generate accurate and reliable precipitation model inputs. The hydrometeorological system used to generate hydrologic model input should provide rainfall distributions that are representative over the watershed. Various methods have been developed to enhance the accuracy of radar rainfall estimates, most of which rely on rain gauge accumulations to adjust the bias inherent in radar measurements. Hydrologic prediction accuracy is directly tied to the accuracy of rainfall measurements distributed over the target watershed. With accurate inputs, it is possible to accurately model the hydraulics and hydrology of runoff, and route the resulting streamflow downstream in the drainage network. Modeled output is formulated for interpretation and use as inundation maps and for warning at critical stream crossings. Reservoir inflow forecasts and urban flood forecasting depend on accurately characterizing soil moisture and rainfall rates, while the model structure effectively represents the runoff processes as demonstrated for basin scales with short and long response times ranging in size from less than 30 km^2 to over 6,000 km^2 at model resolutions ranging from 60-m to 500-m model resolution.

REFERENCES

[1] Vieux, B.E. & Bedient, P.B., Assessing urban hydrologic prediction accuracy through event reconstruction. *Journal of Hydrolofy,* **299**(3–4), pp. 217–236, 2004. http://dx.doi.org/10.1016/S0022-1694(04)00366-X

[2] Vieux, B.E. *Distributed Hydrologic Modeling Using GIS*, 3rd edn, Springer, forthcoming, 2016.

[3] Vieux, B.E. & Moreda, F.G., Ordered physics-based parameter adjustment of a distributed model. *Calibration of Watershed Models*, pp. 267–281, 2003. http://dx.doi.org/10.1029/ws006p0267

[4] V*flo*® User Manual, available at: http://vflo.vieuxinc.com/vflo-guide, (accessed May 2016).

[5] Bedient, P.B., Huber, W.C. & Vieux, B.E., *Hydrology and Floodplain Analysis*. 4th edn, Prentice-Hall: Upper Saddle River, NJ, 2013.

[6] Looper, J.P. & Vieux, B.E., An assessment of distributed flash flood forecasting accuracy using radar and rain gauge input for a physics-based distributed hydrologic model. *Jouranl of Hydrology.* **412**, pp. 114–132, 2012. http://dx.doi.org/10.1016/j.jhydrol.2011.05.046

[7] Vieux, B.E., Kang, B., Park, J.H., On distributed hydrologic prediction: sensitivity to accuracy of initial soil moisture conditions and radar rainfall input. *Journal of Hydrologic Engineering*, **14**(7), pp. 671–689, 2009. http://dx.doi.org/10.1061/(ASCE)HE.1943-5584.0000039

MODELLING INTEGRATED EXTREME HYDROLOGY

FRANCISCO PADILLA[1], J-HORACIO HERNÁNDEZ[2], RICARDO JUNCOSA[1] & PABLO R.-VELLANDO[1]
[1]ETS de Ingenieros de Caminos, University of A Coruña, Spain, and Wessex Institute of Technology, UK.
[2]Ingeniería Geomática e Hidráulica, University of Guanajuato, Mexico.

ABSTRACT

The present research is concerned with some numerical developments and practical application of a physically based numerical model *FreshWaterSheds* that incorporates a finite element solution to the steady/transient problems of the joint ground/surface fresh/salt water flows in inland and coastal regulated watersheds. The proposed model considers surface and groundwater interactions to be 2-D horizontally distributed and depth-averaged through a diffusive wave approach. Infiltration rates, overland flows and evapotranspiration processes are considered by diffuse discharge from surface water, unsaturated subsoil and groundwater table. New improvements also allow for the management of surface water flow control through the capacity of diversion on flooding zones of catchment areas, as well as on river beds, spillways and outflow operations of floodgates in weirs and dams of reservoirs. Practical application regards the flooding hazard of Aznalcóllar toxic spillages. This flooding disaster was caused by the sequential ruptures of the dikes of two mining residual reservoirs of a pyrite mine, releasing about $10{\cdot}10^6$ m^3 of contaminated wastewater and mining sludge onto the Guadiamar River. The numerical model was adapted to the wastewater and sludge properties of both sudden spillages, as well as to the river bed, the flooded zones and the underneath alluvial aquifer. The model simulation and calibration were made during the date of this hydrological hazard to the likely discharges and dual hydrograph produced by the sudden twofold failure of both reservoirs.
Keywords: extreme hydrology, finite elements, flood routing, integrated surface/subsurface flows, numerical modelling, wastewater.

1 INTRODUCTION

Nowadays, there is an increasing need for integrated surface and ground water numerical modelling. The philosophy and role of hydrological models in water resources has been widely described [1, 2.

Recent numerical methodologies were also developed on combined surface and ground-water applications to the whole of the water resources of a particular river basin [3, 4]. MIKE SHE and MIKE BASIN are two examples of numerical and physically based modelling systems developed by DHI [5] for the major flow processes of the hydrological cycle which integrates the Saint-Venant surface equation to a vertical 1-D Richards equation for unsaturated flow and a 3-D finite element solver for saturated flow [6]; Camporese *et al.* [7], apply a catchment's hydrology model that couples a finite element solver for the Richards equation describing variably saturated porous media, and a finite differences solver for the diffusive wave equation describing surface flow.

Standard features of detailed process-based numerical models for water resources management at the scales of the watershed could include: a large variety of numerical conditions and sub-models for the different hydrological components, water uses and managements.

DOI: 10.2495/SAFE-V6-N3-685-696

These characteristics will not be described in any case here; this research will focus instead on some new features and improvements, summarised below, on an integrated surface-subsurface flow numerical model *FreshWaterSheds* that could be also suitable for applications on flooding events of different nature.

In this field, a new methodology of finite element modelling is being recently developed which considers new numerical approaches for the joint surface and ground water regional flows of continental freshwater and coastal saltwater on natural and regulated watersheds [8, 9].

The practical application regards the flooding hazard of Aznalcóllar toxic spillages of 1998. This flooding disaster was caused by the sequential ruptures of the dikes of the two mining waste reservoirs of a pyrite mine, releasing about $10 \cdot 10^6$ m^3 of contaminated wastewater and sludge onto the Guadiamar River. These two sudden spillages of mining wastewater and sludge were caused by the sudden and sequential twofold failure of both reservoirs, spreading downstream over the river bed and margins, flooded zones and underneath alluvial aquifer, producing the resulting flood routing and the recorded dual hydrograph downstream the Guadiamar River.

2 MODEL DESCRIPTION

The *FreshWaterSheds* model for continental and coastal catchment applications couples subsurface and surface regional hydrology, by a diffusive wave approach, with a joint finite element solver for the saturated porous media flows of fresh and saltwater through an immiscible sharp interface [8].

In this sense, this numerical approach (Fig. 1) was mainly developed in order to be able to analyse anisotropic groundwater behaviour as well as surface water flows at a watershed scale. It can also consider other water components such as the evapotranspiration processes, that is, a diffusive discharge from surface water and soils within the unsaturated zone by a root water uptake sub-model, as well as the overland flow by a rainfall-runoff sub-model based on an exponential method for assessing the infiltration rates [10].

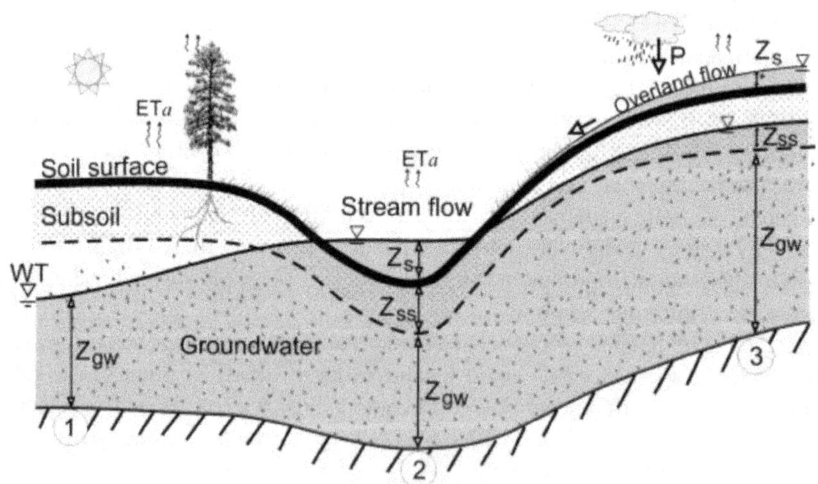

Figure 1: Surface/groundwater flows in *FreshWaterSheds* interactions. Thicknesses of groundwater (z_{gw}), subsoil (z_{ss}) and stream/overland flow (z_s). Water table (WT). Actual evapotranspiration (ETa). [9].

The regulation capacities of diversion works may be considered for withdrawing and bypassing certain flow rates of water from a river bed towards other places, by transferring them mainly through impervious channels, pipelines or tunnels. The place where the water is transferred could be located in the same catchment area or in a different one.

FreshWaterSheds, the current tool used in water resources modelling, is a two dimensional finite element model for regional surface and groundwater flows through drainage basins, developed for a temporal implicit (Eulerian) centred (Crank-Nicholson) and spatially centered (Galerkin) numerical approach.

The inputs of freshwater could be exclusively due to rainfall; however, pumping and injection wells of prescribed flow rates have been also implemented, as well as irrigation and artificial recharges, which could be also prescribed inside the considered domain. Therefore, the boundary conditions of the modelled region, at the scale of the watershed, are mainly those of the outflow or discharge conditions. To this respect, prescribed water heads, outflow face and open boundary conditions are commonly used at the discharge zones of the drainage basin [11].

2.1 Surface-groundwater interactions

FreshWaterSheds numerical model uses a simplification of the equations governing the free surface flows. In particular, as often happens with other commonly used diffusive wave equations describing surface flow approaches in hydrology, only the classical mass conservation or continuity equation is considered for the present 2-D model.

$$\frac{dh}{dt} + \nabla \cdot \mathbf{q} = 0; \text{ which leads to: } \frac{dh}{dt} = \nabla \cdot (T^* \nabla h) \tag{1}$$

In the context of flow diffusion, h represents the hydraulic head, T^* is the diffusivity tensor, and $\mathbf{q}$ is the mean horizontal water flow rate per unit length and defined by the Fick's law as a pure-diffusive process.

When a phenomenon is strongly diffusive, it is necessary to take into account the wave nature of diffusive processes to perform accurate predictions, and the hyperbolic diffusion theory is required. Then, the appropriate equation of continuity is derived substituting in system (1) Fick's law by a more general equation due to Cattaneo [12], namely

$$\frac{dh}{dt} + \nabla \cdot \mathbf{q} = 0; \text{ where: } \mathbf{q} + D\mathbf{q} = \mathbf{q} + \tau \frac{d\mathbf{q}}{dt} = |\mathbf{q}| = -T^* \nabla h$$

$$\text{leading to: } \frac{dh}{dt} = \nabla \cdot (T^* \nabla h) + \nabla \cdot \left(\tau \frac{d\mathbf{q}}{dt} \right) \tag{2}$$

This equation defines the hyperbolic diffusion theory, where τ is the so-called relaxation factor which has dimensions of time.

If we make $\tau/\mathrm{dt} = \mathbf{x}$, and knowing that because of continuity

$$\nabla \cdot (d\mathbf{q}) = \frac{dh}{dt} \tag{3}$$

The following equations for the hyperbolic diffusion theory can be found:

$$(1 - \mathbf{x}) \frac{dh}{dt} = \nabla \cdot (T^* \nabla h); \text{ or otherwise: } (1 - \mathbf{x}) \ \nabla \cdot (d\mathbf{q}) = -\nabla \cdot |\mathbf{q}| \tag{4}$$

where, if we gave solution to $\nabla\cdot(d\mathbf{q})$ with respect to the vertical coordinate of the water depth Z_s, and V_s taken as the water velocity:

$$\frac{\partial(d\mathbf{q})}{\partial z_s}=\frac{d\mathbf{q}}{z_s}=\frac{D\mathbf{q}}{z_s^{\mathbf{x}}}\quad;\quad\frac{\partial\left(dv_s\cdot z_s\right)}{\partial z_s}=dv_s+z_s\frac{\partial(dv_s)}{\partial z_s}=\frac{Dv_s\cdot z_s}{z_s^{\mathbf{x}}}=\frac{\mathbf{x}dv_s\cdot z_s}{z_s^{\mathbf{x}}}\tag{5}$$

which can be rewritten after integration as: (5)

$$|\mathbf{q}|=-dv_s^0\cdot(1-\mathbf{x})\left(e^{\mathbf{x}\frac{z_s^{1-\mathbf{x}}}{1-\mathbf{x}}}-1\right)=-dv_s^0\cdot f\left(z_s,\mathbf{x}\right)$$

$$\left(dv_s\cdot z_s\right)=(d\mathbf{q})=dv_s^0\left(e^{\mathbf{x}\frac{z_s^{1-\mathbf{x}}}{1-\mathbf{x}}}-1\right);$$

That gives the mean horizontal flow rates per unit length $\boldsymbol{q}$ as a function of the diffusivity and the hydraulic gradient.

$$|\mathbf{q}|=-v_s^0\nabla h\cdot f\left(z_s,\mathbf{x}\right)=-K_s f(z_s,\mathbf{x})\cdot\nabla h=-T^*\nabla h\tag{6}$$

Therefore, the $f(z_s, \mathbf{x})$ function, which has here dimensions of thickness, and the water velocity v_s, which can be stated through the hydraulic conductivity K_s multiplied by the hydraulic gradient, both help to define $|\mathbf{q}|$ as the hydraulic gradient and a function that enhances the hydraulic transmissibility (T^*), here behaving like a diffusive parameter adapted to the free surface flow, that in fact depends on the surface water depth z_s and can be adjusted numerically only through the $\mathbf{x}$ parameter.

In order to take into account the diversion capacity of a flooding event or a weir built on a river bed, the following equation has been implemented, defining the velocity of diversion V, in terms of the flow rate being diverted per unit area of flooded surface.

$$V\ (m/s)=\frac{z_s\,(m)\cdot V_0}{10\ (m)}\tag{7}$$

where Z_s is the headwater depth of the overland flow (in meters) or amount of energy available to convey water through diversion, and V_0 is a parameter that defines a particular velocity of the surface water diversion (in terms of flow rate, that is, diverted volume per unit time and area of flooded surface).

Triangular elements of three nodes are used for the analytic integration of the corresponding numerical formulation for steady and transient conditions. The preconditioned iterative algorithm GMRES [13] provides the solution to the system of equations by means of a reduction in the computer memory requirements and allowing for a simple processing of the numerical mesh.

3 MODEL APPLICATION

A large toxic spill occurred in Los Frailes (Aznalcóllar) last April 25 1998, following the rupture of two large tailings reservoirs containing the effluents of a copper and zinc mine that produces pyritic waste rock. Afterwards, it was reported that the failure was caused by a landslide in the base of the joint earth dike of the two mining residual reservoirs, releasing about $10\cdot10^6$ m^3 of contaminated wastewater and mining sludge onto the Guadiamar River, near Sevilla city (South of Spain) (Fig. 2).

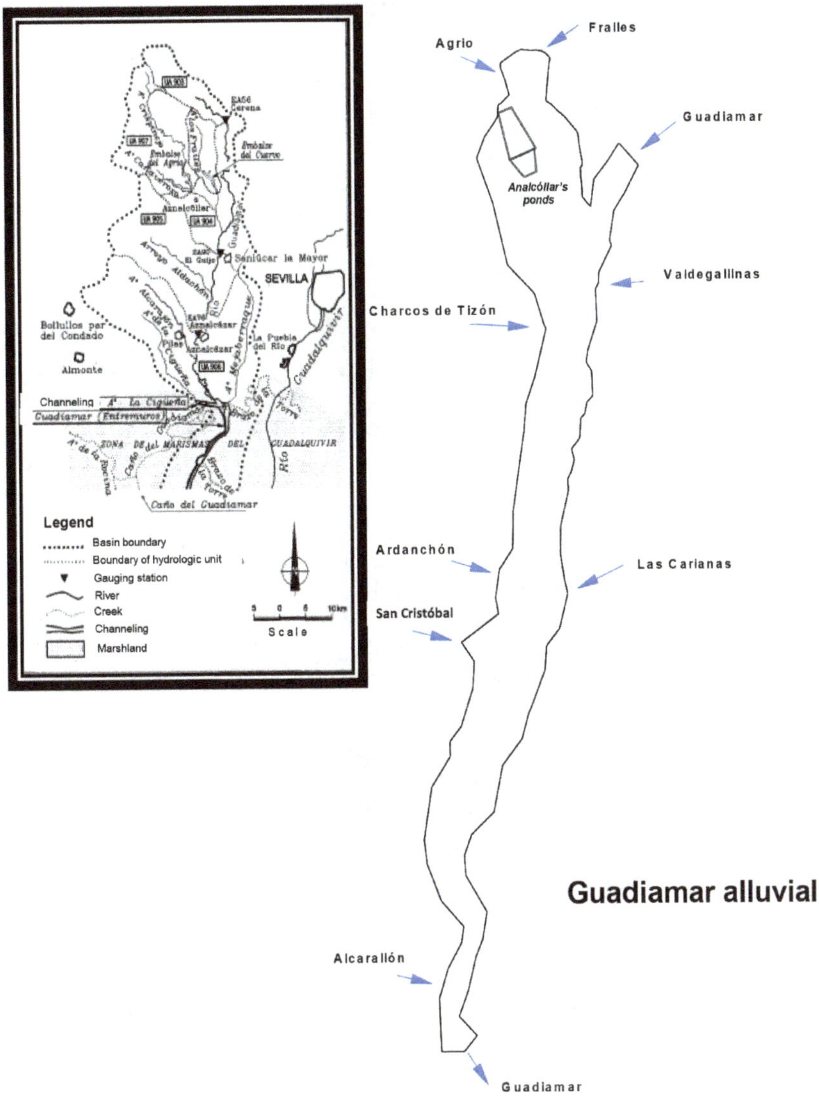

Figure 2: Localisation of Aznalcóllar's ponds, Guadiamar basin and main alluvial, Sevilla (South of Spain).

The aim of this research is to perform the transient simulation of these flooding events by the simultaneous flows of surface and ground waters through the alluvial aquifer and the rivers Agrio and Guadiamar during the 24 h of April 25 1998, the day that lasted the main flooding events caused by the sequential ruptures occurred at the Aznalcóllar wastewater ponds.

The location of the discharges that caused the flooding events should be restricted to the area where the main breakage occurred. The site after rupture responds to an opening in the earth dam of about 45 m wide and 23 m high. This, though relatively small to the scale of the discrete model, can be seen in Fig. 3.

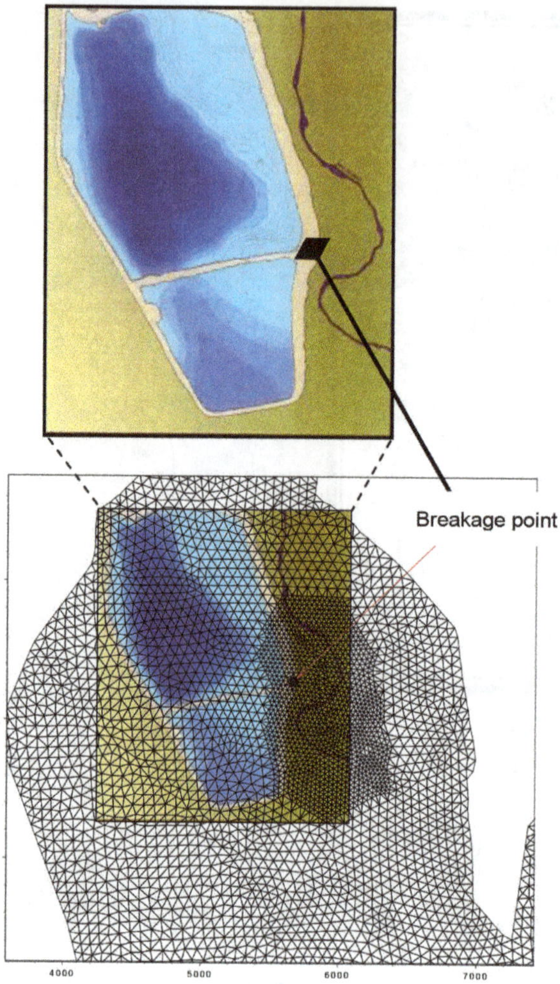

Figure 3: Discrete numerical model in the vicinity of Aznalcóllar's mining waste reservoirs.

The apparently first breakage of the ground retaining wall began in the early hours of Saturday April 25, 1998, between 0:30 am and 1:00 am [14]. The origin of the spillage was then in the area of intersection of the main wall with the dike that divides the reservoir into two ponds: the pyroclastic deposit or pond located to the north, larger, and the pond of pyrites located to the south. Essentially, the general breakage caused firstly the spillage of acidic water from the pyroclastic pond, and secondly and afterwards, a significant amount of sludge mostly from the pyrites pond [15]. This fact was recorded in the limnigraph of the gauging station EA-90 El Guijo. There are two main peaks appreciated in the limnigraph, which are interpreted as corresponding to the two phases in the breaking, one of them mainly from the pyroclastic pond and the other from the pyrites pond [16]; Fig. 4).

The simulation conditions of the model have also been adapted in order to take into account the lateral contributions of surface and groundwater coming from upstream and outside from the watershed that drains its freshwaters throughout the alluvial discrete model of the Agrio and Guadiamar rivers. As a result of the hydrological analysis, lateral contributions are dis-

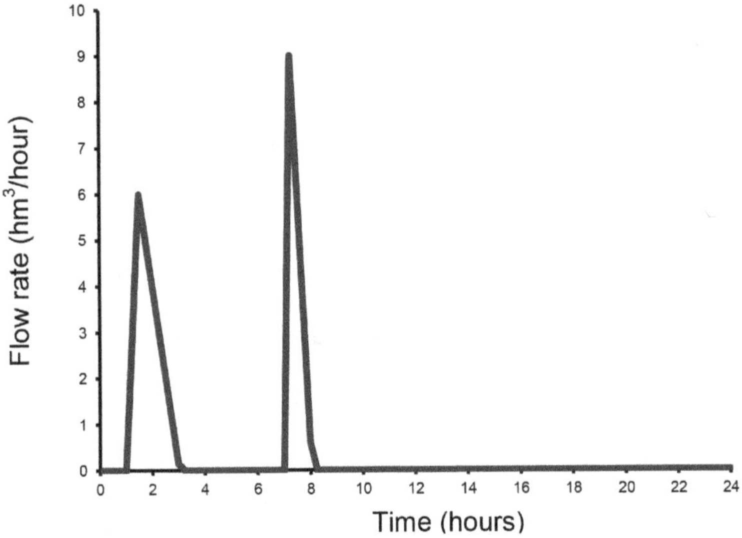

Figure 4: Observed and simulated limnigraphs at the El Guijo's gauging station the 25 April, 1998.

tributed between surface and groundwater. All considered, the dry year adds up to $100{\cdot}10^6$ m^3/year [17]. The distribution of these water contributions were also made up by sub-basins. As a result of all the above hydrological conditions, the simulation is performed for a known averaged water contribution of a dry year that is characteristic of the alluvial of the rivers Agrio and Guadiamar.

As follows up from the analysis of the existing data on the state of the two reservoirs of mining waste, before and after the accidental release (Fig. 3), and the recorded limnigraph of the events in the El Guijo's gauging station (Fig. 4), it seems quite clear that the accidental spillage occurred in two phases.

On the other hand, the volumes and types of materials involved in the breakages of both reservoirs of pyroclasts and pyrites are relatively well known, essentially ***water***, ***sludge*** and

Table 1: Likely distribution of released ***water***, ***sludge*** and ***soils*** in the two spillages produced by the breakage of the Aznalcóllar's waste reservoirs.

Volumes (hm^3)	Pyroclastic pond	Pond of pyrites	Dike	Total materials	*Total*
1st Spillage	6.0 water			6.0 water	6.0
2nd Spillage	1.58 water	0.85 water		2.43 water	
	0.84 sludge	1.05 sludge		1.89 sludge	4.5
			0.18 soils	0.18 soils	
Total materials	7.58 water	0.85 water		8.43 water	
	0.84 sludge	1.05 sludge		1.89 sludge	10.5
			0.18 soils	0.18 soils	
Total	***8.42***	***1.9***	***0.18***	***10.5***	***10.5***

mixed materials from the dikes or ***soils***. Table 1 indicates a likely distribution of these materials in the two different discharges caused by the rupture of the main joint dike of both tailing reservoirs of Aznalcóllar.

The knowledge of the floods caused by the accident, as registered in EA-90 El Guijo gauging station (Fig. 4), as well as the hypothesis of the distribution of the volumes of the materials involved in the discharges caused by the breakages of the ponds of pyroclasts and pyrites (Table 1), allow somehow for an approximation of the evolution of the discharges caused by the accident of Aznalcóllar (Fig. 5).

The two discharges have a temporal separation of 6 hours, and reached approximately the discharged volumes, that is, 6 Hm^3 and 4.5 Hm^3, respectively. In short, the maximum discharge of the first spillage (greater volume) would have been smaller than the maximum of the second spillage (smaller volume).

4 RESULTS AND DISCUSSION

It should first be checked whether the simulated water levels in the vicinity of the gauging station EA-90 El Guijo properly compared with the limnigraph registered after the flooding hazard. Figure 4 provides the comparison between the observed and simulated water depths during the 24 h of the day of the accident.

The drainage simulated layout and the presented outcomes should, however, be interpreted as indicative and approximate (Fig. 6). An overview of the velocity fields generated by this flooding event can also be locally depicted (Fig. 7).

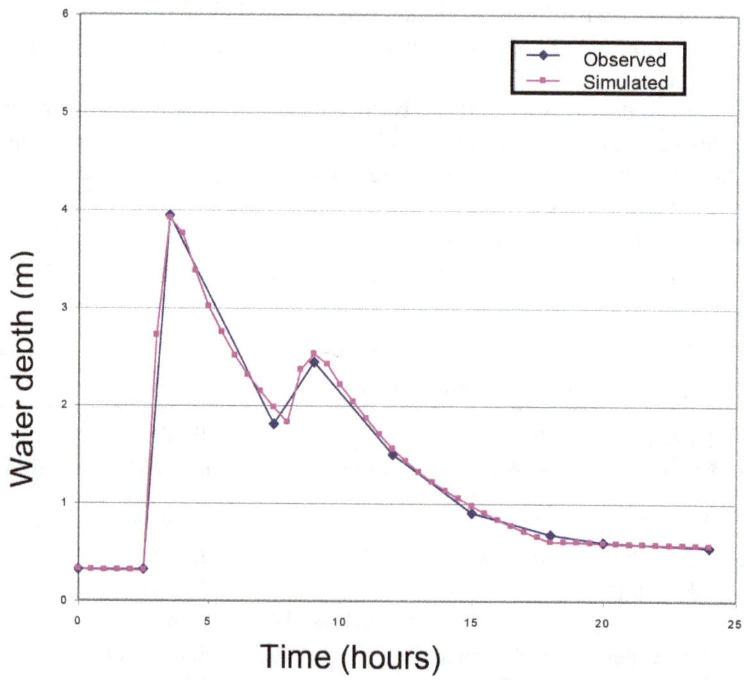

Figure 5: Hypothetical hydrograph of the two discharges produced by the twofold breakage of Aznalcóllar's waste reservoirs.

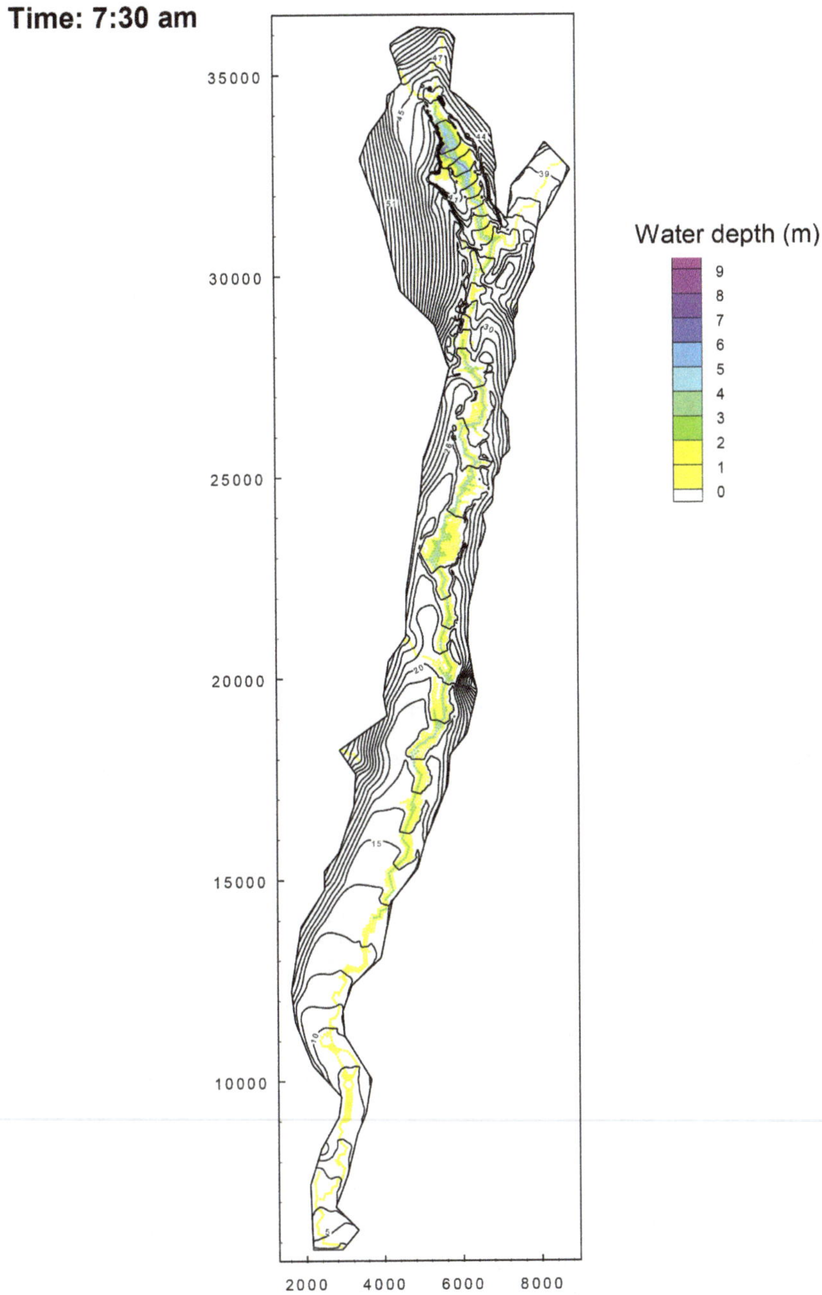

Figure 6: General hydrological layout of groundwater levels and surface water depths in the Guadiamar alluvial at 7:30 am of the 25 April 1998.

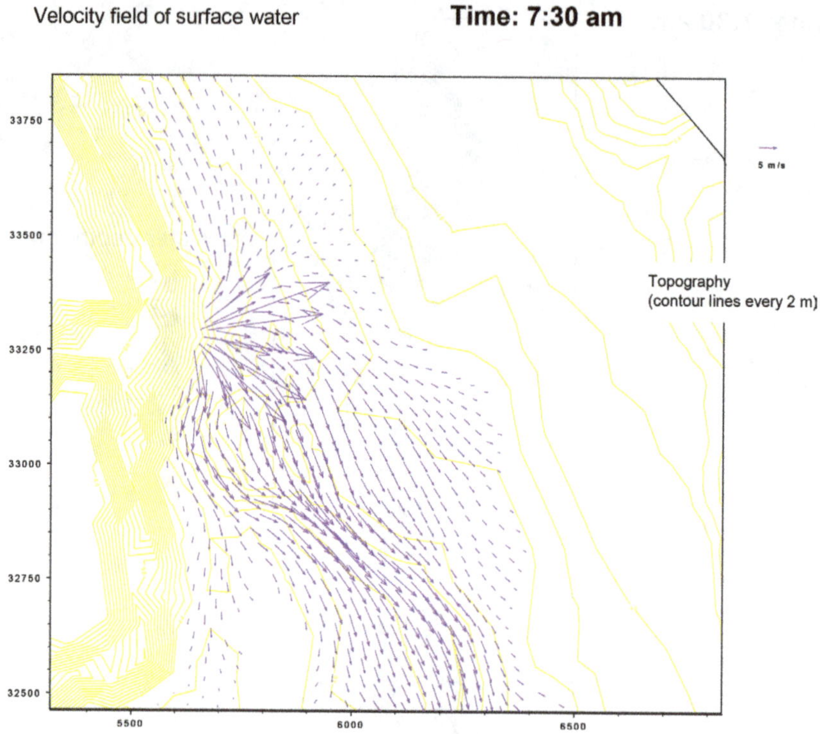

Figure 7: Velocity field of surface water in the vicinity of Aznalcóllar's waste reservoirs at 7:30 am of the 25 April 1998.

Thence this hydrological alternative involving a two phase release of 10.5 hm^3 of waste water, where there are sequentially two discharges both with the same origin, gives rise to two floods spaced in time and two stages of flooding in certain sectors. Thus the general transient hydrological layout, as well as the different sorts of velocity fields, indicate a strong relationship between groundwater and surface water, as there is a repetition of certain previously flooded areas as well as intermediate periods of temporary drying of a given soil surface, including likely flooding periods of the same previously flooded surfaces.

5 CONCLUSIONS

Some new developments have been performed on a process-based integrated numerical model *FreshWaterSheds* that incorporates a finite element solution to the steady/transient problems of the integrated ground/surface fresh/salt water flows in inland and coastal regulated watersheds. The proposed model considers surface and groundwater interactions to be 2-D horizontally distributed and depth-averaged through a new diffusive wave approach.

This research also aims to improve the acquired knowledge about the characteristics of the alluvial basins of the Agrio and Guadiamar rivers during the period of rupture of the Aznalcóllar mining waste repository. The work has conducted to the analysis of a hydrological alternative that releases a volume of waste water of 10.5 Hm^3, but nevertheless distributed in a two-phase breakage of a waste water reservoir.

The simulation conditions are prepared for the times of the reservoir breakages, with an approximate first phase spillage of 6 hm^3 and a second phase spillage of 4.5 hm^3 with a 6 h time in-between. In this sense, a brief calibration of the hydrological parameters of the numerical model were needed in order to better adjust the simulated results to the water depths recorded at the gauging station of El Guijo during the day of the accident.

It is noteworthy that, given the actual knowledge of these flooding events and its very particular hydrological conditions, heretofore the simulated outcomes about the groundwater/surface water relationships, the velocity fields, the flooding surfaces and the general drainage layout seem in general likely enough. It can also be concluded that the *FreshWaterSheds* numerical model can provide useful and quite precise results with regard to the integrated behaviours of groundwater and surface waters of different kinds as related to flood routing and flooding events which are significant in the analysis of extreme hydrology in river basins.

ACKNOWLEDGEMENTS

The work reported in this paper was part of the research conducted on the breakage of Aznalcóllar's dam, funded by DRAGADOS. Our special acknowledgments to José Polimón López and José María Rodríguez Ortiz, directors of the research works.

REFERENCES

[1] Abbot, M.B. & Refsgaard, J.C., *Distributed Hydrological Modelling. Water Science and Technology Library,* Kluver Academic Publishers: Dordrecht, p. 22, 1996. http://dx.doi.org/10.1007/978-94-009-0257-2

[2] Burnash, R.J.C., Ferral, R.L., McGuire, R.A. & McGuire, R.A., A generalized stream flow simulation system. Conceptual modeling for digital computer. Joint developed by Federal-State River Forecast Center and Nacional Weather Service. Dep. Water Resources (California), 1973.

[3] Ross, M., Said, A., Torut, A., Tara, K. & Geurink, J., A new discretization scheme for integrated surface and groundwater modelling. In: *Integrated Water Resource Management*, ed. T.V. Hromadka, American Institute of Hydrology: Las Vegas, Nevada, pp. 143–156, 2005.

[4] Sophocleous, M. & Perkins, S.P., Methodology and application of combined watershed and ground-water models in Kansas. *Journal of Hydrology*, **236**(3–4), pp. 185–201, 2000. http://dx.doi.org/10.1016/S0022-1694(00)00293-6

[5] DHI. MIKE BASIN. A tool river planning and management. Report, Danish Hydr. Inst., Horsholm, Denmark, 1997.

[6] Graham, D.N. & Butts, M.B., Flexible, integrated watershed modelling with MIKE SHE. In: *Watershed Models*, eds. V.P. Singh & D.K. Frevert, CRC Press: Boca Raton, pp. 245–272, 2005.

[7] Camporese, M., Paniconi, C., Putti, M. & Orlandini, S., Surface-subsurface flow modeling with path-based runoff routing, boundary condition-based coupling, and assimilation of multisource observation data. *Water Resources Research*, **46**(2), 2010. http://dx.doi.org/10.1029/2008WR007536

[8] Padilla, F., Méndez, A., Fernández, R. & Vellando, P., Numerical modelling of surface-water/groundwater flows for freshwater/saltwater hydrology: the case of the alluvial coastal aquifer of the Low Guadalhorce River, Malaga, Spain. *Environmental Geology*, **55**(1), pp. 215–226, 2008. http://dx.doi.org/10.1007/s00254-007-0977-2

[9] Padilla, F., Hernández, H., Juncosa, R. & Vellando, P., A numerical solution for the integrated analysis of water resources management: Application to the Mero River watershed – La Coruña, Spain". *Journal of Water Resource and Protection*, **7**, pp. 815–829, 2015. http://dx.doi.org/10.4236/jwarp.2015.710066

[10] Hernández, J.H., Padilla, F., Juncosa, R., R-Vellando, P. & Fernández, A., A numerical solution to integrated water flows: Application to the flooding of an open pit mine at the Barcés River catchment – La Coruña, Spain. *Journal of Hydrology*, **472–473**, pp. 328–339, 2012.
http://dx.doi.org/10.1016/j.jhydrol.2012.09.040

[11] Padilla, F. & Cruz-Sanjulián, J., Modeling seawater intrusion with open boundary conditions. *Ground Water*, **35**(4), pp. 702–712, 1997.

[12] Cattaneo, M.C., Sur une forme de l'équation de la chaleur limitant le paradoxe d'une propagation instantanée. *Comptes Rendus de L'Académie de Sciences: Serie I-Mathematics*, **247**, pp. 431–433, 1958.

[13] Saad, Y. & Schultz, M.H., GMRES: A generalized minimal residual algorithm for solving nonsymmetric linear systems. *SIAM Journal of on Scientific and Statistical Computing*, **7**(3), pp. 856–869, 1986.
http://dx.doi.org/10.1137/0907058

[14] Alonso, E. & Gens, A., Rotura de la balsa de residuos mineros de Aznalcóllar. Universidad Politécnica de Cataluña: Barcelona, 2000 (report).

[15] Mediavilla, C., Arenas, J.M., Carrero, G., Galache, J., Silgado, A. & Vázquez, E.M., Actuaciones realizadas tras el accidente de Aznalcóllar. *Boletín Geológico y Minero*. Volumen especial "Las aguas y los suelo tras el accidente de Aznalcóllar", pp. 35–56, 2001.

[16] Mediavilla, C., Borja, F., López Geta, J.A., Martín Machuca, M., Mantecón, R., del Olmo, P., Palancar, M. & Vives, R., Marco geográfico, geológico e hidrológico regional de la cuenca del Guadiamar. *Boletín Geológico y Minero*. Volumen especial "Las aguas y los suelo tras el accidente de Aznalcóllar", pp. 13–34, 2001.

[17] Carrera, J., Bernet, O., Bolzicco, J., Castro, A., Jaén, M., Padilla, F., Salvany, J.M., Vázquez-Suñé, E. & Vázquez, M., Modelación numérica de los acuíferos potencialmente afectados por la rotura de la balsa de Aznalcóllar. *Boletín Geológico y Minero*. Volumen especial "Las aguas y los suelo tras el accidente de Aznalcóllar", pp. 199–227, 2001.

FLOW INTENSIFICATION INDUCED BY TIDAL OSCILLATIONS IN TRIBUTARIES OF THE MEKONG RIVER

H. TAKAGI[1], V. P. D. TRI[2], T. V. TY[2], N. D. THAO[3], L. T. ANH[3] & G. V. VINH[4]
[1]Tokyo Institute of Technology, Japan.
[2]University of Can Tho, Vietnam.
[3]Ho Chi Minh City University of Technology, Vietnam.
[4]Hydrological Center in the Cuu Long River, Vietnam.

ABSTRACT

Strong currents induced by tidal oscillations have been overlooked in previous flood-risk assessments of the Mekong River. To discuss the potential disaster risk associated with tidal-flow intensification, this study applied ocean tidal modelling to a typical dry season when the tidal regime is predominant compared to the fluvial regime. The model can forecast tidal levels and velocities at a given time and location in the urban area of the Mekong Delta, which is characterized by many tributaries, channels, and low-lying lands, in addition to the main stream of the Mekong River. Numerical simulations revealed that a rapid flow, which could exceed 1 m/s in both the ebb and flood tidal phases, likely occurred along the tributary. Although this study only focused on the tidal regime, such locally intensified flows can be further amplified by a high river discharge or storm surge, having potentially dangerous consequences, such as difficulties in handling ships and small ships being capsized.
Keywords: Can Tho city, ebb tide, flood tide, flow intensification, ocean tidal modelling, rapid flow, the Mekong Delta, the Mekong River, tributary.

1 INTRODUCTION

The Mekong River flows southward, covering a distance of approximately 4,800 km from its source to the sea, and draining a total catchment area of 795,000 km^2 that falls within six different countries: China, Myanmar, Lao PDR, Thailand, Cambodia, and Vietnam. The Mekong River ranks 10th among the rivers of the world, on the basis of its mean annual flow at the river mouth [1]. The Mekong Delta begins at Phnom Penh, Cambodia, where the river divides into its two main distributaries: the Mekong and the Bassac. The Mekong then divides into six main channels and the Bassac divides into three, forming the delta, which is made up of a vast triangular plain of approximately 55,000 km^2 [1].

The present paper highlights Can Tho City, which is the regional capital and has the highest population in the Mekong Delta (approximately 1.2 million). Can Tho is surrounded by a main river (Hau River), a tributary (Can Tho River), and many canals, demonstrating a complicated river system that appears to yield complex hydraulics (Fig. 1). Takagi *et al.* [2] revealed that ocean tides predominantly determine the water elevation, even in Can Tho City, which is 80 km inland from the river mouth. Given the importance of these oceanic influences on deltaic

DOI: 10.2495/SAFE-V6-N3-697-703

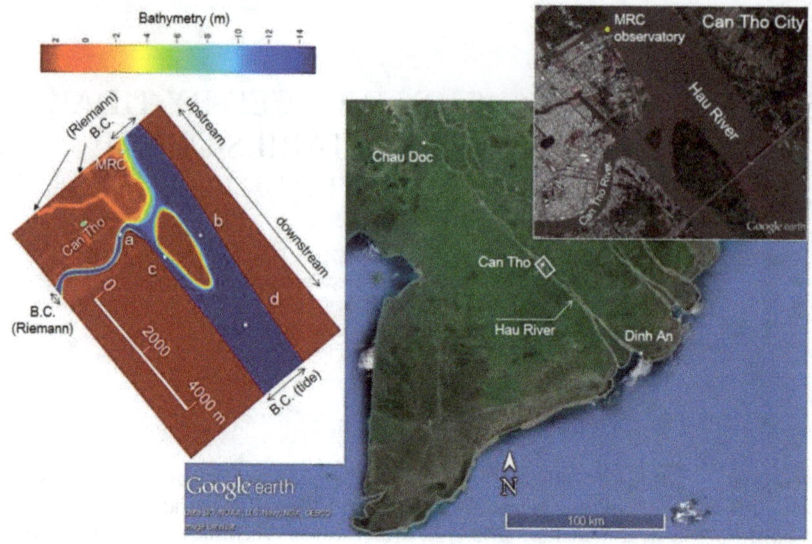

Figure 1: Can Tho, which is the largest city in the Mekong Delta, is located in southern Vietnam. The white rectangle on the satellite image indicates the computational domain used in this study, encompassing the main river (Hau River), the tributary (Can Tho River), and small canals. The frame on the left shows the domain used for the numerical simulation with open boundary conditions.

regions, Takagi *et al.* [3, 4] developed a tidal model and investigated the mechanisms of the tidal propagation and its influence on flooding in the urban areas of the Mekong Delta.

While numerous studies have investigated flood inundations and sea-level rise [5–8], it appears that the flow velocity during a flood was overlooked in previous flood-risk assessments of the Mekong River. Thus, to investigate the potential disaster risk associated with flow intensification, in this study, ocean tidal modelling was applied to a typical dry season when the tidal regime predominates compared to the fluvial regime.

2 METHODOLOGIES

The ocean tides that travel from the deep sea to the shallow waters of the continental shelf – up the Hau River – and eventually flood downtown Can Tho can be simulated by Delft3D-FLOW. Although this model is applicable to a far wider region, the region is limited by the computational domain to encompass only Can Tho City and its surrounding area (Fig. 1) in order to reduce the computational cost and enable a detailed analysis of the local topography and bathymetry, which is characterized by tributaries, channels, and low-lying land. Thus, the present model may have an advantage for evaluating the detailed hydraulic characteristics within the domain.

The simulation was conducted for a typical dry season of the Mekong Delta, including the days of March 10-11, 2012, when the authors performed a field survey (Fig. 2). The computational grid size was set as 10 m throughout the domain. The river bathymetry data used in the computational domain were obtained by the authors by using a portable echo sounder (Fig. 1). The Manning's *n* value was set as 0.02 ($s \cdot m^{-1/3}$) for the riverbed and 0.05 ($s \cdot m^{-1/3}$) for the land area [9].

Figure 2: Photo taken by the author near point (a) on Fig. 1 at 18:30 on March 10, 2012. The tidal elevation was almost the highest for the day, demonstrating that the tidal regime is predominant in Can Tho, which is 80 km inland from the river mouth.

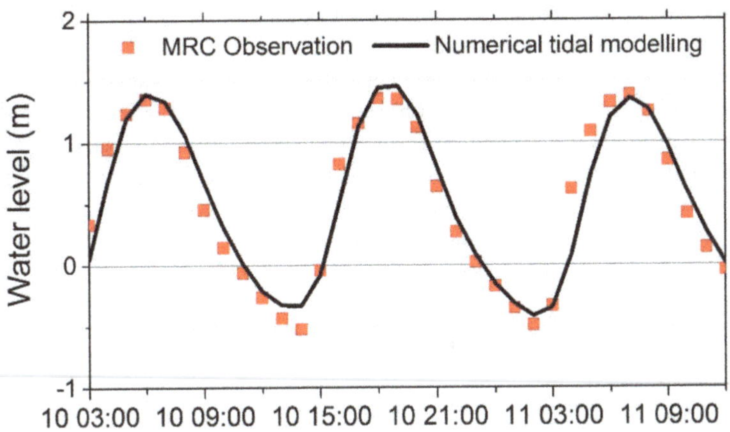

Figure 3: Comparison between the numerically simulated and measured water levels at the observatory operated by the MRC.

The tidal constituents obtained from the data at the Mekong River Commission (MRC) observatory (Fig. 1) were assigned along the downstream boundary to generate ocean tides that propagate across Can Tho. The semidiurnal and diurnal components are considered to be the factors governing the water elevation. Among all the tidal components, the semidiurnal tidal constituents with a period of approximately 12 h appear to have the strongest influence [2, 4]. On the other hand, the upstream boundary and the other tributaries were set as having the Riemann condition, which enables progressive waves to pass through without reflection.

Though there is a distance of several kilometres between the MRC observatory and the downstream boundary, it was confirmed that these obtained constituents can be used without any particular modification [4].

Figure 3 shows the tidal fluctuations at one of the output points, which is indicated as 'MRC' in Fig. 1. The simulated and observed water levels agree well, in terms of both the amplitude and phase.

3 RESULTS AND DISCUSSION

Figure 4 shows the simulated depth-averaged velocities at five different locations in the river basin (Fig. 1). The point referred to as 'MRC' indicates the location of the monitoring station, which is operated by the MRC. The depth-averaged velocity varied largely depending on the location in the river basin. Particularly, the flow velocity at point (a), which reached a maximum of 1.1 m/s, was almost three times that at point (d) in the main stream. Clearly, flow intensification occurred as the tidal current from the downstream river entered the narrower channels and vice versa, causing a rapid flow.

Figure 5 demonstrates that flow intensification occurred during both ebb and flood tides. In the ebb-tide phase, a strong flow developed in the tributary (Can Tho River), impinged on the sandbank, and then split into two streams, merging again approximately 2.5 km downstream (Fig. 5a). Interestingly, in the slack-tide phase, when the tidal current almost ceased, two whirlwinds developed near the edge of the sandbank, indicating that the tide was about to reverse its direction (Fig. 5b). In the flood-tide phase, the tide from the sea propagated upstream, and part of the water departed from the main stream and flowed into the narrower tributary, causing a rapid flow (Fig. 5c).

Typically, the maximum tidal velocity in spring tide may not exceed 1 m/s in estuaries [10]. Considering the general understanding of tidal flow, the tidal current in this particular tributary is considered to be remarkably fast. Thus, the local intensification of the flow velocity can be an important characteristic for considering the flood risk in the urban areas of the Mekong Delta. This phenomenon can be further intensified during unusually high tides or typhoon-induced storm surges [11]. Because numerous local fishermen, tourist guides, and floating-market merchants use small wooden ships (see Fig.2), there is a considerable risk that they may encounter difficulties in handling ships or risk their ships being capsized owing to these unexpected unusually rapid flows.

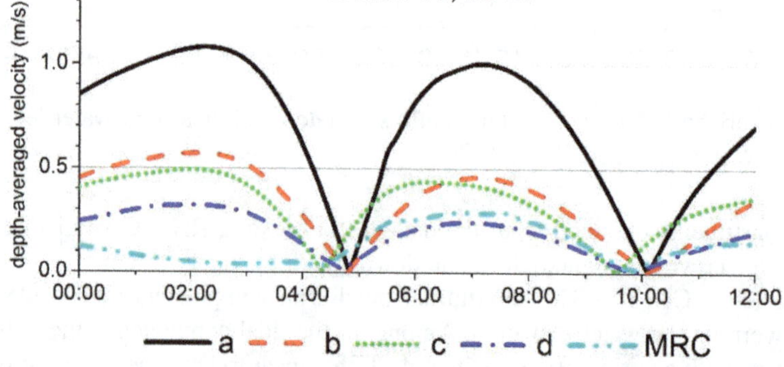

Figure 4: Simulated depth-averaged velocity during a semi-diurnal cycle in a typical dry season. The outputs correspond to the five locations indicated in Fig. 1.

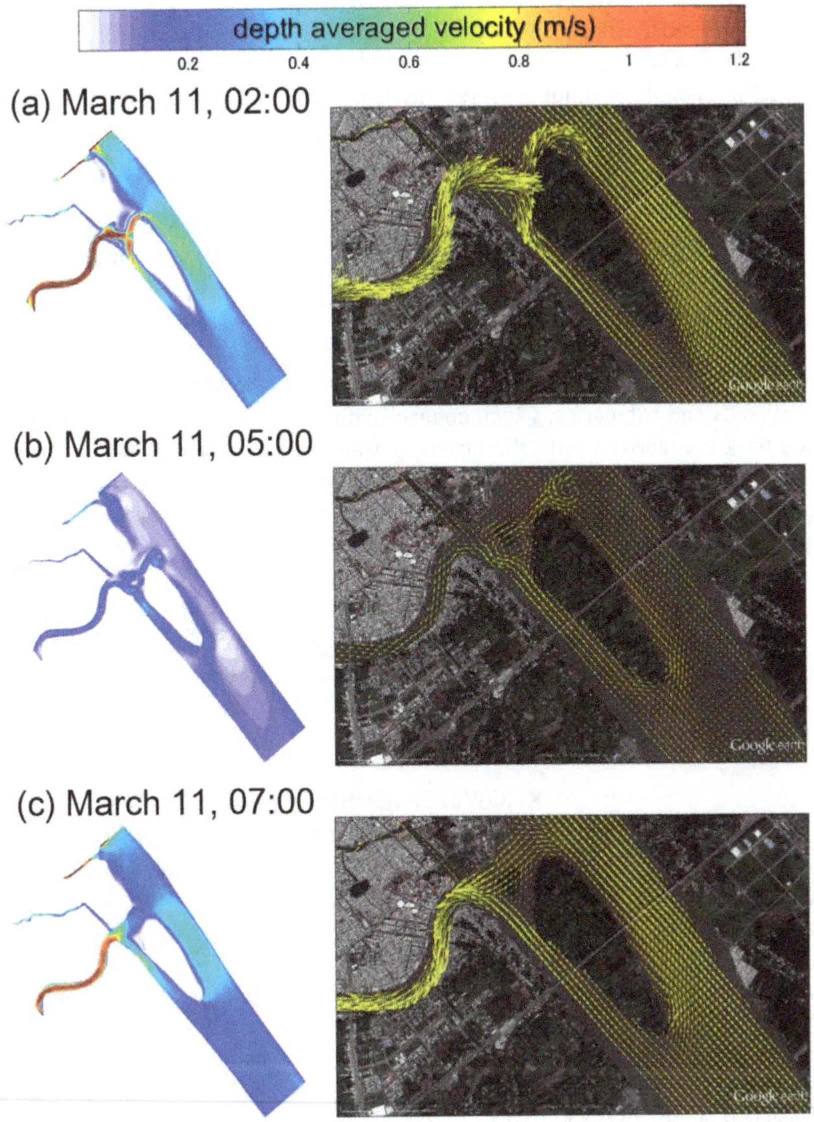

Figure 5: Tidal-flow intensification observed in the numerical simulation: (a) ebb-tide phase, (b) slack-tide phase, (c) flood-tide phase.

It is expected that the tidal currents at a given time and location can be predicted by using the model. However, it should be noted that the field survey was conducted in March, which is a typical dry season, with very limited rainfall. The weather in Vietnam is characterized by two monsoon seasons: the southwest monsoon, from April to September, and the northeast monsoon, from October to late March or early April. Southern Vietnam, including the Mekong Delta, experiences the majority of its rainfall between May and October. Therefore, the water levels in March are not significantly influenced by the discharge from upstream rivers.

For example, in October 2013, at the end of the rainy season, numerous local communities across the delta suffered a historical flood. The water level rose to approximately 50 cm above the predicted tidal levels, indicating significant pluvial and fluvial influences [4]. Although Delft3D-FLOW is capable of simulating river discharge from the upstream boundary in addition to tides, the present simulation was limited to tidal forcing. Therefore, the present model cannot be used to reproduce such an abnormally high water level induced by a high river discharge. Additionally, water levels in deltas are influenced by the sea-level rise and ENSO events [12–14]. The construction of dams will inevitably alter the variability of the seasonal flow in the upper Mekong River. Furthermore, local changes driven by urbanization can significantly influence the prediction of floods in Can Tho.

4 CONCLUSION

Tidal modelling was applied to a complex waterway system in the Mekong Delta that is characterised by canals and tributaries, which connect with the main stream. The model, which was verified by a comparison with the observed water levels, enabled us to examine the mechanisms of tidal propagation. The model analysis indicated that the flow velocity largely varies depending on the location and the flow intensification in the tributary, which exceeds 1 m/s. This kind of local amplification in the tidal current is important for evaluating the flood/inundation risk in the urban areas of the Mekong Delta, as small ships are likely to encounter difficulties in handling or risk being capsized because of the unexpectedly rapid flows that occur during these abnormal high tides and typhoon storm surges. The findings of the present study further our understanding of the mechanisms of tidal propagation in complex waterway systems and are thus useful for identifying potential urban flood risks associated with oceanic tides.

ACKNOWLEDGEMENTS

The present research was funded by the JSPS KAKENHI (Grant Number 26702009) and The Obayashi Foundation. The authors thank, Shintaro Maeda, Yuto Takahashi, and Chiaki Tsurudome for their valuable assistance in accomplishing this study.

REFERENCES

[1] Mekong River Commission (MRC), *Overview of the Hydrology of the Mekong Basin*. ISSN: 1728 3248, 2005.

[2] Takagi, H., Ty, T.V., Thao, N.D. & Esteban, M., Ocean tides and the influence of sea-level rise on floods in urban areas of the Mekong Delta. *Journal of Flood Risk Management*, **8**, pp. 292–300, 2015. http://dx.doi.org/10.1111/jfr3.12094

[3] Takagi, H, Thao, N.D., Esteban, M., Mikami, T., Cong, L.V. & Ca, V.T., Coastal disasters in Vietnam. In *Handbook of Coastal Disaster Mitigation for Engineers and Planners*, eds. Esteban, M., Takagi, H. & Shibayama, T. Elsevier: New York, U.S., pp. 235–255. http://dx.doi.org/10.1016/B978-0-12-801060-0.00012-5

[4] Takagi, H., Tsurudome, C., Nguyen, D.T., Le, T.A., Tran, V.T. & Van P.D.T., Ocean tidal modelling for urban flood risk assessment in the Mekong Delta. *Hydrological Research Letters*, **10**, pp. 21–26, 2016. http://dx.doi.org/10.3178/hrl.10.21

[5] Wassmann, R., Hien, N.X., Hoanh, C.T. & Tuong, T.P., Sea level rise affecting the Vietnamese Mekong Delta: water elevation in the flood season and implications for rice production. *Climatic Change*, **66**, pp. 89–107, 2004.

http://dx.doi.org/10.1023/B:CLIM.0000043144.69736.b7

[6] Carew-Reid, J., Rapid assessment of the extent and impact of sea level rise in Viet Nam. *Climate Change Discussion Paper 1, ICEM –International Centre for Environmental Management,* **82**, 2008.

[7] Takagi, H., Ty, T.V. & Thao, N.D., Investigation on floods in Can Tho City: influence of ocean tides and sea level rise for the Mekong Delta's largest City, eds. N.D. Thao, H. Takagi & M. Esteban. *Coastal Disasters and Climate Change in Vietnam: Engineering and Planning Perspectives*, pp. 257–274, 2014. http://dx.doi.org/10.1016/B978-0-12-800007-6.00012-5

[8] Yamamoto, L., Environmental displacement in Vietnam. In: D.T. Nguyen, H. Takagi & M. Esteban. *Coastal Disasters and Climate Change in Vietnam - Engineering and Planning Perspectives*, Elsevier: New York, pp. 379–393, 2014. http://dx.doi.org/10.1016/b978-0-12-800007-6.00018-6

[9] Takagi, H., Li, S., de Leon, M., Esteban, M., Mikami, T., Matsumaru, R., Shibayama, T. & Nakamura, R., Storm surge and evacuation in urban areas during the peak of a storm. *Coastal Engineering*, **108**, pp. 1–9, 2016. http://dx.doi.org/10.1016/j.coastaleng.2015.11.002

[10] Savenije, H.H.G., A simple analytical expression to describe tidal damping or amplification. *Journal of Hydrology*, **243**, pp. 205–215, 2001. http://dx.doi.org/10.1016/S0022-1694(00)00414-5

[11] Takagi, H., Nguyen, D.T., Esteban, M., Tran, T.T., Hanne, L.K. & Mikami, T., Vulnerability of coastal areas in Southern Vietnam against tropical cyclones and storm surges. *Proceedings of ICEC*, pp. 292–299, 2012.

[12] Nguyen, D.T., Takagi, H. & Esteban, M. (eds), *Coastal Disasters and Climate Change in Vietnam -Engineering and Planning Perspectives*, Elsevier: New York, U.S., p. 424, 2014, ISBN: 978-0128000076

[13] Esteban, M., Takagi, H. & Shibayama, T., *Handbook of Coastal Disaster Mitigation for Engineers and Planners*, Elsevier: New York, U.S., p. 780, 2015, ISBN: 978-0-12-801060-0.

[14] Takagi, H., Esteban, M., Mikami, T. & Fujii, D., *Projection of coastal floods in 2050 Jakarta*, Urban Climate, Vol.17, pp. 135-145, 2016. http://dx.doi.org/10.1016/j.uclim.2016.05.003

Flood Early Warning Systems

Knowledge and Tools for their Critical Assessment

www.ingramcontent.com/pod-product-compliance
Lightning Source LLC
LaVergne TN
LVHW062229200926
848484LV00006B/1506